Inequalities

Inequalities

Special Issue Editor
Shigeru Furuichi

MDPI • Basel • Beijing • Wuhan • Barcelona • Belgrade

Special Issue Editor
Shigeru Furuichi
Nihon University
Japan

Editorial Office
MDPI
St. Alban-Anlage 66
4052 Basel, Switzerland

This is a reprint of articles from the Special Issue published online in the open access journal *Mathematics* (ISSN 2227-7390) from 2018 to 2019 (available at: https://www.mdpi.com/journal/mathematics/special_issues/Inequalities_2018)

For citation purposes, cite each article independently as indicated on the article page online and as indicated below:

LastName, A.A.; LastName, B.B.; LastName, C.C. Article Title. *Journal Name* **Year**, *Article Number*, Page Range.

ISBN 978-3-03928-062-9 (Pbk)
ISBN 978-3-03928-063-6 (PDF)

Contents

About the Special Issue Editor

Shigeru Furuichi (Ph.D.) received the B.S. degree from Department of Mathematics, Tokyo University of Science, and M.S. and Ph.D. from Department of information Science, Tokyo University of Science, Japan. He was an Assistant Professor and Lecturer from 1997 to 2007 in Department of Electronics and Computer Science, Tokyo University of Science, Yamaguchi, Japan. He was an Associate Professor from 2008 to 2012 and has been Professor from 2013 to the present in Department of Information Science, College of Humanities and Sciences, Nihon University. His research interests include inequality, matrices/operators, and entropy.

Preface to "Inequalities"

It is our great pleasure to publish this book. All contents were peer-reviewed by multiple referees and published as papers in the Special Issue "Inequalities" in the journal *Mathematics*. They give new and interesting results in mathematical inequalities so that the readers will be able to obtain the latest developments in the fields of mathematical inequalities.

Shigeru Furuichi
Special Issue Editor

mathematics

MDPI

Article

Coefficient Inequalities of Functions Associated with Petal Type Domains

Sarfraz Nawaz Malik [1,*], **Shahid Mahmood** [2], **Mohsan Raza** [3], **Sumbal Farman** [1] **and Saira Zainab** [4]

1 Department of Mathematics, COMSATS University Islamabad, Wah Campus 47040, Pakistan; sumbalfarman678@gmail.com
2 Department of Mechanical Engineering, Sarhad University of Science and I.T, Ring Road, Peshawar 25000, Pakistan; shahidmahmood757@gmail.com
3 Department of Mathematics, Government College University Faisalabad, Faisalabad 38000, Pakistan; mohsan976@yahoo.com
4 Department of Mathematical Sciences, Fatima Jinnah Women University, Rawalpindi 46000, Pakistan; sairazainab07@yahoo.com
* Correspondence: snmalik110@yahoo.com

Received: 13 November 2018; Accepted: 1 December 2018; Published: 3 December 2018

Abstract: In the theory of analytic and univalent functions, coefficients of functions' Taylor series representation and their related functional inequalities are of major interest and how they estimate functions' growth in their specified domains. One of the important and useful functional inequalities is the Fekete-Szegö inequality. In this work, we aim to analyze the Fekete-Szegö functional and to find its upper bound for certain analytic functions which give parabolic and petal type regions as image domains. Coefficient inequalities and the Fekete-Szegö inequality of inverse functions to these certain analytic functions are also established in this work.

Keywords: analytic functions; starlike functions; convex functions; Fekete-Szegö inequality

MSC: Primary 30C45, 33C10; Secondary 30C20, 30C75

---

1. Introduction and Preliminaries

Let $\mathcal{A}$ be the class of functions f of the form

$$f(z) = z + \sum_{n=2}^{\infty} a_n z^n, \quad (1)$$

which are analytic in the open unit disk $\mathcal{U} = \{z : |z| < 1\}$ and $\mathcal{S}$ be the class of functions from $\mathcal{A}$ which are univalent in $\mathcal{U}$. One of the classical results regarding univalent functions related to coefficients a_n of a function's Taylor series, named as the Fekete-Szegö problem, introduced by Fekete and Szegö [1], is defined as follows:

If $f \in \mathcal{S}$ and is of the form (1), then

$$\left|a_3 - \lambda a_2^2\right| \le \begin{cases} 3 - 4\lambda, & \text{if} \quad \lambda \le 0, \\ 1 + 2\exp\left(\frac{2\lambda}{\lambda - 1}\right), & \text{if} \quad 0 \le \lambda \le 1, \\ 4\lambda - 3, & \text{if} \quad \lambda \ge 1. \end{cases}$$

This result is sharp. The Fekete-Szegö problem has a rich history in literature. Several results dealing with maximizing the non-linear functional $\left|a_3 - \lambda a_2^2\right|$ for various classes and subclasses of univalent functions have been proved. The functional has been examined for λ to be both a real and complex number. Several authors used certain classified techniques to maximize the Fekete-Szegö functional $\left|a_3 - \lambda a_2^2\right|$ for different types of functions having interesting geometric characteristics of image domains. For more details and results, we refer to [1–11]. The function f is said to be subordinate to the function g, written symbolically as $f \prec g$, if there exists a schwarz function w such that

$$f(z) = g(w(z)), \quad z \in \mathcal{U}, \tag{2}$$

where $w(0) = 0$, $|w(z)| < 1$ for $z \in \mathcal{U}$. Let P denote the class of analytic functions p such that $p(0) = 1$ and $p \prec \frac{1+z}{1-z}$, $z \in \mathcal{U}$. For details, see [12].

In 1991, Goodman [13] initiated the concept of a conic domain by introducing generalized convex functions which generated the first parabolic region as an image domain of analytic functions. He introduced and defined the class UCV of uniformly convex functions as follows:

$$UCV = \left\{ f \in \mathcal{A} : \Re\left(1 + (z - \zeta)\frac{f''(z)}{f'(z)}\right) > 0,\ z, \zeta \in \mathcal{U} \right\}.$$

Later on, Rønning [14], and Ma and Minda [7] independently gave the most suitable one variable characterization of the class UCV and defined it as follows:

$$UCV = \left\{ f \in \mathcal{A} : \Re\left(1 + \frac{zf''(z)}{f'(z)}\right) > \left|\frac{zf''(z)}{f'(z)}\right|,\ z \in \mathcal{U} \right\}.$$

This characterization gave birth to the first conic (parabolic) domain

$$\Omega = \{w : \Re w > |w - 1|\}.$$

This domain was then generalized by Kanas and Wiśniowska [15,16] who introduced the domain

$$\Omega_k = \{w : \Re w > k|w - 1|,\ k \geq 0\}.$$

The conic domain Ω_k represents the right half plane for $k = 0$, hyperbolic regions when $0 < k < 1$, parabolic region for $k = 1$ and elliptic regions when $k > 1$. For more details, we refer [15,16]. This conic domain Ω_k has been extensively studied in [17–19]. The domain Ω was also generalized by Noor and Malik [20] by introducing the domain

$$\Omega[A,B] = \left\{ u + iv : \left[\left(B^2 - 1\right)\left(u^2 + v^2\right) - 2(AB - 1)u + \left(A^2 - 1\right)\right]^2 \right.$$
$$\left. > \left(-2(B+1)\left(u^2 + v^2\right) + 2(A + B + 2)u - 2(A+1)\right)^2 + 4(A - B)^2 v^2 \right\}.$$

The domain $\Omega[A,B]$ represents the petal type region, for more details, we refer to [20]. Now, we consider the following class of functions which take all values from the domain $\Omega[A,B]$, $-1 \leq B < A \leq 1$.

Definition 1. *A function $p(z)$ is said to be in the class $UP[A,B]$, if and only if*

$$p(z) \prec \frac{(A+1)\widetilde{p}(z) - (A-1)}{(B+1)\widetilde{p}(z) - (B-1)}, \quad -1 \leq B < A \leq 1, \tag{3}$$

where $\widetilde{p}(z) = 1 + \frac{2}{\pi^2}\left(\log \frac{1+\sqrt{z}}{1-\sqrt{z}}\right)^2$, $z \in \mathcal{U}$.

It can be seen that $\Omega[1,-1] = \Omega_1 = \Omega$. This fact leads us to the following implications of different well-known classes of analytic functions.

1. $UP[A,B] \subset P\left(\frac{3-A}{3-B}\right)$, the well-known class of functions with real part greater than $\frac{3-A}{3-B}$, see [12].

2. $UP[1,-1] = \mathcal{P}(\widetilde{p})$, the well-known class of functions, introduced by Kanas and Wiśniowska [4,21].

Now we consider the following classes $UCV[A,B]$ of uniformly Janowski convex functions and $ST[A,B]$ of corresponding Janowski starlike functions (see [20]) as follows.

Definition 2. *A function $f \in \mathcal{A}$ is said to be in the class $UCV[A,B]$, $-1 \le B < A \le 1$, if and only if*

$$\Re\left(\frac{(B-1)\frac{(zf'(z))'}{f'(z)}-(A-1)}{(B+1)\frac{(zf'(z))'}{f'(z)}-(A+1)}\right) > \left|\frac{(B-1)\frac{(zf'(z))'}{f'(z)}-(A-1)}{(B+1)\frac{(zf'(z))'}{f'(z)}-(A+1)}-1\right|,$$

or equivalently,

$$\frac{(zf'(z))'}{f'(z)} \in UP[A,B]. \tag{4}$$

Definition 3. *A function $f \in \mathcal{A}$ is said to be in the class $ST[A,B]$, $-1 \le B < A \le 1$, if and only if*

$$\Re\left(\frac{(B-1)\frac{zf'(z)}{f(z)}-(A-1)}{(B+1)\frac{zf'(z)}{f(z)}-(A+1)}\right) > \left|\frac{(B-1)\frac{zf'(z)}{f(z)}-(A-1)}{(B+1)\frac{zf'(z)}{f(z)}-(A+1)}-1\right|,$$

or equivalently,

$$\frac{zf'(z)}{f(z)} \in UP[A,B]. \tag{5}$$

It can easily be seen that $f \in UCV[A,B] \iff zf' \in ST[A,B]$. It is clear that $UCV[1,-1] = UCV$ and $ST[1,-1] = ST$, the well-known classes of uniformly convex and corresponding starlike functions respectively, introduced by Goodman [13] and Rønning [22].

In 1994, Ma and Minda [7] found the maximum bound of Fekete-Szegö functional $|a_3 - \lambda a_2^2|$ for uniformly convex functions of class UCV and then Kanas [21] investigated the same for the functions of class $\mathcal{P}(\widetilde{p})$. Our aim is to solve this classical Fekete-Szegö problem for the functions of classes $UP[A,B]$, $UCV[A,B]$ and $ST[A,B]$. We need the following lemmas (see [7]) to prove our results.

Lemma 1. *If $p(z) = 1 + p_1 z + p_2 z^2 + \cdots$ is a function with positive real part in $\mathcal{U}$, then, for any complex number μ,*

$$\left|p_2 - \mu p_1^2\right| \le 2\max\{1, |2\mu - 1|\}$$

and the result is sharp for the functions

$$p_0(z) = \frac{1+z}{1-z} \quad or \quad p_*(z) = \frac{1+z^2}{1-z^2}, \qquad (z \in \mathcal{U}).$$

Lemma 2. *If $p(z) = 1 + p_1 z + p_2 z^2 + \cdots$ is a function with positive real part in $\mathcal{U}$, then, for any real number v,*

$$\left|p_2 - v p_1^2\right| \le \begin{cases} -4v+2, & v \le 0, \\ 2, & 0 \le v \le 1, \\ 4v-2, & v \ge 1. \end{cases}$$

When $v < 0$ or $v > 1$, the equality holds if and only if $p(z)$ is $\frac{1+z}{1-z}$ or one of its rotations. If $0 < v < 1$, then, the equality holds if and only if $p(z) = \frac{1+z^2}{1-z^2}$ or one of its rotations. If $v = 0$, the equality holds if and only if,

$$p(z) = \left(\frac{1+\eta}{2}\right)\frac{1+z}{1-z} + \left(\frac{1-\eta}{2}\right)\frac{1-z}{1+z} \quad (0 \le \eta \le 1),$$

or one of its rotations. If $v = 1$, then, the equality holds if and only if $p(z)$ is reciprocal of one of the function such that equality holds in the case of $v = 0$. Although the above upper bound is sharp, when $0 < v < 1$, it can be improved as follows:

$$\left|p_2 - vp_1^2\right| + |p_1|^2 \leq 2 \qquad \left(0 < v \leq \frac{1}{2}\right)$$

and

$$\left|p_2 - vp_1^2\right| + (1-v)|p_1|^2 \leq 2 \quad \left(\frac{1}{2} < v \leq 1\right).$$

2. Main Results

Theorem 1. *Let $p \in UP[A,B]$, $-1 \leq B < A \leq 1$ and of the form $p(z) = 1 + \sum_{n=1}^{\infty} p_n z^n$. Then, for a complex number μ, we have*

$$\left|p_2 - \mu p_1^2\right| \leq \frac{4}{\pi^2}(A-B) . \max\left(1, \left|\frac{4}{\pi^2}(B+1) - \frac{2}{3} + 4\mu\left(\frac{A-B}{\pi^2}\right)\right|\right) \tag{6}$$

and for a real number μ, we have

$$\left|p_2 - \mu p_1^2\right| \leq \frac{2(A-B)}{\pi^2}\begin{cases} \frac{4}{3} - \frac{8}{\pi^2}(B+1) - \frac{8}{\pi^2}(A-B)\mu, & \mu \leq -\frac{\pi^2}{12(A-B)} - \frac{B+1}{A-B}, \\ 2, & -\frac{\pi^2}{12(A-B)} - \frac{B+1}{A-B} \leq \mu \leq \frac{5\pi^2}{12(A-B)} - \frac{B+1}{A-B}, \\ -\frac{4}{3} + \frac{8}{\pi^2}(B+1) + \frac{8}{\pi^2}(A-B)\mu, & \mu \geq \frac{5\pi^2}{12(A-B)} - \frac{B+1}{A-B}. \end{cases} \tag{7}$$

These results are sharp and the equality in (6) holds for the functions

$$p_1(z) = \frac{\frac{2(A+1)}{\pi^2}\left(\log\frac{1+\sqrt{z}}{1-\sqrt{z}}\right)^2 + 2}{\frac{2(B+1)}{\pi^2}\left(\log\frac{1+\sqrt{z}}{1-\sqrt{z}}\right)^2 + 2} \tag{8}$$

or

$$p_2(z) = \frac{\frac{2(A+1)}{\pi^2}\left(\log\frac{1+z}{1-z}\right)^2 + 2}{\frac{2(B+1)}{\pi^2}\left(\log\frac{1+z}{1-z}\right)^2 + 2}. \tag{9}$$

When $\mu < -\frac{\pi^2}{12(A-B)} - \frac{B+1}{A-B}$ or $\mu > \frac{5\pi^2}{12(A-B)} - \frac{B+1}{A-B}$, the equality in (7) holds for the function $p_1(z)$ or one of its rotations. If $-\frac{\pi^2}{12(A-B)} - \frac{B+1}{A-B} < \mu < \frac{5\pi^2}{12(A-B)} - \frac{B+1}{A-B}$, then, the equality in (7) holds for the function $p_2(z)$ or one of its rotations. If $\mu = -\frac{\pi^2}{12(A-B)} - \frac{B+1}{A-B}$, the equality in (7) holds for the function

$$p_3(z) = \left(\frac{1+\eta}{2}\right)p_1(z) + \left(\frac{1-\eta}{2}\right)p_1(-z), \quad (0 \leq \eta \leq 1), \tag{10}$$

or one of its rotations. If $\mu = \frac{5\pi^2}{12(A-B)} - \frac{B+1}{A-B}$, then, the equality in (7) holds for the functions $p(z)$ which is reciprocal of one of the function such that equality holds in the case for $\mu = -\frac{\pi^2}{12(A-B)} - \frac{B+1}{A-B}$.

Proof. For $h \in P$ and of the form $h(z) = 1 + \sum_{n=1}^{\infty} c_n z^n$, we consider

$$h(z) = \frac{1 + w(z)}{1 - w(z)},$$

where $w(z)$ is such that $w(0) = 0$ and $|w(z)| < 1$. It follows easily that

$$\begin{aligned} w(z) &= \frac{h(z)-1}{h(z)+1} \\ &= \frac{(1+c_1z+c_2z^2+c_3z^3+\cdots)-1}{(1+c_1z+c_2z^2+c_3z^3+\cdots)+1} \\ &= \frac{1}{2}c_1z+\left(\frac{1}{2}c_2-\frac{1}{4}c_1^2\right)z^2+\left(\frac{1}{2}c_3-\frac{1}{2}c_2c_1+\frac{1}{8}c_1^3\right)z^3+\cdots. \end{aligned} \tag{11}$$

Now, if $\widetilde{p}(z) = 1 + R_1z + R_2z^2 + \cdots$, then from (11), one may have

$$\begin{aligned} \widetilde{p}(w(z)) &= 1+R_1w(z)+R_2(w(z))^2+R_3(w(z))^3+\cdots \\ &= 1+R_1\left(\frac{1}{2}c_1z+\left(\frac{1}{2}c_2-\frac{1}{4}c_1^2\right)z^2+\left(\frac{1}{2}c_3-\frac{1}{2}c_2c_1+\frac{1}{8}c_1^3\right)z^3+\cdots\right) \\ &\quad +R_2\left(\frac{1}{2}c_1z+\left(\frac{1}{2}c_2-\frac{1}{4}c_1^2\right)z^2+\left(\frac{1}{2}c_3-\frac{1}{2}c_2c_1+\frac{1}{8}c_1^3\right)z^3+\cdots\right)^2 \\ &\quad +R_3\left(\frac{1}{2}c_1z+\left(\frac{1}{2}c_2-\frac{1}{4}c_1^2\right)z^2+\left(\frac{1}{2}c_3-\frac{1}{2}c_2c_1+\frac{1}{8}c_1^3\right)z^3+\cdots\right)^3+\cdots, \end{aligned}$$

where $R_1 = \frac{8}{\pi^2}$, $R_2 = \frac{16}{3\pi^2}$ and $R_3 = \frac{184}{45\pi^2}$, see [21]. Using these, the above series reduces to

$$\widetilde{p}(w(z)) = 1+\frac{4}{\pi^2}c_1z+\frac{4}{\pi^2}\left(c_2-\frac{1}{6}c_1^2\right)z^2+\frac{4}{\pi^2}\left(c_3-\frac{1}{3}c_2c_1+\frac{2}{45}c_1^3\right)z^3+\cdots. \tag{12}$$

Since $p \in UP[A,B]$, so from relations (2), (3) and (12), one may have

$$\begin{aligned} p(z) &= \frac{(A+1)\widetilde{p}(w(z))-(A-1)}{(B+1)\widetilde{p}(w(z))-(B-1)} \\ &= \frac{2+(A+1)\frac{4}{\pi^2}c_1z+(A+1)\frac{4}{\pi^2}\left(c_2-\frac{1}{6}c_1^2\right)z^2+\cdots}{2+(B+1)\frac{4}{\pi^2}c_1z+(B+1)\frac{4}{\pi^2}\left(c_2-\frac{1}{6}c_1^2\right)z^2+\cdots}. \end{aligned}$$

This implies that

$$\begin{aligned} p(z) &= 1+(A-B)\frac{2}{\pi^2}c_1z+(A-B)\frac{2}{\pi^2}\left(c_2-\frac{1}{6}c_1^2-\frac{2}{\pi^2}(B+1)c_1^2\right)z^2+ \\ &\quad (A-B)\frac{8}{\pi^2}\left(\left(\frac{(B+1)^2}{\pi^4}+\frac{B+1}{6\pi^2}+\frac{1}{90}\right)c_1^3-\left(\frac{B+1}{\pi^2}+\frac{1}{12}\right)c_2c_1+\frac{1}{4}c_3\right)z^3+\cdots. \end{aligned} \tag{13}$$

If $p(z) = 1 + \sum_{n=1}^{\infty} p_nz^n$, then equating coefficients of z and z^2, one may have

$$\begin{aligned} p_1 &= \frac{2}{\pi^2}(A-B)c_1, \\ p_2 &= \frac{2}{\pi^2}(A-B)\left(c_2-\frac{1}{6}c_1^2-\frac{2}{\pi^2}(B+1)c_1^2\right). \end{aligned}$$

Now for a complex number μ, consider

$$p_2-\mu p_1^2 = \frac{2}{\pi^2}(A-B)\left[c_2-c_1^2\left(\frac{1}{6}+\frac{2}{\pi^2}(B+1)+\mu\frac{2}{\pi^2}(A-B)\right)\right].$$

This implies that

$$\left|p_2 - \mu p_1^2\right| = \frac{2}{\pi^2}(A-B)\left|c_2 - \left(\frac{1}{6} + \frac{2}{\pi^2}(B+1) + \mu\frac{2}{\pi^2}(A-B)\right)c_1^2\right|. \tag{14}$$

Using Lemma 1, one may have

$$\left|p_2 - \mu p_1^2\right| \leq \frac{2}{\pi^2}(A-B).2\max(1, |2v-1|),$$

where

$$v = \frac{1}{6} + \frac{2}{\pi^2}(B+1) + \mu\frac{2}{\pi^2}(A-B).$$

This leads us to the required inequality (6) and applying Lemma 2 to the expression (14) for real number μ, we get the required inequality (7). Sharpness follows from the functions $p_i(z)$; $i = 1,2,3$, defined by (8)–(10), and the following series form.

$$\begin{aligned} p_1(z) &= 1 + \frac{4(A-B)}{\pi^2}z + \frac{8(A-B)}{\pi^2}\left(\frac{1}{3} - \frac{2(B+1)}{\pi^2}\right)z^2 + \\ &\frac{16(A-B)}{\pi^2}\left(4\left(\frac{(B+1)^2}{\pi^4} + \frac{B+1}{6\pi^2} + \frac{1}{90}\right) - 2\left(\frac{B+1}{\pi^2} + \frac{1}{12}\right) + \frac{1}{4}\right)z^3 + \cdots, \end{aligned}$$

$$\begin{aligned} p_2(z) &= 1 + \frac{4(A-B)}{\pi^2}z^2 + \frac{8(A-B)}{\pi^2}\left(\frac{1}{3} - \frac{2(B+1)}{\pi^2}\right)z^4 + \\ &\frac{16(A-B)}{\pi^2}\left(4\left(\frac{(B+1)^2}{\pi^4} + \frac{B+1}{6\pi^2} + \frac{1}{90}\right) - 2\left(\frac{B+1}{\pi^2} + \frac{1}{12}\right) + \frac{1}{4}\right)z^6 + \cdots. \end{aligned}$$

□

Corollary 1. *Let $p \in UP[1,-1] = \mathcal{P}(p_1) = \mathcal{P}(\tilde{p})$ and of the form $p(z) = 1 + \sum_{n=1}^{\infty} p_n z^n$. Then, for a complex number μ, we have*

$$\left|p_2 - \mu p_1^2\right| \leq \frac{8}{\pi^2}.\max\left(1, \left|\frac{8\mu}{\pi^2} - \frac{2}{3}\right|\right) \tag{15}$$

and for real number μ, we have

$$\left|p_2 - \mu p_1^2\right| \leq \frac{4}{\pi^2}\begin{cases} \frac{4}{3} - \frac{16}{\pi^2}\mu, & \mu \leq -\frac{\pi^2}{24}, \\ 2, & -\frac{\pi^2}{24} \leq \mu \leq \frac{5\pi^2}{24}, \\ -\frac{4}{3} + \frac{16}{\pi^2}\mu, & \mu \geq \frac{5\pi^2}{24}. \end{cases} \tag{16}$$

These inequalities are sharp.

In [4,21], Kanas studied the class $\mathcal{P}(p_k)$ which consists of functions who take all values from the conic domain Ω_k. Kanas [21] found the bound of Fekete-Szegö functional for the class $\mathcal{P}(p_k)$ whose particular case for $k = 1$ is as follows:

Let $p(z) = 1 + b_1 z + b_2 z^2 + b_3 z^3 + \cdots \in \mathcal{P}(p_1)$. Then, for real number μ, we have

$$\left|b_2 - \mu b_1^2\right| \leq \frac{8}{\pi^2}\begin{cases} 1 - \frac{8}{\pi^2}\mu, & \mu \leq 0, \\ 1, & \mu \in (0,1], \\ 1 + \frac{8}{\pi^2}(\mu - 1), & \mu \geq 1. \end{cases} \tag{17}$$

We observe that Corollary 1 improves the bounds of the Fekete-Szegö functional $\left|p_2 - \mu p_1^2\right|$ for the functions of class $\mathcal{P}\left(p_1\right)$.

Theorem 2. *Let $f \in UCV\left[A, B\right]$, $-1 \leq B < A \leq 1$ and of the form (1). Then, for a real number μ, we have*

$$\left|a_3 - \mu a_2^2\right| \leq \frac{A-B}{3\pi^2} \begin{cases} \frac{4}{3} - \frac{8}{\pi^2}\left(B+1\right) + \frac{4}{\pi^2}\left(A-B\right)\left(2-3\mu\right), & \mu \leq \frac{2}{3} - \frac{\pi^2}{18(A-B)} - \frac{2(B+1)}{3(A-B)}, \\ 2, & \frac{2}{3} - \frac{\pi^2}{18(A-B)} - \frac{2(B+1)}{3(A-B)} \leq \mu \leq \frac{2}{3} + \frac{5\pi^2}{18(A-B)} - \frac{2(B+1)}{3(A-B)}, \\ -\frac{4}{3} + \frac{8}{\pi^2}\left(B+1\right) - \frac{4}{\pi^2}\left(A-B\right)\left(2-3\mu\right), & \mu \geq \frac{2}{3} + \frac{5\pi^2}{18(A-B)} - \frac{2(B+1)}{3(A-B)}. \end{cases} \tag{18}$$

This result is sharp.

Proof. If $f \in UCV\left[A, B\right]$, $-1 \leq B < A \leq 1$, then it follows from relations (2)–(4),

$$\frac{\left(zf'\left(z\right)\right)'}{f'\left(z\right)} = \frac{\left(A+1\right)\widetilde{p}\left(w\left(z\right)\right) - \left(A-1\right)}{\left(B+1\right)\widetilde{p}\left(w\left(z\right)\right) - \left(B-1\right)},$$

where $w\left(z\right)$ is such that $w\left(0\right) = 0$ and $\left|w\left(z\right)\right| < 1$. The right hand side of above expression gets its series form from (13) and reduces to

$$\begin{aligned} \frac{\left(zf'\left(z\right)\right)'}{f'\left(z\right)} &= 1 + \left(A-B\right)\frac{2}{\pi^2}c_1 z + \left(A-B\right)\frac{2}{\pi^2}\left(c_2 - \frac{1}{6}c_1^2 - \frac{2}{\pi^2}\left(B+1\right)c_1^2\right)z^2 + \\ & \left(A-B\right)\frac{8}{\pi^2}\left(\left(\frac{\left(B+1\right)^2}{\pi^4} + \frac{B+1}{6\pi^2} + \frac{1}{90}\right)c_1^3 - \left(\frac{B+1}{\pi^2} + \frac{1}{12}\right)c_2 c_1 + \frac{1}{4}c_3\right)z^3 + \cdots. \end{aligned} \tag{19}$$

If $f\left(z\right) = z + \sum_{n=2}^{\infty} a_n z^n$, then one may have

$$\frac{\left(zf'\left(z\right)\right)'}{f'\left(z\right)} = 1 + 2a_2 z + \left(6a_3 - 4a_2^2\right)z^2 + \left(12a_4 - 18a_2 a_3 + 8a_2^3\right)z^3 + \cdots. \tag{20}$$

From (19) and (20), comparison of coefficients of z and z^2 gives

$$a_2 = \frac{1}{\pi^2}\left(A-B\right)c_1 \tag{21}$$

and

$$6a_3 - 4a_2^2 = \left(A-B\right)\frac{2}{\pi^2}\left(c_2 - \frac{1}{6}c_1^2 - \frac{2}{\pi^2}\left(B+1\right)c_1^2\right).$$

This implies, by using (21), that

$$a_3 = \frac{1}{3\pi^2}\left(A-B\right)\left(c_2 - \frac{1}{6}c_1^2 - \frac{2}{\pi^2}\left(B+1\right)c_1^2 + \frac{2}{\pi^2}\left(A-B\right)c_1^2\right). \tag{22}$$

Now, for a real number μ, consider

$$\begin{aligned}\left|a_3-\mu a_2^2\right| &= \left|(A-B)\frac{1}{3\pi^2}\left(c_2-\frac{1}{6}c_1^2-\frac{2}{\pi^2}(B+1)c_1^2\right)+\frac{2}{3\pi^4}(A-B)^2c_1^2-\mu\frac{1}{\pi^4}(A-B)^2c_1^2\right| \\ &= \frac{A-B}{3\pi^2}\left|c_2-c_1^2\left(\frac{1}{6}+\frac{2}{\pi^2}(B+1)-\frac{2}{\pi^2}(A-B)+\frac{3\mu}{\pi^2}(A-B)\right)\right| \\ &= \frac{A-B}{3\pi^2}\left|c_2-vc_1^2\right|,\end{aligned}$$

where

$$v=\frac{1}{6}+\frac{2}{\pi^2}(B+1)-\frac{1}{\pi^2}(A-B)(2-3\mu).$$

Applying Lemma 2 leads us to the required result. The inequality (18) is sharp and equality holds for $\mu<\frac{2}{3}-\frac{\pi^2}{18(A-B)}-\frac{2(B+1)}{3(A-B)}$ or $\mu>\frac{2}{3}+\frac{5\pi^2}{18(A-B)}-\frac{2(B+1)}{3(A-B)}$ when $f(z)$ is $f_1(z)$ or one of its rotations, where $f_1(z)$ is defined such that $\frac{(zf_1'(z))'}{f_1'(z)}=p_1(z)$. If $\frac{2}{3}-\frac{\pi^2}{18(A-B)}-\frac{2(B+1)}{3(A-B)}<\mu<\frac{2}{3}+\frac{5\pi^2}{18(A-B)}-\frac{2(B+1)}{3(A-B)}$, then, the equality holds for the function $f_2(z)$ or one of its rotations, where $f_2(z)$ is defined such that $\frac{(zf_2'(z))'}{f_2'(z)}=p_2(z)$. If $\mu=\frac{2}{3}-\frac{\pi^2}{18(A-B)}-\frac{2(B+1)}{3(A-B)}$, the equality holds for the function $f_3(z)$ or one of its rotations, where $f_3(z)$ is defined such that $\frac{(zf_3'(z))'}{f_3'(z)}=p_3(z)$. If $\mu=\frac{2}{3}+\frac{5\pi^2}{18(A-B)}-\frac{2(B+1)}{3(A-B)}$, then, the equality holds for $f(z)$, which is such that $\frac{(zf'(z))'}{f'(z)}$ is reciprocal of one of the function such that equality holds in the case of $\mu=\frac{2}{3}-\frac{\pi^2}{18(A-B)}-\frac{2(B+1)}{3(A-B)}$. □

For $A=1$, $B=-1$, the above result takes the following form which is proved by Ma and Minda [8].

Corollary 2. *Let $f\in UCV[1,-1]=UCV$ and of the form (1). Then, for a real number μ,*

$$\left|a_3-\mu a_2^2\right|\le\frac{2}{3\pi^2}\begin{cases}\frac{4}{3}+\frac{8}{\pi^2}(2-3\mu), & \mu\le\frac{2}{3}-\frac{\pi^2}{36}, \\ 2, & \frac{2}{3}-\frac{\pi^2}{36}\le\mu\le\frac{2}{3}+\frac{5\pi^2}{36}, \\ -\frac{4}{3}-\frac{8}{\pi^2}(2-3\mu), & \mu\ge\frac{2}{3}+\frac{5\pi^2}{36}.\end{cases}$$

This result is sharp.

Theorem 3. *Let $f\in ST[A,B]$, $-1\le B<A\le1$ and of the form (1). Then, for a real number μ,*

$$\left|a_3-\mu a_2^2\right|\le\frac{A-B}{\pi^2}\begin{cases}\frac{4}{3}-\frac{8}{\pi^2}(B+1)+\frac{8}{\pi^2}(A-B)(1-2\mu), & \mu\le\frac{1}{2}-\frac{\pi^2}{24(A-B)}-\frac{B+1}{2(A-B)}, \\ 2, & \frac{1}{2}-\frac{\pi^2}{24(A-B)}-\frac{B+1}{2(A-B)}\le\mu\le\frac{1}{2}+\frac{5\pi^2}{24(A-B)}-\frac{B+1}{2(A-B)}, \\ -\frac{4}{3}+\frac{8}{\pi^2}(B+1)-\frac{8}{\pi^2}(A-B)(1-2\mu), & \mu\ge\frac{1}{2}+\frac{5\pi^2}{24(A-B)}-\frac{B+1}{2(A-B)}.\end{cases} \quad (23)$$

This result is sharp.

Proof. The proof follows similarly as in Theorem 2. □

For $A=1$, $B=-1$, the above result reduces to the following form.

Corollary 3. *Let $f \in ST[1,-1]$ and of the form (1). Then, for a real number μ,*

$$\left|a_3-\mu a_2^2\right| \le \frac{2}{\pi^2}\begin{cases} \frac{4}{3}+\frac{16}{\pi^2}(1-2\mu), & \mu \le \frac{1}{2}-\frac{\pi^2}{48}, \\ 2, & \frac{1}{2}-\frac{\pi^2}{48} \le \mu \le \frac{1}{2}+\frac{5\pi^2}{48}, \\ -\frac{4}{3}-\frac{16}{\pi^2}(1-2\mu), & \mu \ge \frac{1}{2}+\frac{5\pi^2}{48}. \end{cases}$$

Now we consider the inverse function $\mathcal{F}$ which maps petal type regions to the open unit disk $\mathcal{U}$, defined as $\mathcal{F}(w)=\mathcal{F}(f(z))=z, z \in \mathcal{U}$ and we find the following coefficient bound for inverse functions. As the classes $UCV[A,B]$ and $ST[A,B]$ are the subclasses of $\mathcal{S}$. Thus the existence of such inverse functions to the functions from $UCV[A,B]$ and $ST[A,B]$ is assured.

Theorem 4. *Let $w=f(z) \in UCV[A,B], -1 \le B < A \le 1$ and $\mathcal{F}(w)=f^{-1}(w)=w+\sum_{n=2}^{\infty} d_n w^n$. Then,*

$$|d_n| \le \frac{4(A-B)}{n(n-1)\pi^2} \qquad (n=2,3,4).$$

Proof. Since $\mathcal{F}(w)=\mathcal{F}(f(z))=z$, so it is easy to see that

$$d_2=-a_2, \qquad d_3=2a_2^2-a_3, \qquad d_4=-a_4+5a_2a_3-5a_2^3.$$

By using (21) and (22), one can have

$$d_2=\frac{-1}{\pi^2}(A-B)c_1 \tag{24}$$

and

$$d_3=\frac{A-B}{3\pi^2}\left[\left(\frac{1}{6}+\frac{2}{\pi^2}(B+1)+\frac{4}{\pi^2}(A-B)\right)c_1^2-c_2\right]. \tag{25}$$

From (19) and (20), comparison of z^3 gives

$$\begin{aligned} a_4 = {} & \frac{A-B}{3\pi^2}\left[\left(\frac{1}{45}+\frac{1}{\pi^2}\left(\frac{1}{3}(B+1)-\frac{1}{4}(A-B)\right)+\frac{1}{\pi^4}\left(2(B+1)^2-3(A-B)(B+1)+(A-B)^2\right)\right)c_1^3\right. \\ & \left.-\left(\frac{1}{6}+\frac{1}{\pi^2}\left(2(B+1)-\frac{3}{2}(A-B)\right)\right)c_2c_1+\frac{1}{2}c_3\right]. \end{aligned}$$

Using the values of $a_n;\ n=2,3,4$, we get

$$\begin{aligned} d_4 = {} & -\frac{A-B}{3\pi^2}\left[\left(\frac{1}{45}+\frac{1}{3\pi^2}\left(B+1+\frac{7}{4}(A-B)\right)+\frac{1}{\pi^4}\left(2(B+1)^2+7(A-B)(B+1)+6(A-B)^2\right)\right)c_1^3\right. \\ & \left.-\left(\frac{1}{6}+\frac{2}{\pi^2}\left(B+1+\frac{7}{4}(A-B)\right)\right)c_2c_1+\frac{1}{2}c_3\right]. \end{aligned} \tag{26}$$

Now, from (24) and (25), one can have

$$|d_2| \le \frac{2}{\pi^2}(A-B)$$

and

$$\begin{aligned} |d_3| \le {} & \frac{A-B}{3\pi^2}\left|\frac{1}{6}+\frac{2}{\pi^2}(B+1)+\frac{4}{\pi^2}(A-B)\right|\left|c_2-c_1^2\right| \\ & +\frac{A-B}{3\pi^2}\left|\frac{5}{6}-\frac{2}{\pi^2}(B+1)-\frac{4}{\pi^2}(A-B)\right||c_2|. \end{aligned}$$

Application of the bounds $\left|c_2 - c_1^2\right| \leq 2$ and $|c_2| \leq 2$ (see Lemma 2 for $v = 1$ and $v = 0$) gives $|d_3| \leq \frac{2(A-B)}{3\pi^2}$. Lastly, (26) reduces to

$$|d_4| \leq \frac{A-B}{3\pi^2}\left[|\lambda_1|\left|c_3 - 2c_2c_1 + c_1^3\right| + |\lambda_2|\left|c_3 - c_2c_1\right| + |\lambda_3|\left|c_3\right|\right], \tag{27}$$

where

$$\lambda_1 = \frac{1}{45} + \frac{1}{3\pi^2}\left(B+1+\frac{7}{4}(A-B)\right) + \frac{1}{\pi^4}\left(2(B+1)^2 + 7(A-B)(B+1) + 6(A-B)^2\right),$$

$$\lambda_2 = \frac{11}{90} + \frac{4}{3\pi^2}\left(B+1+\frac{7}{4}(A-B)\right) - \frac{2}{\pi^4}\left(2(B+1)^2 + 7(A-B)(B+1) + 6(A-B)^2\right)$$

and

$$\lambda_3 = \frac{16}{45} - \frac{5}{3\pi^2}\left(B+1+\frac{7}{4}(A-B)\right) + \frac{1}{\pi^4}\left(2(B+1)^2 + 7(A-B)(B+1) + 6(A-B)^2\right).$$

Applying the bounds $\left|c_3 - 2c_2c_1 + c_1^3\right| \leq 2$, see [23], $|c_3 - c_2c_1| \leq 2$ and $|c_3| \leq 2$, see [7] to the right hand side of (27) and using the fact that $\lambda_i \geq 0$; $i = 1, 2, 3$, we have $|d_4| \leq \frac{A-B}{3\pi^2}$ and this completes the proof. □

For $A = 1$, $B = -1$, the above result takes the following form which is proved by Ma and Minda [8].

Corollary 4. *Let $w = f(z) \in UCV$ and $\mathcal{F}(w) = f^{-1}(w) = w + \sum_{n=2}^{\infty} d_n w^n$. Then,*

$$|d_n| \leq \frac{8}{n(n-1)\pi^2} \qquad (n = 2, 3, 4).$$

Theorem 5. *Let $w = f(z) \in UCV[A, B]$, $-1 \leq B < A \leq 1$ and $\mathcal{F}(w) = f^{-1}(w) = w + \sum_{n=2}^{\infty} d_n w^n$. Then, for a real number μ, we have*

$$\left|d_3 - \mu d_2^2\right| \leq \frac{A-B}{3\pi^2}\begin{cases} \frac{4}{3} - \frac{8}{\pi^2}(B+1) - \frac{4}{\pi^2}(A-B)(4-3\mu), & \mu \geq \frac{4}{3} + \frac{\pi^2}{18(A-B)} + \frac{2(B+1)}{3(A-B)}, \\ 2, & \frac{4}{3} - \frac{5\pi^2}{18(A-B)} + \frac{2(B+1)}{3(A-B)} \leq \mu \leq \frac{4}{3} + \frac{\pi^2}{18(A-B)} + \frac{2(B+1)}{3(A-B)}, \\ -\frac{4}{3} + \frac{8}{\pi^2}(B+1) + \frac{4}{\pi^2}(A-B)(4-3\mu), & \mu \leq \frac{4}{3} - \frac{5\pi^2}{18(A-B)} + \frac{2(B+1)}{3(A-B)}. \end{cases}$$

This result is sharp.

Proof. The proof follows directly from (24), (25) and Lemma 2. □

Author Contributions: Conceptualization, M.R.; Formal analysis, M.R.; Funding acquisition, S.M.; Investigation, S.F.; Methodology, S.N.M. and S.F.; Supervision, S.N.M.; Validation, S.M.; Visualization, S.Z.; Writing—original draft, S.Z.; Writing—review and editing, S.Z.

Funding: This research is partially supported by Sarhad University of Science and I.T, Ring Road, Peshawar 25000.

Conflicts of Interest: The authors declare no conflict of interest.

References

1. Fekete, M.; Szegö, G. Eine bemerkung uber ungerade schlichte funktionen. *J. Lond. Math. Soc.* **1933**, *8*, 85–89. [CrossRef]
2. Agrawal, S. Coefficient estimates for some classes of functions associated with q-function theory. *Bull. Aust. Math. Soc.* **2017**, *95*, 446–456. [CrossRef]
3. Ahuja, O.P.; Jahangiri, M. Fekete–Szegö problem for a unified class of analytic functions. *Panam. Math. J.* **1997**, *7*, 67–78.
4. Kanas, S. An unified approach to the Fekete–Szegö problem. *Appl. Math. Comput.* **2012**, *218*, 8453–8461. [CrossRef]
5. Keogh, F.R.; Merkes, E.P. A coefficient inequality for certain classes of analytic functions. *Proc. Am. Math. Soc.* **1969**, *20*, 8–12. [CrossRef]
6. Koepf, W. On the Fekete-Szegö problem for close to convex functions I and II. *Proc. Am. Math. Soc.* **1987**, *101*, 89–95; reprinted in *Arch. Math. (Basel)* **1987**, *49*, 420–433.
7. Ma, W.; Minda, D. A unified treatment of some special classes of univalent functions. In Proceedings of the Conference on Complex Analysis, Tianjin, China, 19–23 June 1992; Conference on Proceedings Lecture Notes for Analysis; International Press: Cambridge, MA, USA, 1994; pp. 157–169.
8. Ma, W.; Minda, D. Uniformly convex functions II. *Ann. Polon. Math.* **1993**, *8*, 275–285. [CrossRef]
9. Raza, M.; Malik, S.M. Upper bound of the third Hankel determinant for a class of analytic functions related with lemniscate of Bernoulli. *J. Inequal. Appl.* **2013**, *2013*, 412. [CrossRef]
10. Sokół J.; Darwish, H.E. Fekete–Szegö problem for starlike and convex functions of complex order. *Appl. Math. Lett.* **2010**, *23*, 777–782.
11. Thomas, D.K.; Verma, S. Invaience of the coefficients of strongly convex functions. *Bull. Aust. Math. Soc.* **2017**, *95*, 436–345. [CrossRef]
12. Goodman, A.W. *Univalent Functions*; Mariner Publishing Company: Tempa, FL, USA, 1983; Volumes I–II.
13. Goodman, A.W. On uniformly convex functions. *Ann. Polon. Math.* **1991**, *56*, 87–92. [CrossRef]
14. Rønning, F. On starlike functions associated with parabolic regions. *Ann. Univ. Mariae Curie-Sklodowska Sect. A* **1991**, *45*, 117–122.
15. Kanas, S.; Wiśniowska, A. Conic regions and k-uniform convexity. *J. Comput. Appl. Math.* **1999**, *105*, 327–336. [CrossRef]
16. Kanas, S.; Wiśniowska, A. Conic domains and starlike functions. *Rev. Roum. Math. Pures Appl.* **2000**, *45*, 647–657.
17. Malik, S.N. *Some topics in Geometric Function Theory*; LAP LAMBERT Academic Publishing: Saarbrucken, Germany, 2017.
18. Malik, S.N.; Raza, M.; Arif, M.; Hussain, S. Coefficient estimates of some subclasses of analytic functions related with conic domains. *Anal. Univ. Ovidius Const. Ser. Mat.* **2013**, *21*, 181–188. [CrossRef]
19. Noor, K.I.; Malik, S.N. On generalized bounded Mocanu variation associated with conic domain. *Math. Comput. Model.* **2012**, *55*, 844–852. [CrossRef]
20. Noor, K.I.; Malik, S.N. On coefficient inequalities of functions associated with conic domains. *Comput. Math. Appl.* **2011**, *62*, 2209–2217. [CrossRef]
21. Kanas, S. Coefficient estimates in subclasses of the Caratheodory class related to conical domains. *Acta Math. Univ. Comen.* **2005**, *74*, 149–161.
22. Rønning, F. Uniformly convex functions and a corresponding class of starlike functions. *Proc. Am. Math. Soc.* **1993**, *118*, 189–196. [CrossRef]
23. Libera, R.J.; Złotkiewcz, E.J. Early coefficients of the inverse of a regular convex function. *Proc. Amer. Math. Soc.* **1982**, *85*, 225–230. [CrossRef]

mathematics

MDPI

Article

Coefficient Inequalities of Functions Associated with Hyperbolic Domains

Sarfraz Nawaz Malik [1], Shahid Mahmood [2], Mohsan Raza [3,*], Sumbal Farman [1], Saira Zainab [4] and Nazeer Muhammad [1]

1 Department of Mathematics, COMSATS University Islamabad, Wah Campus 47040, Pakistan; snmalik110@yahoo.com (S.N.M.); sumbalfarman678@gmail.com (S.F.); nazeermuhammad@ciitwah.edu.pk (N.M.)
2 Department of Mechanical Engineering, Sarhad University of Science and I.T, Ring Road, Peshawar 25000, Pakistan; shahidmahmood757@gmail.com
3 Department of Mathematics, Government College University Faisalabad, Faisalabad 38000, Pakistan
4 Department of Mathematical Sciences, Fatima Jinnah Women University, Rawalpindi 46000, Pakistan; sairazainab07@yahoo.com
* Correspondence: mohsan976@yahoo.com

Received: 4 December 2018; Accepted: 8 January 2019; Published: 16 January 2019

Abstract: In this work, our focus is to study the Fekete-Szegö functional in a different and innovative manner, and to do this we find its upper bound for certain analytic functions which give hyperbolic regions as image domain. The upper bounds obtained in this paper give refinement of already known results. Moreover, we extend our work by calculating similar problems for the inverse functions of these certain analytic functions for the sake of completeness.

Keywords: analytic functions; starlike functions; convex functions; Fekete-Szegö inequality

MSC: 30C45, 33C10; Secondary: 30C20, 30C75

1. Introduction and Preliminaries

We consider the class of analytic functions f in the open unit disk $\mathcal{U} = \{z : |z| < 1\}$, defined as

$$f(z) = z + \sum_{n=2}^{\infty} a_n z^n. \tag{1}$$

We also consider $\mathcal{S}$, the class of those functions from $\mathcal{A}$ which are univalent in $\mathcal{U}$. Fekete-Szegö problem may be considered as one of the most important results about univalent functions, which is related to coefficients a_n of a function's taylor series and was introduced by Fekete and Szegö [1]. We state it as:

If $f \in \mathcal{S}$ and is of the form (1), then

$$\left|a_3 - \lambda a_2^2\right| \leq \begin{cases} 3 - 4\lambda, & \text{if} \quad \lambda \leq 0, \\ 1 + 2\exp\left(\frac{2\lambda}{\lambda - 1}\right), & \text{if} \quad 0 \leq \lambda \leq 1, \\ 4\lambda - 3, & \text{if} \quad \lambda \geq 1. \end{cases}$$

The problem of maximizing the absolute value of the functional $a_3 - \lambda a_2^2$ is called Fekete-Szegö problem. This result is sharp and is studied thoroughly by many researchers. The equality holds true for Koebe function. The case $0 < \lambda < 1$ provides an example of an extremal problem over $\mathcal{S}$ in which Koebe fails to be extremal. In this regard, one can find a number of results related to the maximization of the non-linear functional $\left|a_3 - \lambda a_2^2\right|$ for various classes and subclasses of univalent functions. Moreover, this functional has also been studied for λ as real as well as complex number. To maximize Fekete-Szegö functional $\left|a_3 - \lambda a_2^2\right|$ for different types of functions, showing interesting geometric characteristics of image domains, several authors used certain classified techniques. For in-depth understanding and more details, we refer the interested readers to study [1–11].

Subordination of two functions f and g is written symbolically as $f \prec g$, and is defined with respect to a schwarz function w such that $w(0) = 0$, $|w(z)| < 1$ for $z \in \mathcal{U}$, as

$$f(z) = g(w(z)), \quad z \in \mathcal{U}. \tag{2}$$

We now include P, the class of analytic functions p such that $p(0) = 1$ and $p \prec \frac{1+z}{1-z}$, $z \in \mathcal{U}$. For details, see [12].

Goodman [13] opened an altogether new area of research with the initiation of the concept of conic domain. He did it in 1991, by introducing parabolic region as image domain of analytic functions. Related to the same, he introduced the class UCV of uniformly convex functions and defined it as follows:

$$UCV = \left\{ f \in \mathcal{A} : \Re\left(1 + (z - \zeta)\frac{f''(z)}{f'(z)}\right) > 0, \ z, \zeta \in \mathcal{U} \right\}.$$

The most suitable one variable characterization of the above defined class UCV of Goodman was independently given by Rønning [14], and Ma and Minda [6]. They defined it as follows:

$$UCV = \left\{ f \in \mathcal{A} : \Re\left(1 + \frac{zf''(z)}{f'(z)}\right) > \left|\frac{zf''(z)}{f'(z)}\right|, \ z \in \mathcal{U} \right\}.$$

It proved its importance by giving birth to a domain, ever first of its kind, that is, conic (parabolic) domain, given as $\Omega = \{w : \Re w > |w - 1|\}$. Later on, $\beta-$uniformly convex functions were introduced by Kanas and Wiśniowska [15], which are defined as:

$$\beta - UCV = \left\{ f \in \mathcal{A} : \Re\left(1 + \frac{zf''(z)}{f'(z)}\right) > \beta\left|\frac{zf''(z)}{f'(z)}\right|, \ z \in \mathcal{U} \right\}.$$

This proved to be a remarkable innovation in this area since it gave the most general conic domain Ω_β, given as under, which covers parabolic as well as hyperbolic and elliptic regions.

$$\Omega_\beta = \{w : \Re w > \beta |w - 1|, \ \beta \geq 0\}.$$

For different values of β, the conic domain Ω_β, represents different image domains. For $\beta = 0$, this represents the right half plane, whereas hyperbolic regions when $0 < \beta < 1$, parabolic region for $\beta = 1$ and elliptic regions when $\beta > 1$. For further investigation, we refer to [15,16]. Another breakthrough occurred in this field when Noor and Malik [17] further generalized this domain Ω_β. They introduced the domain

$$\begin{aligned} \Omega_\beta[A,B] = \Big\{ u + iv : & \left[(B^2-1)(u^2+v^2) - 2(AB-1)u + (A^2-1)\right]^2 \\ & > \beta^2\left[\left(-2(B+1)(u^2+v^2) + 2(A+B+2)u - 2(A+1)\right)^2 + 4(A-B)^2 v^2\right] \Big\}. \end{aligned} \tag{3}$$

The class of functions given in the following definition takes all values from the above domain $\Omega_\beta[A,B]$, $-1 \leq B < A \leq 1$, $\beta \geq 0$. For more details, we refer to [17].

Definition 1. *A function $p(z)$ is said to be in the class $\beta - P[A,B]$, if and only if,*

$$p(z) \prec \frac{(A+1)\widetilde{p}_\beta(z) - (A-1)}{(B+1)\widetilde{p}_\beta(z) - (B-1)}, \quad -1 \leq B < A \leq 1, \ \beta \geq 0, \tag{4}$$

where $\widetilde{p}_\beta(z)$ *is defined by*

$$\widetilde{p}_\beta(z) = \begin{cases} \frac{1+z}{1-z}, & \beta = 0, \\ 1 + \frac{2}{\pi^2}\left(\log \frac{1+\sqrt{z}}{1-\sqrt{z}}\right)^2, & \beta = 1, \\ 1 + \frac{2}{1-\beta^2}\sinh^2\left[\left(\frac{2}{\pi}\arccos\beta\right)\operatorname{arctanh}\sqrt{z}\right], & 0 < \beta < 1, \\ 1 + \frac{1}{\beta^2-1}\sin\left(\frac{\pi}{2R(t)}\int_0^{\frac{u(z)}{\sqrt{t}}}\frac{1}{\sqrt{1-x^2}\sqrt{1-(tx)^2}}dx\right) + \frac{1}{\beta^2-1}, & \beta > 1, \end{cases} \tag{5}$$

where $u(z) = \frac{z-\sqrt{t}}{1-\sqrt{tz}}$, $t \in (0,1)$, $z \in \mathcal{U}$ *and* z *is chosen such that* $\beta = \cosh\left(\frac{\pi R'(t)}{4R(t)}\right)$, $R(t)$ *is the Legendre's complete elliptic integral of the first kind, and* $R'(t)$ *is complementary integral of* $R(t)$. *For more details about the function* $\widetilde{p}_\beta(z)$, *we refer the readers to [15,16].*

It may be noted that if we restrict the domain as $\Omega_\beta[1,-1] = \Omega_\beta$, then it becomes the conic domain defined by Kanas and Wiśniowska [15,16]. With the help of this important fact, we notice the following important connections of different well-known classes of analytic functions.

1. $\beta - P[A,B] \subset P\left(\frac{2\beta+1-A}{2\beta+1-B}\right)$, the class of functions with real part greater than $\frac{2\beta+1-A}{2\beta+1-B}$.
2. $\beta - P[1,-1] = \mathcal{P}(\widetilde{p}_\beta)$, the well-known class introduced by Kanas and Wiśniowska [15,16].
3. $0 - P[A,B] = P[A,B]$, the well-known class introduced by Janowski [18].

We now include the two very important classes $\beta - UCV[A,B]$ of $\beta-$uniformly Janowski functions and $\beta - ST[A,B]$ of corresponding $\beta-$Janowski starlike functions which are used in Section 2 of this paper. These are introduced in [17] and defined as follows.

Definition 2. *A function* $f \in \mathcal{A}$ *is said to be in the class* $\beta - UCV[A,B]$, $\beta \geq 0$, $-1 \leq B < A \leq 1$, *if and only if*,

$$\Re\left(\frac{(B-1)\frac{(zf'(z))'}{f'(z)} - (A-1)}{(B+1)\frac{(zf'(z))'}{f'(z)} - (A+1)}\right) > \beta\left|\frac{(B-1)\frac{(zf'(z))'}{f'(z)} - (A-1)}{(B+1)\frac{(zf'(z))'}{f'(z)} - (A+1)} - 1\right|,$$

or equivalently,

$$\frac{(zf'(z))'}{f'(z)} \in \beta - P[A,B]. \tag{6}$$

Definition 3. *A function* $f \in \mathcal{A}$ *is said to be in the class* $\beta - ST[A,B]$, $\beta \geq 0$, $-1 \leq B < A \leq 1$, *if and only if*,

$$\Re\left(\frac{(B-1)\frac{zf'(z)}{f(z)} - (A-1)}{(B+1)\frac{zf'(z)}{f(z)} - (A+1)}\right) > \beta\left|\frac{(B-1)\frac{zf'(z)}{f(z)} - (A-1)}{(B+1)\frac{zf'(z)}{f(z)} - (A+1)} - 1\right|,$$

or equivalently,

$$\frac{zf'(z)}{f(z)} \in \beta - P[A,B]. \tag{7}$$

It can easily be seen that $f(z) \in \beta - UCV[A,B] \iff zf'(z) \in \beta - ST[A,B]$. It is clear that $\beta - UCV[1,-1] = \beta - UCV$ and $\beta - ST[1,-1] = \beta - ST$, the well-known classes of β-uniformly convex and corresponding β-starlike functions respectively, introduced by Kanas and Wiśniowska [15,16].

As it is mentioned earlier that a number of well known researchers contributed in the development of this area of study, to mark the importance of our work in this stream of work, we take a quick review of what is done so far. In 1994, Ma and Minda [6] found the maximum bound of Fekete-Szegö functional $|a_3 - \lambda a_2^2|$ for the class UCV of uniformly convex functions whereas Kanas [19] solved the

Fekete-Szegö problem for the functions of class $\mathcal{P}\left(\widetilde{p}_{\beta}\right)$. Further, for the functions of classes $\beta - UCV$ and $\beta - ST$, the same problem was studies by Mishra and Gochhayat [20]. Keeping in view the ongoing research, our aim for this paper is to solve the classical Fekete-Szegö problem for the functions of classes $\beta - P\left[A,B\right]$, $\beta - UCV\left[A,B\right]$ and $\beta - ST\left[A,B\right]$. To prove our results, we need the following lemmas. For the proofs, one may study the reference [6].

Lemma 1. *If $p\left(z\right) = 1 + p_1 z + p_2 z^2 + \cdots$ is a function with positive real part in $\mathcal{U}$, then, for any complex number μ,*

$$\left|p_2 - \mu p_1^2\right| \leq 2\max\left\{1, \left|2\mu - 1\right|\right\}$$

and the result is sharp for the functions

$$p_0\left(z\right) = \frac{1+z}{1-z} \quad \text{or} \quad p_*\left(z\right) = \frac{1+z^2}{1-z^2}, \qquad \left(z \in \mathcal{U}\right).$$

Lemma 2. *If $p\left(z\right) = 1 + p_1 z + p_2 z^2 + \cdots$ is a function with positive real part in $\mathcal{U}$, then, for any real number v,*

$$\left|p_2 - v p_1^2\right| \leq \begin{cases} -4v + 2, & v \leq 0, \\ 2, & 0 \leq v \leq 1, \\ 4v - 2, & v \geq 1. \end{cases}$$

When $v < 0$ or $v > 1$, the equality holds if and only if $p\left(z\right)$ is $\frac{1+z}{1-z}$ or one of its rotations. If $0 < v < 1$, then, the equality holds if and only if $p\left(z\right) = \frac{1+z^2}{1-z^2}$ or one of its rotations. If $v = 0$, the equality holds if and only if,

$$p\left(z\right) = \left(\frac{1+\eta}{2}\right)\frac{1+z}{1-z} + \left(\frac{1-\eta}{2}\right)\frac{1-z}{1+z} \quad \left(0 \leq \eta \leq 1\right),$$

or one of its rotations. If $v = 1$, then, the equality holds if and only if $p\left(z\right)$ is reciprocal of one of the function such that equality holds in the case of $v = 0$. Although the above upper bound is sharp, when $0 < v < 1$, it can be improved as follows:

$$\left|p_2 - v p_1^2\right| + \left|p_1\right|^2 \leq 2 \qquad \left(0 < v \leq \frac{1}{2}\right)$$

and

$$\left|p_2 - v p_1^2\right| + \left(1 - v\right)\left|p_1\right|^2 \leq 2 \quad \left(\frac{1}{2} < v \leq 1\right).$$

2. Main Results

Theorem 1. *Let $p \in \beta - P\left[A,B\right], -1 \leq B < A \leq 1, 0 < \beta < 1$, and of the form $p\left(z\right) = 1 + \sum_{n=1}^{\infty} p_n z^n$. Then, for a complex number μ, we have*

$$\left|p_2 - \mu p_1^2\right| \leq \frac{\left(A-B\right)T^2}{1-\beta^2} \cdot \max\left(1, \left|\frac{\left(B+1\right)T^2}{\left(1-\beta^2\right)} + \mu\frac{\left(A-B\right)T^2}{\left(1-\beta^2\right)} - \frac{T^2}{3} - \frac{2}{3}\right|\right) \tag{8}$$

and for real number μ, we have

$$\left|p_2 - \mu p_1^2\right| \leq \frac{(A-B)T^2}{1-\beta^2} \begin{cases} \frac{2}{3} + \frac{T^2}{3} - \frac{(B+1)T^2}{1-\beta^2} - \frac{\mu(A-B)T^2}{1-\beta^2}, & \mu \leq -\frac{1-\beta^2}{3(A-B)T^2} - \frac{B+1}{A-B} + \frac{1-\beta^2}{3(A-B)}, \\ 1, & -\frac{1-\beta^2}{3(A-B)T^2} - \frac{B+1}{A-B} + \frac{1-\beta^2}{3(A-B)} \leq \mu \\ & \leq \frac{5\left(1-\beta^2\right)}{3(A-B)T^2} - \frac{B+1}{A-B} + \frac{1-\beta^2}{3(A-B)}, \\ -\frac{2}{3} - \frac{T^2}{3} + \frac{(B+1)T^2}{1-\beta^2} + \frac{\mu(A-B)T^2}{1-\beta^2}, & \mu \geq \frac{5\left(1-\beta^2\right)}{3(A-B)T^2} - \frac{B+1}{A-B} + \frac{1-\beta^2}{3(A-B)}, \end{cases} \tag{9}$$

where $T = T(\beta) = \frac{2}{\pi}\arccos(\beta)$ *and the equality in* (8) *holds for the functions*

$$p_1(z) = \frac{\frac{A+1}{1-\beta^2}\sinh^2\left[\left(\frac{2}{\pi}\arccos\beta\right)\operatorname{arctanh}\sqrt{z}\right]+1}{\frac{B+1}{1-\beta^2}\sinh^2\left[\left(\frac{2}{\pi}\arccos\beta\right)\operatorname{arctanh}\sqrt{z}\right]+1} \tag{10}$$

or

$$p_2(z) = \frac{\frac{A+1}{1-\beta^2}\sinh^2\left[\left(\frac{2}{\pi}\arccos\beta\right)\operatorname{arctanh}(z)\right]+1}{\frac{B+1}{1-\beta^2}\sinh^2\left[\left(\frac{2}{\pi}\arccos\beta\right)\operatorname{arctanh}(z)\right]+1}. \tag{11}$$

When $\mu < -\frac{1-\beta^2}{3(A-B)T^2} - \frac{B+1}{A-B} + \frac{1-\beta^2}{3(A-B)}$ *or* $\mu > \frac{5(1-\beta^2)}{3(A-B)T^2} - \frac{B+1}{A-B} + \frac{1-\beta^2}{3(A-B)}$*, the equality in* (9) *for the function* $p_1(z)$ *or one of its rotations. If* $-\frac{1-\beta^2}{3(A-B)T^2} - \frac{B+1}{A-B} + \frac{1-\beta^2}{3(A-B)} < \mu < \frac{5(1-\beta^2)}{3(A-B)T^2} - \frac{B+1}{A-B} + \frac{1-\beta^2}{3(A-B)}$*, then the equality in* (9) *holds for the function* $p_2(z)$ *or one of its rotations. If* $\mu = -\frac{1-\beta^2}{3(A-B)T^2} - \frac{B+1}{A-B} + \frac{1-\beta^2}{3(A-B)}$*, the equality in* (9) *holds for the function*

$$p_3(z) = \left(\frac{1+\eta}{2}\right)p_1(z) + \left(\frac{1-\eta}{2}\right)p_1(-z), \quad (0 \le \eta \le 1), \tag{12}$$

or one of its rotations. If $\mu = \frac{5(1-\beta^2)}{3(A-B)T^2} - \frac{B+1}{A-B} + \frac{1-\beta^2}{3(A-B)}$*, then, the equality in* (9) *holds for the function* $p(z)$ *which is reciprocal of one of the function such that equality holds in the case for* $\mu = -\frac{1-\beta^2}{3(A-B)T^2} - \frac{B+1}{A-B} + \frac{1-\beta^2}{3(A-B)}$.

Proof. For $h \in P$ and of the form $h(z) = 1 + \sum_{n=1}^{\infty} c_n z^n$, we consider

$$h(z) = \frac{1+w(z)}{1-w(z)},$$

where $w(z)$ is such that $w(0) = 0$ and $|w(z)| < 1$. It follows easily that

$$\begin{aligned} w(z) &= \frac{h(z)-1}{h(z)+1} \\ &= \frac{(1+c_1z+c_2z^2+c_3z^3+\cdots)-1}{(1+c_1z+c_2z^2+c_3z^3+\cdots)+1} \\ &= \frac{1}{2}c_1z + \left(\frac{1}{2}c_2 - \frac{1}{4}c_1^2\right)z^2 + \left(\frac{1}{2}c_3 - \frac{1}{2}c_2c_1 + \frac{1}{8}c_1^3\right)z^3 + \cdots. \end{aligned} \tag{13}$$

Now, if $\widetilde{p}_\beta(w(z)) = 1 + R_1(\beta)w(z) + R_2(\beta)w^2(z) + R_3(\beta)w^3(z) + \cdots$, then from (13), one may have

$$\begin{aligned} \widetilde{p}_\beta(w(z)) &= 1 + R_1(\beta)w(z) + R_2(\beta)w^2(z) + R_3(\beta)w^3(z) + \cdots, \\ &= 1 + R_1(\beta)\left(\frac{1}{2}c_1z + \left(\frac{1}{2}c_2 - \frac{1}{4}c_1^2\right)z^2 + \left(\frac{1}{2}c_3 - \frac{1}{2}c_2c_1 + \frac{1}{8}c_1^3\right)z^3 + \cdots\right) + \\ &\quad R_2(\beta)\left(\frac{1}{2}c_1z + \left(\frac{1}{2}c_2 - \frac{1}{4}c_1^2\right)z^2 + \left(\frac{1}{2}c_3 - \frac{1}{2}c_2c_1 + \frac{1}{8}c_1^3\right)z^3 + \cdots\right)^2 + \\ &\quad R_3(\beta)\left(\frac{1}{2}c_1z + \left(\frac{1}{2}c_2 - \frac{1}{4}c_1^2\right)z^2 + \left(\frac{1}{2}c_3 - \frac{1}{2}c_2c_1 + \frac{1}{8}c_1^3\right)z^3 + \cdots\right)^3 + \cdots, \end{aligned}$$

where $R_1(\beta)$, $R_2(\beta)$ and $R_3(\beta)$ are given by

$$R_1(\beta) = \frac{2T^2}{1-\beta^2},$$

$$R_2(\beta) = \frac{2T^2}{3(1-\beta^2)}\left(2+T^2\right),$$

$$R_3(\beta) = \frac{2T^2}{9(1-\beta^2)}\left(\frac{23}{5}+4T^2+\frac{2}{5}T^4\right),$$

and $T = T(\beta) = \frac{2}{\pi}\arccos(\beta)$, $0 < \beta < 1$, see [19]. Using these, the above series reduces to

$$\begin{aligned} \widetilde{p}_\beta(w(z)) &= 1+\frac{T^2}{1-\beta^2}c_1z+\frac{T^2}{1-\beta^2}\left(\left(T^2-1\right)\frac{1}{6}c_1^2+c_2\right)z^2+ \\ & \frac{T^2}{1-\beta^2}\left(\frac{1}{9}\left(\frac{2}{5}-\frac{1}{2}T^2+\frac{1}{10}T^4\right)c_1^3-\frac{1}{3}\left(1-T^2\right)c_2c_1+c_3\right)z^3+\cdots. \end{aligned} \tag{14}$$

Since $p \in \beta - P[A,B]$, $0<\beta<1$, so from relations (2), (4) and (14), one may have

$$\begin{aligned} p(z) &= \frac{(A+1)\widetilde{p}_\beta(w(z))-(A-1)}{(B+1)\widetilde{p}_\beta(w(z))-(B-1)} \\ &= 1+\frac{(A-B)}{2}\frac{T^2}{1-\beta^2}c_1z+\frac{(A-B)}{2}\frac{T^2}{1-\beta^2}\left(\frac{T^2c_1^2}{6}-\frac{1}{6}c_1^2-\frac{(B+1)T^2}{2(1-\beta^2)}c_1^2+c_2\right)z^2+\cdots. \end{aligned} \tag{15}$$

If $p(z) = 1+\sum_{n=1}^{\infty} p_nz^n$, then equating coefficients of like powers of z, we have

$$\begin{aligned} p_1 &= \frac{(A-B)}{2}\frac{T^2}{1-\beta^2}c_1, \\ p_2 &= \frac{(A-B)}{2}\frac{T^2}{1-\beta^2}\left(\frac{T^2c_1^2}{6}-\frac{1}{6}c_1^2-\frac{(B+1)T^2}{2(1-\beta^2)}c_1^2+c_2\right). \end{aligned}$$

Now for complex number μ, consider

$$p_2-\mu p_1^2 = \frac{(A-B)}{2}\frac{T^2}{1-\beta^2}\left(\frac{T^2c_1^2}{6}-\frac{1}{6}c_1^2-\frac{(B+1)T^2}{2(1-\beta^2)}c_1^2+c_2\right)-\mu\frac{(A-B)^2T^4}{4(1-\beta^2)^2}c_1^2.$$

This implies that

$$\left|p_2-\mu p_1^2\right| = \frac{(A-B)T^2}{2(1-\beta^2)}\left|c_2-c_1^2\left(\frac{1}{6}-\frac{T^2}{6}+\frac{(B+1)T^2}{2(1-\beta^2)}+\mu\frac{(A-B)T^2}{2(1-\beta^2)}\right)\right|. \tag{16}$$

Now using Lemma 1, we have

$$\left|p_2-\mu p_1^2\right| \leq \frac{(A-B)T^2}{2(1-\beta^2)}\cdot 2\max(1,|2v-1|),$$

where

$$v = \frac{1}{6}-\frac{T^2}{6}+\frac{(B+1)T^2}{2(1-\beta^2)}+\mu\frac{(A-B)T^2}{2(1-\beta^2)}.$$

This leads us to the required inequality (8) and applying Lemma 2 to the expression (16) for real number μ, we get the required inequality (9). □

For $A = 1$, $B = -1$, the above result reduces to the following form.

Corollary 1. *Let $p \in \beta - P[1,-1] = \mathcal{P}(\widetilde{p}_\beta)$, $0 < \beta < 1$, and of the form $p(z) = 1 + \sum_{n=1}^{\infty} p_n z^n$. Then, for a complex number μ, we have*

$$\left|p_2 - \mu p_1^2\right| \le \frac{2T^2}{1-\beta^2} \cdot \max\left(1, \left|\mu \frac{2T^2}{(1-\beta^2)} - \frac{T^2}{3} - \frac{2}{3}\right|\right) \tag{17}$$

and for real number μ, we have

$$\left|p_2 - \mu p_1^2\right| \le \frac{T^2}{1-\beta^2} \begin{cases} \frac{4}{3} + \frac{2}{3}T^2 - \frac{4\mu T^2}{1-\beta^2}, & \mu < -\frac{1-\beta^2}{6T^2} + \frac{(1-\beta^2)}{6}, \\ 2, & -\frac{(1-\beta^2)}{6T^2} + \frac{(1-\beta^2)}{6} \le \mu \le \frac{5(1-\beta^2)}{6T^2} + \frac{1-\beta^2}{6}, \\ -\frac{4}{3} - \frac{2}{3}T^2 + \frac{4\mu T^2}{1-\beta^2}, & \mu > \frac{5(1-\beta^2)}{6T^2} + \frac{1-\beta^2}{6}. \end{cases} \tag{18}$$

These results are sharp.

In [3,19], Kanas studied the class $\mathcal{P}(\widetilde{p}_\beta)$ which consists of functions who take all values from the conic domain Ω_β. Kanas [19] found the bound of Fekete-Szegö functional for the class $\mathcal{P}(\widetilde{p}_\beta)$ whose particular case for $0 < \beta < 1$ is as follows:

Let $p(z) = 1 + b_1 z + b_2 z^2 + b_3 z^3 + \cdots \in \mathcal{P}(\widetilde{p}_\beta)$, $0 < \beta < 1$. Then, for real number μ, we have

$$\left|b_2 - \mu b_1^2\right| \le \frac{2T^2}{1-\beta^2} \begin{cases} 1 - \mu \frac{2T^2}{1-\beta^2}, & \mu \le 0, \\ 1, & \mu \in (0,1], \\ 1 + (\mu - 1)\frac{2T^2}{1-\beta^2}, & \mu \ge 1. \end{cases} \tag{19}$$

For certain values of β and μ, we have the following bounds for $\left|p_2 - \mu p_1^2\right|$, shown in Table 1.

Table 1. Comparison of Fekete-Szegö inequalities.

β	μ	Bound from (18)	Bound from (19)
0.3	3	4.8652	5.51463
0.3	2	2.82267	3.47193
0.5	2	1.84841	2.5939
0.5	−1	2.37422	2.5939
0.7	3	2.28155	3.03221
0.7	−1	1.7698	2.01932

We observe that Corollary 1 gives more refined bounds of Fekete-Szegö functional $\left|p_2 - \mu p_1^2\right|$ for the functions of class $\mathcal{P}(\widetilde{p}_\beta)$, $0 < \beta < 1$ as compared to that from (19) as can be seen from above table.

Theorem 2. *Let $f \in \beta - UCV[A,B]$, $-1 \le B < A \le 1$, $0 \le \beta < 1$ and of the form (1), then for a real number μ, we have*

$$\left|a_3 - \mu a_2^2\right| \le \frac{(A-B)T^2}{12(1-\beta^2)} \begin{cases} \frac{4}{3} + \frac{2T^2}{3} - \frac{2(B+1)T^2}{1-\beta^2} + (2-3\mu)\frac{(A-B)T^2}{1-\beta^2}, & \mu \le \frac{2}{3} - \frac{2(1-\beta^2)}{9(A-B)T^2} - \frac{2(B+1)}{3(A-B)} + \frac{2(1-\beta^2)}{9(A-B)}, \\ 2, & \frac{2}{3} - \frac{2(1-\beta^2)}{9(A-B)T^2} - \frac{2(B+1)}{3(A-B)} + \frac{2(1-\beta^2)}{9(A-B)} \le \mu \\ & \le \frac{2}{3} + \frac{10(1-\beta^2)}{9(A-B)T^2} - \frac{2(B+1)}{3(A-B)} + \frac{2(1-\beta^2)}{9(A-B)}, \\ -\frac{4}{3} - \frac{2T^2}{3} + \frac{2(B+1)T^2}{1-\beta^2} - (2-3\mu)\frac{(A-B)T^2}{1-\beta^2}, & \mu \ge \frac{2}{3} + \frac{10(1-\beta^2)}{9(A-B)T^2} - \frac{2(B+1)}{3(A-B)} + \frac{2(1-\beta^2)}{9(A-B)}. \end{cases} \tag{20}$$

Proof. If $f(z) \in \beta - UCV[A,B]$, $-1 \le B < A \le 1$, $0 \le \beta < 1$, then it follows from relations (2), (4), and (6) that

$$\frac{(zf'(z))'}{f'(z)} = \frac{(A+1)\widetilde{p}_\beta(w(z)) - (A-1)}{(B+1)\widetilde{p}_\beta(w(z)) - (B-1)}.$$

This implies by using (15) that

$$\frac{(zf'(z))'}{f'(z)} = 1 + \frac{(A-B)}{2}\frac{T^2}{1-\beta^2}c_1 z + \frac{(A-B)}{2}\frac{T^2}{1-\beta^2}\left(\frac{T^2c_1^2}{6} - \frac{1}{6}c_1^2 - \frac{(B+1)T^2}{2(1-\beta^2)}c_1^2 + c_2\right)z^2 + \cdots. \quad (21)$$

If $f(z) = z + \sum_{n=2}^{\infty} a_n z^n$, then one may have

$$\frac{(zf'(z))'}{f'(z)} = 1 + 2a_2 z + \left(6a_3 - 4a_2^2\right)z^2 + \left(12a_4 - 18a_2a_3 + 8a_2^3\right)z^3 + \cdots. \quad (22)$$

From (21) and (22), comparison of like powers of z gives

$$a_2 = \frac{(A-B)T^2}{4(1-\beta^2)}c_1, \quad (23)$$

and

$$a_3 = \frac{(A-B)T^2}{12(1-\beta^2)}\left(c_2 - \left(\frac{1}{6} - \frac{T^2}{6} + \frac{(B+1)T^2}{2(1-\beta^2)} - \frac{(A-B)T^2}{2(1-\beta^2)}\right)c_1^2\right). \quad (24)$$

Now, for a real number μ, we consider

$$\begin{aligned}\left|a_3 - \mu a_2^2\right| &= \frac{(A-B)T^2}{12(1-\beta^2)}\left|c_2 - \left(\frac{1}{6} - \frac{T^2}{6} + \frac{(B+1)T^2}{2(1-\beta^2)} - \frac{(A-B)T^2}{2(1-\beta^2)}\right)c_1^2 - \mu\frac{3(A-B)}{4}\frac{T^2}{1-\beta^2}c_1^2\right| \\ &= \frac{(A-B)T^2}{12(1-\beta^2)}\left|c_2 - \left(\frac{1}{6} - \frac{T^2}{6} + \frac{(B+1)T^2}{2(1-\beta^2)} - \frac{(A-B)T^2}{2(1-\beta^2)} + \mu\frac{3(A-B)T^2}{4(1-\beta^2)}\right)c_1^2\right|.\end{aligned}$$

Now applying Lemma 2, we have the required result. The inequality (20) is sharp and equality holds for $\mu < \frac{2}{3} - \frac{2(1-\beta^2)}{9(A-B)T^2} - \frac{2(B+1)}{3(A-B)} + \frac{2(1-\beta^2)}{9(A-B)}$ or $\mu > \frac{2}{3} + \frac{10(1-\beta^2)}{9(A-B)T^2} - \frac{2(B+1)}{3(A-B)} + \frac{2(1-\beta^2)}{9(A-B)}$ when $f(z)$ is $f_1(z)$ or one of its rotations, where $f_1(z)$ is defined such that $\frac{(zf_1'(z))'}{f_1'(z)} = p_1(z)$. If $\frac{2}{3} - \frac{2(1-\beta^2)}{9(A-B)T^2} - \frac{2(B+1)}{3(A-B)} + \frac{2(1-\beta^2)}{9(A-B)} < \mu < \frac{2}{3} + \frac{10(1-\beta^2)}{9(A-B)T^2} - \frac{2(B+1)}{3(A-B)} + \frac{2(1-\beta^2)}{9(A-B)}$, then, the equality holds for the function $f_2(z)$ or one of its rotations, where $f_2(z)$ is defined such that $\frac{(zf_2'(z))'}{f_2'(z)} = p_2(z)$. If $\mu = \frac{2}{3} - \frac{2(1-\beta^2)}{9(A-B)T^2} - \frac{2(B+1)}{3(A-B)} + \frac{2(1-\beta^2)}{9(A-B)}$, the equality holds for the function $f_3(z)$ or one of its rotations, where $f_3(z)$ is defined such that $\frac{(zf_3'(z))'}{f_3'(z)} = p_3(z)$. If $\mu = \frac{2}{3} + \frac{10(1-\beta^2)}{9(A-B)T^2} - \frac{2(B+1)}{3(A-B)} + \frac{2(1-\beta^2)}{9(A-B)}$, then, the equality holds for $f(z)$, which is such that $\frac{(zf'(z))'}{f'(z)}$ is reciprocal of one of the function such that equality holds in the case of $\mu = \frac{2}{3} - \frac{2(1-\beta^2)}{9(A-B)T^2} - \frac{2(B+1)}{3(A-B)} + \frac{2(1-\beta^2)}{9(A-B)}$. □

For $A = 1$, $B = -1$, the above result takes the following form which is proved by Mishra and Gochhayat [20].

Corollary 2. *Let $f \in \beta - UCV[1,-1] = \beta - UCV$, $0 \le \beta < 1$ and of the form (1), then*

$$\left|a_3 - \mu a_2^2\right| \le \frac{T^2}{6(1-\beta^2)}\begin{cases} \frac{4}{3} + \frac{2T^2}{3} + (4-6\mu)\frac{T^2}{1-\beta^2}, & \mu \le \frac{2}{3} - \frac{1-\beta^2}{9T^2} + \frac{1-\beta^2}{9}, \\ 2, & \frac{2}{3} - \frac{1-\beta^2}{9T^2} + \frac{1-\beta^2}{9} \le \mu \le \frac{2}{3} + \frac{5(1-\beta^2)}{9T^2} + \frac{1-\beta^2}{9}, \\ -\frac{4}{3} - \frac{2T^2}{3} - (4-6\mu)\frac{T^2}{1-\beta^2}, & \mu \ge \frac{2}{3} + \frac{5(1-\beta^2)}{9T^2} + \frac{1-\beta^2}{9}. \end{cases}$$

Theorem 3. *If* $f(z) \in \beta - ST[A,B]$, $-1 \le B < A \le 1$, $0 < \beta < 1$ *and of the form* (1), *then for a real number* μ, *we have*

$$\left|a_3 - \mu a_2^2\right| \le \frac{(A-B)T^2}{2(1-\beta^2)} \begin{cases} \frac{2}{3} + \frac{T^2}{3} - \frac{(B+1)T^2}{1-\beta^2} + (1-2\mu)\frac{(A-B)T^2}{1-\beta^2}, & \mu \le \frac{1}{2} - \frac{1-\beta^2}{6T^2(A-B)} - \frac{B+1}{2(A-B)} + \frac{1-\beta^2}{6(A-B)}, \\ 1, & \frac{1}{2} - \frac{1-\beta^2}{6T^2(A-B)} - \frac{B+1}{2(A-B)} + \frac{1-\beta^2}{6(A-B)} \le \mu \\ & \le \frac{1}{2} + \frac{5(1-\beta^2)}{6(A-B)T^2} - \frac{B+1}{2(A-B)} + \frac{1-\beta^2}{6(A-B)}, \\ -\frac{2}{3} - \frac{T^2}{3} + \frac{(B+1)T^2}{1-\beta^2} - (1-2\mu)\frac{(A-B)T^2}{1-\beta^2}, & \mu \ge \frac{1}{2} + \frac{5(1-\beta^2)}{6(A-B)T^2} - \frac{B+1}{2(A-B)} + \frac{1-\beta^2}{6(A-B)}. \end{cases}$$

This result is sharp.

Proof. The proof follows similarly as in Theorem 2. □

For $A = 1$, $B = -1$, the above result takes the following form which is proved by Mishra and Gochhayat [20].

Corollary 3. *Let* $f \in \beta - ST[1,-1] = \beta - ST$, $0 < \beta < 1$ *and of the form* (1). *Then, for a real number* μ,

$$\left|a_3 - \mu a_2^2\right| \le \frac{T^2}{1-\beta^2} \begin{cases} \frac{2}{3} + \frac{T^2}{3} + (1-2\mu)\frac{2T^2}{1-\beta^2}, & \mu \le \frac{1}{2} - \frac{1-\beta^2}{12T^2} + \frac{1-\beta^2}{12}, \\ 1, & \frac{1}{2} - \frac{1-\beta^2}{12T^2} + \frac{1-\beta^2}{12} \le \mu \le \frac{1}{2} + \frac{5(1-\beta^2)}{12T^2} + \frac{1-\beta^2}{12}, \\ -\frac{2}{3} - \frac{T^2}{3} - (1-2\mu)\frac{2T^2}{1-\beta^2}, & \mu \ge \frac{1}{2} + \frac{5(1-\beta^2)}{12T^2} + \frac{1-\beta^2}{12}. \end{cases}$$

Now we consider the inverse function $\mathcal{F}$ which maps regions presented by (3) to the open unit disk $\mathcal{U}$, defined as $\mathcal{F}(w) = \mathcal{F}(f(z)) = z$, $z \in \mathcal{U}$ and we find the following coefficient bound for inverse functions. The functions of classes $\beta - UCV[A,B]$ and $\beta - ST[A,B]$ have inverses as they are univalent too.

Theorem 4. *Let* $w = f(z) \in \beta - UCV[A,B]$, $-1 \le B < A \le 1$, $0 \le \beta < 1$ *and* $\mathcal{F}(w) = f^{-1}(w) = w + \sum_{n=2}^{\infty} d_n w^n$. *Then,*

$$|d_n| \le \frac{(A-B)T^2}{2(1-\beta^2)} \qquad (n = 2,3).$$

Proof. Since $\mathcal{F}(w) = \mathcal{F}(f(z)) = z$, so it is easy to see that

$$d_2 = -a_2, \qquad d_3 = 2a_2^2 - a_3, \qquad d_4 = -a_4 + 5a_2a_3 - 5a_2^3.$$

By using (23) and (24), one can have

$$d_2 = -\frac{(A-B)T^2}{4(1-\beta^2)}c_1 \tag{25}$$

and

$$\begin{aligned} d_3 &= \frac{(A-B)T^2}{12(1-\beta^2)}\left[\left(\frac{1}{6} - \frac{T^2}{6} + \frac{(B+1)T^2}{2(1-\beta^2)} + \frac{(A-B)T^2}{1-\beta^2}\right)c_1^2 - c_2\right] \\ &= \frac{(A-B)T^2}{12(1-\beta^2)}\left(\frac{1}{6} - \frac{T^2}{6} + \frac{(B+1)T^2}{2(1-\beta^2)} + \frac{(A-B)T^2}{1-\beta^2}\right)(c_1^2 - c_2) \\ &\quad - \frac{(A-B)T^2}{12(1-\beta^2)}\left(\frac{11}{6} + \frac{T^2}{6} - \frac{(B+1)T^2}{2(1-\beta^2)} - \frac{(A-B)T^2}{1-\beta^2}\right)c_2 + \frac{(A-B)T^2}{12(1-\beta^2)}c_2. \end{aligned} \tag{26}$$

Now, from (25) and (26), one can have

$$|d_2| \leq \frac{(A-B)\,T^2}{2\,(1-\beta^2)}$$

and

$$\begin{aligned} |d_3| &\leq \frac{(A-B)\,T^2}{12\,(1-\beta^2)} \left| \frac{1}{6} - \frac{T^2}{6} + \frac{(B+1)\,T^2}{2\,(1-\beta^2)} + \frac{(A-B)\,T^2}{1-\beta^2} \right| \left| c_2 - c_1^2 \right| \\ &\quad + \frac{(A-B)\,T^2}{12\,(1-\beta^2)} \left| \frac{11}{6} + \frac{T^2}{6} - \frac{(B+1)\,T^2}{2\,(1-\beta^2)} - \frac{(A-B)\,T^2}{1-\beta^2} \right| |c_2| + \frac{(A-B)\,T^2}{12\,(1-\beta^2)} |c_2| \\ &= \frac{(A-B)\,T^2}{12\,(1-\beta^2)} \left\{ |\lambda_1| \left| c_2 - c_1^2 \right| + |\lambda_2|\,|c_2| + |c_2| \right\}, \end{aligned}$$

where $\lambda_1 = \frac{1}{6} - \frac{T^2}{6} + \frac{(B+1)T^2}{2(1-\beta^2)} + \frac{(A-B)T^2}{(1-\beta^2)}$ and $\lambda_2 = \frac{11}{6} + \frac{T^2}{6} - \frac{(B+1)T^2}{2(1-\beta^2)} - \frac{(A-B)T^2}{(1-\beta^2)}$. We see that $\lambda_i \geq 0$; $i = 1, 2$ for $-1 \leq B < A \leq 1, 0 \leq \beta < 1$. Thus, the application of bounds $\left| c_2 - c_1^2 \right| \leq 2$ and $|c_2| \leq 2$ (see Lemma 2 for $v = 1$ and $v = 0$) gives

$$\begin{aligned} |d_3| &\leq \frac{(A-B)\,T^2}{6\,(1-\beta^2)} \{\lambda_1 + \lambda_2 + 1\} \\ &= \frac{(A-B)\,T^2}{2\,(1-\beta^2)} \end{aligned}$$

□

Theorem 5. *Let $w = f(z) \in \beta - UCV\,[A, B]$, $-1 \leq B < A \leq 1, 0 \leq \beta < 1$ and $\mathcal{F}(w) = f^{-1}(w) = w + \sum_{n=2}^{\infty} d_n w^n$. Then, for a real number μ, we have*

$$\left| d_3 - \mu d_2^2 \right| \leq \frac{(A-B)\,T^2}{12\,(1-\beta^2)} \begin{cases} \frac{4}{3} + \frac{2T^2}{3} - \frac{2(B+1)T^2}{1-\beta^2} - (4-3\mu)\,\frac{(A-B)T^2}{1-\beta^2}, & \mu \geq \frac{4}{3} + \frac{2(1-\beta^2)}{9(A-B)T^2} - \frac{2(1-\beta^2)}{9(A-B)} + \frac{2(B+1)}{3(A-B)}, \\ 2, & \frac{4}{3} - \frac{10(1-\beta^2)}{9(A-B)T^2} - \frac{2(1-\beta^2)}{9(A-B)} + \frac{2(B+1)}{3(A-B)} \leq \mu \\ & \leq \frac{4}{3} + \frac{2(1-\beta^2)}{9(A-B)T^2} - \frac{2(1-\beta^2)}{9(A-B)} + \frac{2(B+1)}{3(A-B)}, \\ -\frac{4}{3} - \frac{2T^2}{3} + \frac{2(B+1)T^2}{1-\beta^2} + (4-3\mu)\,\frac{(A-B)T^2}{1-\beta^2}, & \mu \leq \frac{4}{3} - \frac{10(1-\beta^2)}{9(A-B)T^2} - \frac{2(1-\beta^2)}{9(A-B)} + \frac{2(B+1)}{3(A-B)}. \end{cases}$$

This result is sharp.

Proof. The proof follows directly from (25), (26), and Lemma 2. □

Author Contributions: Conceptualization, M.R.; Formal analysis, S.N.M. and M.R.; Funding acquisition, S.M.; Investigation, S.F.; Methodology, S.N.M., M.R. and S.F.; Supervision, S.N.M.; Validation, S.M., S.Z. and N.M.; Visualization, S.Z.; Writing—original draft, S.Z.; Writing—review & editing, S.Z.

Funding: This research received no external funding.

Acknowledgments: The authors are grateful to the referees for their valuable comments and suggestions which improved the presentation of paper and quality of work.

Conflicts of Interest: The authors declare no conflict of interest.

References

1. Fekete, M.; Szegö, G. Eine bemerkung uber ungerade schlichte funktionen. *J. Lond. Math. Soc.* **1933**, *8*, 85–89. [CrossRef]
2. Ahuja, O.P.; Jahangiri, M. Fekete–Szegö problem for a unified class of analytic functions, Panamer. *Math. J.* **1997**, *7*, 67–78.
3. Kanas, S. An unified approach to the Fekete–Szegö problem. *Appl. Math. Comput.* **2012**, *218*, 8453–8461. [CrossRef]
4. Keogh, F.R.; Merkes, E.P. A coefficient inequality for certain classes of analytic functions. *Proc. Am. Math. Soc.* **1969**, *20*, 8–12. [CrossRef]
5. Koepf, W. On the Fekete-Szegö problem for close to convex functions I and II. *Arch. Math. (Basel)* **1987**, *49*, 420–433. [CrossRef]
6. Ma, W.; Minda, D. A unified treatment of some special classes of univalent functions. In Proceedings of the Conference On Complex Analysis, Tianjin, China, 19–23 June 1992; pp. 157–169.
7. Ma, W.; Minda, D. Uniformly convex functions II. *Ann. Polon. Math.* **1993**, *8*, 275–285. [CrossRef]
8. Raza, M.; Malik, S.M. Upper bound of the third Hankel determinant for a class of analytic functions related with lemniscate of Bernoulli. *J. Inequal. Appl.* **2013**, *2013*, 412. [CrossRef]
9. Raina, R.K.; Sokół, J. On coefficient estimates for a certain class of starlike functions. *Hacett. J. Math. Stat.* **2015**, *44*, 1427–1433. [CrossRef]
10. Sokół, J.; Darwish, H.E. Fekete–Szegö problem for starlike and convex functions of complex order. *Appl. Math. Lett.* **2010**, *23*, 777–782.
11. Sokół, J.; Raina, R.K.; Özgür, N.Y. Applications of k-Fibonacci numbers for the starlike analytic functions. *Hacett. J. Math. Stat.* **2015**, *44*, 121–127.
12. Goodman, A.W. *Univalent Functions*; vols. I–II; Mariner Publishing Company: Tempa, FL, USA, 1983.
13. Goodman, A.W. On uniformly convex functions. *Ann. Polon. Math.* **1991**, *56*, 87–92. [CrossRef]
14. Rønning, F. On starlike functions associated with parabolic regions. *Ann. Univ. Mariae Curie-Sklodowska Sect A* **1991**, *45*, 117–122.
15. Kanas, S.; Wiśniowska, A. Conic regions and k-uniform convexity. *J. Comput. Appl. Math.* **1999**, *105*, 327–336. [CrossRef]
16. Kanas, S.; Wiśniowska, A. Conic domains and starlike functions. *Rev. Roum. Math. Pures Appl.* **2000**, *45*, 647–657.
17. Noor, K.I.; Malik, S.N. On coefficient inequalities of functions associated with conic domains. *Comput. Math. Appl.* **2011**, *62*, 2209–2217. [CrossRef]
18. Janowski, W. Some extremal problems for certain families of analytic functions. *Ann. Polon. Math.* **1973**, *28*, 297–326. [CrossRef]
19. Kanas, S. Coefficient estimates in subclasses of the Caratheodory class related to conical domains. *Acta Math. Univ. Comen.* **2005**, *74*, 149–161.
20. Mishra, A.K.; Gochhayat, P. The Fekete–Szegö problem for k-uniformly convex functions and for a class defined by the Owa–Srivastava operator. *J. Math. Anal. Appl.* **2008**, *347*, 563–572. [CrossRef]

mathematics

MDPI

Article

Some Inequalities for g-Frames in Hilbert C^*-Modules

Zhong-Qi Xiang

College of Mathematics and Computer Science, Shangrao Normal University, Shangrao 334001, China; lxsy20110927@163.com; Tel.: +86-793-815-9108

Received: 8 November 2018; Accepted: 26 December 2018 ; Published: 27 December 2018

Abstract: In this paper, we obtain new inequalities for g-frames in Hilbert C^*-modules by using operator theory methods, which are related to a scalar $\lambda \in \mathbb{R}$ and an adjointable operator with respect to two g-Bessel sequences. It is demonstrated that our results can lead to several known results on this topic when suitable scalars and g-Bessel sequences are chosen.

Keywords: Hilbert C^*-module; g-frame; g-Bessel sequence; adjointable operator

MSC: 46L08; 42C15; 47B48; 46H25

---

1. Introduction

Since their appearance in the literature [1] on nonharmonic Fourier series, frames for Hilbert spaces have been a useful tool and applied to different branches of mathematics and other fields. For details on frames, the reader can refer to the papers [2–11]. The author in [12] extended the concept of frames to bounded linear operators and thus gave us the notion of g-frames, which possess some properties that are quite different from those of frames (see [13,14]).

In the past decade, much attention has been paid to the extension of frame and g-frame theory from Hilbert spaces to Hilbert C^*-modules, and some significant results have been presented (see [15–23]). It should be pointed out that, due to the essential differences between Hilbert spaces and Hilbert C^*-modules and the complex structure of the C^*-algebra involved in a Hilbert C^*-module, the problems on frames and g-frames for Hilbert C^*-modules are expected to be more complicated than those for Hilbert spaces. Also, increasingly more evidence is indicating that there is a close relationship between the theory of wavelets and frames and Hilbert C^*-modules in many aspects. This suggests that the discussion of frame and g-frame theory in Hilbert C^*-modules is interesting and important.

The authors in [24] provided a surprising inequality while further discussing the remarkable identity for Parseval frames derived from their research on effective algorithms to compute the reconstruction of a signal, which was later generalized to the situation of general frames and dual frames [25]. Those inequalities have already been extended to several generalized versions of frames in Hilbert spaces [26–28]. Moreover, the authors in [29–31] showed that g-frames in Hilbert C^*-modules have their inequalities based on the work in [24,25]; it is worth noting that the inequalities given in [30] are associated with a scalar in $[0,1]$ or $[\frac{1}{2},1]$. In this paper, we establish several new inequalities for g-frames in Hilbert C^*-modules, where a scalar λ in $\mathbb{R}$, the real number set, and an adjointable operator with respect to two g-Bessel sequences are involved. Also, we show that some corresponding results in [29,31] can be considered a special case of our results.

We continue with this section for a review of some notations and definitions.

This paper adopts the following notations: $\mathbb{J}$ and $\mathcal{A}$ are, respectively, a finite or countable index set and a unital C^*-algebra; $\mathcal{H}$, $\mathcal{K}$, and $\mathcal{K}_j$'s ($j \in \mathbb{J}$) are Hilbert C^*-modules over $\mathcal{A}$ (or simply Hilbert $\mathcal{A}$-modules), setting $\langle f,f\rangle = |f|^2$ for any $f \in \mathcal{H}$. The family of all adjointable operators from $\mathcal{H}$ to $\mathcal{K}$ is designated $\mathrm{End}^*_{\mathcal{A}}(\mathcal{H},\mathcal{K})$, which is abbreviated to $\mathrm{End}^*_{\mathcal{A}}(\mathcal{H})$ if $\mathcal{K} = \mathcal{H}$.

A sequence $\Lambda = \{\Lambda_j \in \mathrm{End}^*_{\mathcal{A}}(\mathcal{H}, \mathcal{K}_j)\}_{j\in\mathbb{J}}$ denotes a g-frame for $\mathcal{H}$ with respect to $\{\mathcal{K}_j\}_{j\in\mathbb{J}}$ if there are real numbers $0 < C \le D < \infty$ satisfying

$$C\langle f, f\rangle \le \sum_{j\in\mathbb{J}} \langle \Lambda_j f, \Lambda_j f\rangle \le D\langle f, f\rangle, \quad \forall f \in \mathcal{H}. \tag{1}$$

If only the second inequality in Equation (1) is required, then Λ is said to be a g-Bessel sequence.

For a given g-frame $\Lambda = \{\Lambda_j \in \mathrm{End}^*_{\mathcal{A}}(\mathcal{H}, \mathcal{K}_j)\}_{j\in\mathbb{J}}$, there is always a positive, invertible, and self-adjoint operator in $\mathrm{End}^*_{\mathcal{A}}(\mathcal{H})$, which we call the g-frame operator of Λ, defined by

$$S_\Lambda : \mathcal{H} \to \mathcal{H}, \quad S_\Lambda f = \sum_{j\in\mathbb{J}} \Lambda_j^* \Lambda_j f. \tag{2}$$

For any $\mathbb{I} \subset \mathbb{J}$, let $\mathbb{I}^c$ be the complement of $\mathbb{I}$. We define a positive and self-adjoint operator in $\mathrm{End}^*_{\mathcal{A}}(\mathcal{H})$ related to $\mathbb{I}$ and a g-frame $\Lambda = \{\Lambda_j \in \mathrm{End}^*_{\mathcal{A}}(\mathcal{H}, \mathcal{K}_j)\}_{j\in\mathbb{J}}$ in the following form

$$S^\Lambda_{\mathbb{I}} : \mathcal{H} \to \mathcal{H}, \quad S^\Lambda_{\mathbb{I}} f = \sum_{j\in\mathbb{I}} \Lambda_j^* \Lambda_j f. \tag{3}$$

Recall that a g-Bessel $\Gamma = \{\Gamma_j \in \mathrm{End}^*_{\mathcal{A}}(\mathcal{H}, \mathcal{K}_j)\}_{j\in\mathbb{J}}$ is an alternate dual g-frame of Λ if, for every $f \in \mathcal{H}$, we have $f = \sum_{j\in\mathbb{J}} \Lambda_j^* \Gamma_j f$.

Let $\Lambda = \{\Lambda_j\}_{j\in\mathbb{J}}$ and $\Gamma = \{\Gamma_j\}_{j\in\mathbb{J}}$ be g-Bessel sequences for $\mathcal{H}$ with respect to $\{\mathcal{K}_j\}_{j\in\mathbb{J}}$. We observe from the Cauchy–Schwarz inequality that the operator

$$S_{\Gamma\Lambda} : \mathcal{H} \to \mathcal{H}, \quad S_{\Gamma\Lambda} f = \sum_{j\in\mathbb{J}} \Gamma_j^* \Lambda_j f \tag{4}$$

is well defined, and a direct calculation shows that $S_{\Gamma\Lambda} \in \mathrm{End}^*_{\mathcal{A}}(\mathcal{H})$.

2. The Main Results

The following result for operators is used to prove our main results.

Lemma 1. *Suppose that $U, V, L \in \mathrm{End}^*_{\mathcal{A}}(\mathcal{H})$ and that $U + V = L$. Then, for any $\lambda \in \mathbb{R}$, we have*

$$U^*U + \frac{\lambda}{2}(V^*L + L^*V) = V^*V + (1 - \frac{\lambda}{2})(U^*L + L^*U) + (\lambda - 1)L^*L \ge (\lambda - \frac{\lambda^2}{4})L^*L.$$

Proof. On the one hand, we obtain

$$U^*U + \frac{\lambda}{2}(V^*L + L^*V) = U^*U + \frac{\lambda}{2}((L^* - U^*)L + L^*(L - U)) = U^*U - \frac{\lambda}{2}(U^*L + L^*U) + \lambda L^*L.$$

On the other hand, we have

$$\begin{aligned}
&V^*V + (1 - \frac{\lambda}{2})(U^*L + L^*U) + (\lambda - 1)L^*L \\
&= (L^* - U^*)(L - U) + (U^*L + L^*U) - \frac{\lambda}{2}(U^*L + L^*U) + (\lambda - 1)L^*L \\
&= L^*L - (U^*L + L^*U) + U^*U + (U^*L + L^*U) - \frac{\lambda}{2}(U^*L + L^*U) + (\lambda - 1)L^*L \\
&= U^*U - \frac{\lambda}{2}(U^*L + L^*U) + \lambda L^*L = (U - \frac{\lambda}{2}L)^*(U - \frac{\lambda}{2}L) + (\lambda - \frac{\lambda^2}{4})L^*L \ge (\lambda - \frac{\lambda^2}{4})L^*L.
\end{aligned}$$

This completes the proof. □

Theorem 1. *Let $\Lambda = \{\Lambda_j\}_{j\in\mathbb{J}}$ be a g-frame for $\mathcal{H}$ with respect to $\{\mathcal{K}_j\}_{j\in\mathbb{J}}$. Suppose that $\Gamma = \{\Gamma_j\}_{j\in\mathbb{J}}$ and $\Theta = \{\Theta_j\}_{j\in\mathbb{J}}$ are two g-Bessel sequences for $\mathcal{H}$ with respect to $\{\mathcal{K}_j\}_{j\in\mathbb{J}}$, and that the operator $S_{\Gamma\Lambda}$ is defined in Equation (4). Then, for any $\lambda \in \mathbb{R}$ and any $f \in \mathcal{H}$, we have*

$$\begin{aligned}\Big|\sum_{j\in\mathbb{J}}(\Gamma_j-\Theta_j)^*\Lambda_j f\Big|^2 + \sum_{j\in\mathbb{J}}\langle\Lambda_j f, \Theta_j S_{\Gamma\Lambda} f\rangle &= \Big|\sum_{j\in\mathbb{J}}\Theta_j^*\Lambda_j f\Big|^2 + \sum_{j\in\mathbb{J}}\langle(\Gamma_j-\Theta_j)S_{\Gamma\Lambda} f, \Lambda_j f\rangle \\ &\geq (\lambda - \frac{\lambda^2}{4})\sum_{j\in\mathbb{J}}\langle\Lambda_j f, (\Gamma_j-\Theta_j)S_{\Gamma\Lambda} f\rangle + (1+\frac{\lambda}{2}-\frac{\lambda^2}{4})\sum_{j\in\mathbb{J}}\langle\Lambda_j f, \Theta_j S_{\Gamma\Lambda} f\rangle \\ &\quad - \frac{\lambda}{2}\sum_{j\in\mathbb{J}}\langle\Theta_j S_{\Gamma\Lambda} f, \Lambda_j f\rangle. \end{aligned} \tag{5}$$

Proof. We let

$$Uf = \sum_{j\in\mathbb{J}}(\Gamma_j-\Theta_j)^*\Lambda_j f \quad \text{and} \quad Vf = \sum_{j\in\mathbb{J}}\Theta_j^*\Lambda_j f \tag{6}$$

for each $f \in \mathcal{H}$. Then, $U, V \in \mathrm{End}^*_{\mathcal{A}}(\mathcal{H})$ and, further,

$$Uf + Vf = \sum_{j\in\mathbb{J}}(\Gamma_j-\Theta_j)^*\Lambda_j f + \sum_{j\in\mathbb{J}}\Theta_j^*\Lambda_j f = \sum_{j\in\mathbb{J}}\Gamma_j^*\Lambda_j f = S_{\Gamma\Lambda} f.$$

By Lemma 1, we get

$$\begin{aligned}&|Uf|^2 + \frac{\lambda}{2}(\langle Vf, S_{\Gamma\Lambda} f\rangle + \langle S_{\Gamma\Lambda} f, Vf\rangle) \\ &\quad = |Vf|^2 + (1-\frac{\lambda}{2})(\langle Uf, S_{\Gamma\Lambda} f\rangle + \langle S_{\Gamma\Lambda} f, Uf\rangle) + (\lambda-1)|S_{\Gamma\Lambda} f|^2.\end{aligned}$$

Hence,

$$\begin{aligned}|Uf|^2 &= |Vf|^2 + (1-\frac{\lambda}{2})(\langle Uf, S_{\Gamma\Lambda} f\rangle + \langle S_{\Gamma\Lambda} f, Uf\rangle) + (\lambda-1)|S_{\Gamma\Lambda} f|^2 \\ &\quad - \frac{\lambda}{2}(\langle Vf, S_{\Gamma\Lambda} f\rangle + \langle S_{\Gamma\Lambda} f, Vf\rangle) \\ &= |Vf|^2 + \langle Uf, S_{\Gamma\Lambda} f\rangle + \langle S_{\Gamma\Lambda} f, Uf\rangle - \frac{\lambda}{2}(\langle Uf, S_{\Gamma\Lambda} f\rangle + \langle S_{\Gamma\Lambda} f, Uf\rangle) \\ &\quad - \frac{\lambda}{2}(\langle Vf, S_{\Gamma\Lambda} f\rangle + \langle S_{\Gamma\Lambda} f, Vf\rangle) + (\lambda-1)|S_{\Gamma\Lambda} f|^2 \\ &= |Vf|^2 + \langle Uf, S_{\Gamma\Lambda} f\rangle + \langle S_{\Gamma\Lambda} f, Uf\rangle - \frac{\lambda}{2}(\langle Uf, S_{\Gamma\Lambda} f\rangle + \langle Vf, S_{\Gamma\Lambda} f\rangle) \\ &\quad - \frac{\lambda}{2}(\langle S_{\Gamma\Lambda} f, Uf\rangle + \langle S_{\Gamma\Lambda} f, Vf\rangle) + (\lambda-1)|S_{\Gamma\Lambda} f|^2 \\ &= |Vf|^2 + \langle Uf, S_{\Gamma\Lambda} f\rangle + \langle S_{\Gamma\Lambda} f, Uf\rangle - \lambda|S_{\Gamma\Lambda} f|^2 + (\lambda-1)|S_{\Gamma\Lambda} f|^2 \\ &= |Vf|^2 + \langle Uf, S_{\Gamma\Lambda} f\rangle + \langle S_{\Gamma\Lambda} f, Uf\rangle - \langle Uf, S_{\Gamma\Lambda} f\rangle - \langle Vf, S_{\Gamma\Lambda} f\rangle.\end{aligned}$$

It follows that

$$|Uf|^2 + \langle Vf, S_{\Gamma\Lambda} f\rangle = |Vf|^2 + \langle S_{\Gamma\Lambda} f, Uf\rangle, \tag{7}$$

from which we arrive at

$$\Big|\sum_{j\in\mathbb{J}}(\Gamma_j-\Theta_j)^*\Lambda_j f\Big|^2 + \sum_{j\in\mathbb{J}}\langle\Lambda_j f, \Theta_j S_{\Gamma\Lambda} f\rangle = \Big|\sum_{j\in\mathbb{J}}\Theta_j^*\Lambda_j f\Big|^2 + \sum_{j\in\mathbb{J}}\langle(\Gamma_j-\Theta_j)S_{\Gamma\Lambda} f, \Lambda_j f\rangle.$$

We are now in a position to prove the inequality in Equation (5).

Again by Lemma 1,

$$
\begin{aligned}
|Uf|^2 &\geq (\lambda - \frac{\lambda^2}{4})|S_{\Gamma\Lambda}f|^2 - \frac{\lambda}{2}(\langle Vf, S_{\Gamma\Lambda}f\rangle + \langle S_{\Gamma\Lambda}f, Vf\rangle) \\
&= (\lambda - \frac{\lambda^2}{4})\langle Uf, S_{\Gamma\Lambda}f\rangle + (\lambda - \frac{\lambda^2}{4})\langle Vf, S_{\Gamma\Lambda}f\rangle - \frac{\lambda}{2}\langle Vf, S_{\Gamma\Lambda}f\rangle - \frac{\lambda}{2}\langle S_{\Gamma\Lambda}f, Vf\rangle \\
&= (\lambda - \frac{\lambda^2}{4})\langle Uf, S_{\Gamma\Lambda}f\rangle + (\frac{\lambda}{2} - \frac{\lambda^2}{4})\langle Vf, S_{\Gamma\Lambda}f\rangle - \frac{\lambda}{2}\langle S_{\Gamma\Lambda}f, Vf\rangle. \qquad (8)
\end{aligned}
$$

Therefore,

$$
\begin{aligned}
&\left|\sum_{j\in\mathbb{J}}(\Gamma_j - \Theta_j)^*\Lambda_j f\right|^2 + \sum_{j\in\mathbb{J}}\langle \Lambda_j f, \Theta_j S_{\Gamma\Lambda}f\rangle = |Uf|^2 + \langle Vf, S_{\Gamma\Lambda}f\rangle \\
&\geq (\lambda - \frac{\lambda^2}{4})\langle Uf, S_{\Gamma\Lambda}f\rangle + (1 + \frac{\lambda}{2} - \frac{\lambda^2}{4})\langle Vf, S_{\Gamma\Lambda}f\rangle - \frac{\lambda}{2}\langle S_{\Gamma\Lambda}f, Vf\rangle \\
&= (\lambda - \frac{\lambda^2}{4})\sum_{j\in\mathbb{J}}\langle \Lambda_j f, (\Gamma_j - \Theta_j)S_{\Gamma\Lambda}f\rangle + (1 + \frac{\lambda}{2} - \frac{\lambda^2}{4})\sum_{j\in\mathbb{J}}\langle \Lambda_j f, \Theta_j S_{\Gamma\Lambda}f\rangle - \frac{\lambda}{2}\sum_{j\in\mathbb{J}}\langle \Theta_j S_{\Gamma\Lambda}f, \Lambda_j f\rangle
\end{aligned}
$$

for any $f \in \mathcal{H}$. □

Corollary 1. *Suppose that $\Lambda = \{\Lambda_j\}_{j\in\mathbb{J}}$ is a g-frame for $\mathcal{H}$ with respect to $\{\mathcal{K}_j\}_{j\in\mathbb{J}}$ with g-frame operator S_Λ and that $\tilde{\Lambda}_j = \Lambda_j S_\Lambda^{-1}$ for each $j \in \mathbb{J}$. Then, for any $\lambda \in \mathbb{R}$, for all $\mathbb{I} \subset \mathbb{J}$ and all $f \in \mathcal{H}$, we have*

$$
\begin{aligned}
\sum_{j\in\mathbb{I}}\langle \Lambda_j f, \Lambda_j f\rangle + \sum_{j\in\mathbb{J}}\langle \tilde{\Lambda}_j S_{\mathbb{I}^c}^{\Lambda} f, \tilde{\Lambda}_j S_{\mathbb{I}^c}^{\Lambda} f\rangle &= \sum_{j\in\mathbb{I}^c}\langle \Lambda_j f, \Lambda_j f\rangle + \sum_{j\in\mathbb{J}}\langle \tilde{\Lambda}_j S_{\mathbb{I}}^{\Lambda} f, \tilde{\Lambda}_j S_{\mathbb{I}}^{\Lambda} f\rangle \\
&\geq (\lambda - \frac{\lambda^2}{4})\sum_{j\in\mathbb{I}^c}\langle \Lambda_j f, \Lambda_j f\rangle + (1 - \frac{\lambda^2}{4})\sum_{j\in\mathbb{I}}\langle \Lambda_j f, \Lambda_j f\rangle.
\end{aligned}
$$

Proof. Taking $\Gamma_j = \Lambda_j S_\Lambda^{-\frac{1}{2}}$ for any $j \in \mathbb{J}$, then it is easy to see that $S_{\Gamma\Lambda} = S_\Lambda^{\frac{1}{2}}$. For each $j \in \mathbb{J}$, let

$$
\Theta_j = \begin{cases} \Gamma_j, & j \in \mathbb{I}, \\ 0, & j \in \mathbb{I}^c. \end{cases}
$$

Now, for each $f \in \mathcal{H}$,

$$
\begin{aligned}
\left|\sum_{j\in\mathbb{J}}(\Gamma_j - \Theta_j)^*\Lambda_j f\right|^2 &= \left|\sum_{j\in\mathbb{I}^c} S_\Lambda^{-\frac{1}{2}}\Lambda_j^*\Lambda_j f\right|^2 = |S_\Lambda^{-\frac{1}{2}} S_{\mathbb{I}^c}^{\Lambda} f|^2 = \langle S_\Lambda^{-\frac{1}{2}} S_{\mathbb{I}^c}^{\Lambda} f, S_\Lambda^{-\frac{1}{2}} S_{\mathbb{I}^c}^{\Lambda} f\rangle \\
&= \langle S_{\mathbb{I}^c}^{\Lambda} f, S_\Lambda^{-1} S_{\mathbb{I}^c}^{\Lambda} f\rangle = \langle S_\Lambda S_\Lambda^{-1} S_{\mathbb{I}^c}^{\Lambda} f, S_\Lambda^{-1} S_{\mathbb{I}^c}^{\Lambda} f\rangle \\
&= \sum_{j\in\mathbb{J}}\langle \Lambda_j S_\Lambda^{-1} S_{\mathbb{I}^c}^{\Lambda} f, \Lambda_j S_\Lambda^{-1} S_{\mathbb{I}^c}^{\Lambda} f\rangle = \sum_{j\in\mathbb{J}}\langle \tilde{\Lambda}_j S_{\mathbb{I}^c}^{\Lambda} f, \tilde{\Lambda}_j S_{\mathbb{I}^c}^{\Lambda} f\rangle. \qquad (9)
\end{aligned}
$$

Since $|\sum_{j\in\mathbb{J}}\Theta_j^*\Lambda_j f|^2 = |\sum_{j\in\mathbb{I}}\Gamma_j^*\Lambda_j f|^2 = |\sum_{j\in\mathbb{I}} S_\Lambda^{-\frac{1}{2}}\Lambda_j^*\Lambda_j f|^2$, a replacement of $\mathbb{I}^c$ by $\mathbb{I}$ in the last item of Equation (9) leads to

$$
\left|\sum_{j\in\mathbb{J}}\Theta_j^*\Lambda_j f\right|^2 = \sum_{j\in\mathbb{J}}\langle \tilde{\Lambda}_j S_{\mathbb{I}}^{\Lambda} f, \tilde{\Lambda}_j S_{\mathbb{I}}^{\Lambda} f\rangle. \qquad (10)
$$

We also have

$$
\sum_{j\in\mathbb{J}}\langle \Lambda_j f, \Theta_j S_{\Gamma\Lambda}f\rangle = \sum_{j\in\mathbb{I}}\langle \Lambda_j f, \Lambda_j f\rangle, \quad \sum_{j\in\mathbb{J}}\langle (\Gamma_j - \Theta_j)S_{\Gamma\Lambda}f, \Lambda_j f\rangle = \sum_{j\in\mathbb{I}^c}\langle \Lambda_j f, \Lambda_j f\rangle. \qquad (11)
$$

Hence, the conclusion follows from Theorem 1. □

Let $\Lambda = \{\Lambda_j\}_{j\in\mathbb{J}}$ be a Parseval g-frame for $\mathcal{H}$ with respect to $\{\mathcal{K}_j\}_{j\in\mathbb{J}}$; then, $S_\Lambda = \mathrm{Id}_{\mathcal{H}}$. Thus, for any $\mathbb{I} \subset \mathbb{J}$,

$$\sum_{j\in\mathbb{J}}\langle\tilde{\Lambda}_j S^{\Lambda}_{\mathbb{I}^c} f, \tilde{\Lambda}_j S^{\Lambda}_{\mathbb{I}^c} f\rangle = \sum_{j\in\mathbb{J}}\langle\Lambda_j S^{\Lambda}_{\mathbb{I}^c} f, \Lambda_j S^{\Lambda}_{\mathbb{I}^c} f\rangle = |S^{\Lambda}_{\mathbb{I}^c} f|^2 = \left|\sum_{j\in\mathbb{I}^c}\Lambda_j^*\Lambda_j f\right|^2.$$

Similarly,

$$\sum_{j\in\mathbb{J}}\langle\tilde{\Lambda}_j S^{\Lambda}_{\mathbb{I}} f, \tilde{\Lambda}_j S^{\Lambda}_{\mathbb{I}} f\rangle = \left|\sum_{j\in\mathbb{I}}\Lambda_j^*\Lambda_j f\right|^2.$$

This fact, together with Corollary 1, yields

Corollary 2. *Suppose that $\Lambda = \{\Lambda_j\}_{j\in\mathbb{J}}$ is a Parseval g-frame for $\mathcal{H}$ with respect to $\{\mathcal{K}_j\}_{j\in\mathbb{J}}$. Then, for any $\lambda \in \mathbb{R}$, for all $\mathbb{I} \subset \mathbb{J}$ and all $f \in \mathcal{H}$, we have*

$$\begin{aligned}\sum_{j\in\mathbb{I}}\langle\Lambda_j f, \Lambda_j f\rangle + \left|\sum_{j\in\mathbb{I}^c}\Lambda_j^*\Lambda_j f\right|^2 &= \sum_{j\in\mathbb{I}^c}\langle\Lambda_j f, \Lambda_j f\rangle + \left|\sum_{j\in\mathbb{I}}\Lambda_j^*\Lambda_j f\right|^2 \\ &\geq (\lambda - \frac{\lambda^2}{4})\sum_{j\in\mathbb{I}^c}\langle\Lambda_j f, \Lambda_j f\rangle + (1 - \frac{\lambda^2}{4})\sum_{j\in\mathbb{I}}\langle\Lambda_j f, \Lambda_j f\rangle.\end{aligned}$$

Corollary 3. *Suppose that $\Lambda = \{\Lambda_j\}_{j\in\mathbb{J}}$ is a g-frame for $\mathcal{H}$ with respect to $\{\mathcal{K}_j\}_{j\in\mathbb{J}}$ with an alternate dual g-frame $\Gamma = \{\Gamma_j\}_{j\in\mathbb{J}}$. Then, for any $\lambda \in \mathbb{R}$, for all $\mathbb{I} \subset \mathbb{J}$ and all $f \in \mathcal{H}$, we have*

$$\begin{aligned}\left|\sum_{j\in\mathbb{I}}\Gamma_j^*\Lambda_j f\right|^2 + \sum_{j\in\mathbb{I}^c}\langle\Lambda_j f, \Gamma_j f\rangle &= \left|\sum_{j\in\mathbb{I}^c}\Gamma_j^*\Lambda_j f\right|^2 + \sum_{j\in\mathbb{I}}\langle\Gamma_j f, \Lambda_j f\rangle \\ &\geq (\lambda - \frac{\lambda^2}{4})\sum_{j\in\mathbb{I}}\langle\Lambda_j f, \Gamma_j f\rangle + (1 + \frac{\lambda}{2} - \frac{\lambda^2}{4})\sum_{j\in\mathbb{I}^c}\langle\Lambda_j f, \Gamma_j f\rangle - \frac{\lambda}{2}\sum_{j\in\mathbb{I}^c}\langle\Gamma_j f, \Lambda_j f\rangle.\end{aligned}$$

Proof. We conclude first that $S_{\Gamma\Lambda} = \mathrm{Id}_{\mathcal{H}}$. Now, the result follows immediately from Theorem 1 if, for any $\mathbb{I} \subset \mathbb{J}$, we take $\Theta_j = \begin{cases}\Gamma_j, & j \in \mathbb{I}^c, \\ 0, & j \in \mathbb{I}.\end{cases}$ □

Remark 1. *Theorems 4.1 and 4.2 in [31] can be obtained if we take $\lambda = 1$, respectively, in Corollaries 1 and 2.*

Theorem 2. *Let $\Lambda = \{\Lambda_j\}_{j\in\mathbb{J}}$ be a g-frame for $\mathcal{H}$ with respect to $\{\mathcal{K}_j\}_{j\in\mathbb{J}}$. Suppose that $\Gamma = \{\Gamma_j\}_{j\in\mathbb{J}}$ and $\Theta = \{\Theta_j\}_{j\in\mathbb{J}}$ are two g-Bessel sequences for $\mathcal{H}$ with respect to $\{\mathcal{K}_j\}_{j\in\mathbb{J}}$ and that the operator $S_{\Gamma\Lambda}$ is defined in Equation (4). Then, for any $\lambda \in \mathbb{R}$ and any $f \in \mathcal{H}$, we have*

$$\begin{aligned}\left|\sum_{j\in\mathbb{J}}(\Gamma_j - \Theta_j)^*\Lambda_j f\right|^2 + \left|\sum_{j\in\mathbb{J}}\Theta_j^*\Lambda_j f\right|^2 \geq (\lambda - \frac{\lambda^2}{2})\left|\sum_{j\in\mathbb{J}}\Gamma_j^*\Lambda_j f\right|^2 &- (1-\lambda)\sum_{j\in\mathbb{J}}\langle(\Gamma_j - \Theta_j)S_{\Gamma\Lambda} f, \Lambda_j f\rangle \\ &+ (1-\lambda)\sum_{j\in\mathbb{J}}\langle\Lambda_j f, \Theta_j S_{\Gamma\Lambda} f\rangle.\end{aligned}$$

Moreover, if U^*V is positive, where U and V are given in Equation (6), then

$$\left|\sum_{j\in\mathbb{J}}(\Gamma_j-\Theta_j)^*\Lambda_j f\right|^2+\left|\sum_{j\in\mathbb{J}}\Theta_j^*\Lambda_j f\right|^2 \leq \sum_{j\in\mathbb{J}}\langle(\Gamma_j-\Theta_j)S_{\Gamma\Lambda}f,\Lambda_j f\rangle+\sum_{j\in\mathbb{J}}\langle\Lambda_j f,\Theta_j S_{\Gamma\Lambda}f\rangle. \tag{12}$$

Proof. Combining Equation (7) with Lemma 1, we obtain

$$\begin{aligned}
&\left|\sum_{j\in\mathbb{J}}(\Gamma_j-\Theta_j)^*\Lambda_j f\right|^2+\left|\sum_{j\in\mathbb{J}}\Theta_j^*\Lambda_j f\right|^2\\
&= |Uf|^2+|Vf|^2 = 2|Vf|^2+\langle S_{\Gamma\Lambda}f,Uf\rangle-\langle Vf,S_{\Gamma\Lambda}f\rangle\\
&\geq (2-\frac{\lambda^2}{2})|S_{\Gamma\Lambda}f|^2-(2-\lambda)(\langle S_{\Gamma\Lambda}f,Uf\rangle+\langle Uf,S_{\Gamma\Lambda}f\rangle)+\langle S_{\Gamma\Lambda}f,Uf\rangle-\langle Vf,S_{\Gamma\Lambda}f\rangle\\
&= (2-\frac{\lambda^2}{2})|S_{\Gamma\Lambda}f|^2-(2-\lambda)\langle S_{\Gamma\Lambda}f,Uf\rangle-(2-\lambda)\langle Uf,S_{\Gamma\Lambda}f\rangle\\
&\quad -(2-\lambda)\langle Vf,S_{\Gamma\Lambda}f\rangle+(1-\lambda)\langle Vf,S_{\Gamma\Lambda}f\rangle+\langle S_{\Gamma\Lambda}f,Uf\rangle\\
&= (2-\frac{\lambda^2}{2})|S_{\Gamma\Lambda}f|^2-(1-\lambda)\langle S_{\Gamma\Lambda}f,Uf\rangle-(2-\lambda)|S_{\Gamma\Lambda}f|^2+(1-\lambda)\langle Vf,S_{\Gamma\Lambda}f\rangle\\
&= (\lambda-\frac{\lambda^2}{2})|S_{\Gamma\Lambda}f|^2-(1-\lambda)\langle S_{\Gamma\Lambda}f,Uf\rangle+(1-\lambda)\langle Vf,S_{\Gamma\Lambda}f\rangle\\
&= (\lambda-\frac{\lambda^2}{2})\left|\sum_{j\in\mathbb{J}}\Gamma_j^*\Lambda_j f\right|^2-(1-\lambda)\sum_{j\in\mathbb{J}}\langle(\Gamma_j-\Theta_j)S_{\Gamma\Lambda}f,\Lambda_j f\rangle+(1-\lambda)\sum_{j\in\mathbb{J}}\langle\Lambda_j f,\Theta_j S_{\Gamma\Lambda}f\rangle
\end{aligned}$$

for any $f\in\mathcal{H}$. We next prove Equation (12). Since U^*V is positive, we see from Equation (7) that

$$|Uf|^2=|Vf|^2+\langle S_{\Gamma\Lambda}f,Uf\rangle-\langle Vf,S_{\Gamma\Lambda}f\rangle=\langle S_{\Gamma\Lambda}f,Uf\rangle-\langle Vf,Uf\rangle\leq\langle S_{\Gamma\Lambda}f,Uf\rangle$$

for each $f\in\mathcal{H}$. A similar discussion gives $|Vf|^2\leq\langle Vf,S_{\Gamma\Lambda}f\rangle$. Thus,

$$\begin{aligned}
\left|\sum_{j\in\mathbb{J}}(\Gamma_j-\Theta_j)^*\Lambda_j f\right|^2+\left|\sum_{j\in\mathbb{J}}\Theta_j^*\Lambda_j f\right|^2 &= |Uf|^2+|Vf|^2\leq\langle S_{\Gamma\Lambda}f,Uf\rangle+\langle Vf,S_{\Gamma\Lambda}f\rangle\\
&= \sum_{j\in\mathbb{J}}\langle(\Gamma_j-\Theta_j)S_{\Gamma\Lambda}f,\Lambda_j f\rangle+\sum_{j\in\mathbb{J}}\langle\Lambda_j f,\Theta_j S_{\Gamma\Lambda}f\rangle.
\end{aligned}$$

□

Corollary 4. *Let $\Lambda=\{\Lambda_j\}_{j\in\mathbb{J}}$ be a g-frame for $\mathcal{H}$ with respect to $\{\mathcal{K}_j\}_{j\in\mathbb{J}}$ with g-frame operator S_Λ, and $\tilde{\Lambda}_j=\Lambda_j S_\Lambda^{-1}$ for each $j\in\mathbb{J}$. Then, for any $\lambda\in\mathbb{R}$, for all $\mathbb{I}\subset\mathbb{J}$ and all $f\in\mathcal{H}$, we have*

$$\begin{aligned}
&(\lambda-\frac{\lambda^2}{2})\sum_{j\in\mathbb{J}}\langle\Lambda_j f,\Lambda_j f\rangle-(1-\lambda)\sum_{j\in\mathbb{I}^c}\langle\Lambda_j f,\Lambda_j f\rangle+(1-\lambda)\sum_{j\in\mathbb{I}}\langle\Lambda_j f,\Lambda_j f\rangle\\
&\leq\sum_{j\in\mathbb{J}}\langle\tilde{\Lambda}_j S_{\mathbb{I}}^{\Lambda}f,\tilde{\Lambda}_j S_{\mathbb{I}}^{\Lambda}f\rangle+\sum_{j\in\mathbb{J}}\langle\tilde{\Lambda}_j S_{\mathbb{I}^c}^{\Lambda}f,\tilde{\Lambda}_j S_{\mathbb{I}^c}^{\Lambda}f\rangle\leq\sum_{j\in\mathbb{J}}\langle\Lambda_j f,\Lambda_j f\rangle.
\end{aligned}$$

Proof. For every $j\in\mathbb{J}$, taking $\Gamma_j=\Lambda_j S_\Lambda^{-\frac{1}{2}}$ and $\Theta_j=\begin{cases}\Gamma_j, & j\in\mathbb{I},\\ 0, & j\in\mathbb{I}^c,\end{cases}$ then the operators U and V defined in Equation (6) can be expressed as $U=S_\Lambda^{-\frac{1}{2}}S_{\mathbb{I}^c}^{\Lambda}$ and $V=S_\Lambda^{-\frac{1}{2}}S_{\mathbb{I}}^{\Lambda}$, respectively. Hence, $U^*V=S_{\mathbb{I}^c}^{\Lambda}S_\Lambda^{-1}S_{\mathbb{I}}^{\Lambda}$. Since $S_\Lambda^{-\frac{1}{2}}S_{\mathbb{I}}^{\Lambda}S_\Lambda^{-\frac{1}{2}}$ and $S_\Lambda^{-\frac{1}{2}}S_{\mathbb{I}^c}^{\Lambda}S_\Lambda^{-\frac{1}{2}}$ are positive and commutative, it follows that

$$0 \leq S_{\Lambda}^{-\frac{1}{2}} S_{\mathbb{I}^c}^{\Lambda} S_{\Lambda}^{-\frac{1}{2}} S_{\Lambda}^{-\frac{1}{2}} S_{\mathbb{I}}^{\Lambda} S_{\Lambda}^{-\frac{1}{2}} = S_{\Lambda}^{-\frac{1}{2}} S_{\mathbb{I}^c}^{\Lambda} S_{\Lambda}^{-1} S_{\mathbb{I}}^{\Lambda} S_{\Lambda}^{-\frac{1}{2}},$$

and, consequently, $S_{\mathbb{I}^c}^{\Lambda} S_{\Lambda}^{-1} S_{\mathbb{I}}^{\Lambda} \geq 0$. Note also that

$$\Big|\sum_{j\in\mathbb{J}} \Gamma_j^* \Lambda_j f\Big|^2 = \Big|S_{\Lambda}^{-\frac{1}{2}} \sum_{j\in\mathbb{J}} \Lambda_j^* \Lambda_j f\Big|^2 = |S_{\Lambda}^{\frac{1}{2}} f|^2 = \langle S_{\Lambda} f, f\rangle = \sum_{j\in\mathbb{J}} \langle \Lambda_j f, \Lambda_j f\rangle.$$

Now, the result follows by combining Theorem 2 and Equations (9)–(11). □

Theorem 3. *Let $\Lambda = \{\Lambda_j\}_{j\in\mathbb{J}}$ be a g-frame for $\mathcal{H}$ with respect to $\{\mathcal{K}_j\}_{j\in\mathbb{J}}$ with g-frame operator S_{Λ}. Suppose that $\Gamma = \{\Gamma_j\}_{j\in\mathbb{J}}$ and $\Theta = \{\Theta_j\}_{j\in\mathbb{J}}$ are two g-Bessel sequences for $\mathcal{H}$ with respect to $\{\mathcal{K}_j\}_{j\in\mathbb{J}}$ and that the operator $S_{\Gamma\Lambda}$ is defined in Equation (4). Then, for any $\lambda \in \mathbb{R}$ and any $f \in \mathcal{H}$, we have*

$$\begin{aligned}\sum_{j\in\mathbb{J}} \langle \Lambda_j f, \Theta_j S_{\Lambda}^{\frac{1}{2}} f\rangle - \Big|\sum_{j\in\mathbb{J}} \Theta_j^* \Lambda_j f\Big|^2 &\leq \sum_{j\in\mathbb{J}} \langle \Lambda_j f, \Theta_j (S_{\Lambda}^{\frac{1}{2}} - S_{\Gamma\Lambda}) f\rangle - \frac{\lambda}{2} \sum_{j\in\mathbb{J}} \langle \Lambda_j f, (\Gamma_j - \Theta_j) S_{\Gamma\Lambda} f\rangle \\ &\quad + (1 - \frac{\lambda}{2}) \sum_{j\in\mathbb{J}} \langle (\Gamma_j - \Theta_j) S_{\Gamma\Lambda} f, \Lambda_j f\rangle + \frac{\lambda^2}{4} \Big|\sum_{j\in\mathbb{J}} \Gamma_j^* \Lambda_j f\Big|^2.\end{aligned}$$

Moreover, if U^*V is positive, where U and V are given in Equation (6), then

$$\sum_{j\in\mathbb{J}} \langle \Lambda_j f, \Theta_j S_{\Lambda}^{\frac{1}{2}} f\rangle - \Big|\sum_{j\in\mathbb{J}} \Theta_j^* \Lambda_j f\Big|^2 \geq \sum_{j\in\mathbb{J}} \langle \Lambda_j f, \Theta_j (S_{\Lambda}^{\frac{1}{2}} - S_{\Gamma\Lambda}) f\rangle.$$

Proof. Combining Equations (7) and (8) leads to

$$\begin{aligned}\sum_{j\in\mathbb{J}} \langle \Lambda_j f, \Theta_j S_{\Lambda}^{\frac{1}{2}} f\rangle - \Big|\sum_{j\in\mathbb{J}} \Theta_j^* \Lambda_j f\Big|^2 &= \langle S_{\Lambda}^{\frac{1}{2}} V f, f\rangle - |Vf|^2 \\ &\leq \langle S_{\Lambda}^{\frac{1}{2}} V f, f\rangle - (\lambda - \frac{\lambda^2}{4}) \langle Uf, S_{\Gamma\Lambda} f\rangle - (\frac{\lambda}{2} - \frac{\lambda^2}{4}) \langle Vf, S_{\Gamma\Lambda} f\rangle \\ &\quad + \frac{\lambda}{2} \langle S_{\Gamma\Lambda} f, Vf\rangle - \langle Vf, S_{\Gamma\Lambda} f\rangle + \langle S_{\Gamma\Lambda} f, Uf\rangle \\ &= \langle Vf, (S_{\Lambda}^{\frac{1}{2}} - S_{\Gamma\Lambda}) f\rangle - (\frac{\lambda}{2} - \frac{\lambda^2}{4}) (\langle Uf, S_{\Gamma\Lambda} f\rangle + \langle Vf, S_{\Gamma\Lambda} f\rangle) - \frac{\lambda}{2} \langle Uf, S_{\Gamma\Lambda} f\rangle \\ &\quad + \frac{\lambda}{2} (\langle S_{\Gamma\Lambda} f, Vf\rangle + \langle S_{\Gamma\Lambda} f, Uf\rangle) + (1 - \frac{\lambda}{2}) \langle S_{\Gamma\Lambda} f, Uf\rangle \\ &= \langle Vf, (S_{\Lambda}^{\frac{1}{2}} - S_{\Gamma\Lambda}) f\rangle - (\frac{\lambda}{2} - \frac{\lambda^2}{4}) |S_{\Gamma\Lambda} f|^2 \\ &\quad - \frac{\lambda}{2} \langle Uf, S_{\Gamma\Lambda} f\rangle + \frac{\lambda}{2} |S_{\Gamma\Lambda} f|^2 + (1 - \frac{\lambda}{2}) \langle S_{\Gamma\Lambda} f, Uf\rangle \\ &= \langle Vf, (S_{\Lambda}^{\frac{1}{2}} - S_{\Gamma\Lambda}) f\rangle + \frac{\lambda^2}{4} |S_{\Gamma\Lambda} f|^2 - \frac{\lambda}{2} \langle Uf, S_{\Gamma\Lambda} f\rangle + (1 - \frac{\lambda}{2}) \langle S_{\Gamma\Lambda} f, Uf\rangle \\ &= \sum_{j\in\mathbb{J}} \langle \Lambda_j f, \Theta_j (S_{\Lambda}^{\frac{1}{2}} - S_{\Gamma\Lambda}) f\rangle - \frac{\lambda}{2} \sum_{j\in\mathbb{J}} \langle \Lambda_j f, (\Gamma_j - \Theta_j) S_{\Gamma\Lambda} f\rangle \\ &\quad + (1 - \frac{\lambda}{2}) \sum_{j\in\mathbb{J}} \langle (\Gamma_j - \Theta_j) S_{\Gamma\Lambda} f, \Lambda_j f\rangle + \frac{\lambda^2}{4} \Big|\sum_{j\in\mathbb{J}} \Gamma_j^* \Lambda_j f\Big|^2, \quad \forall f \in \mathcal{H}.\end{aligned}$$

Suppose that U^*V is positive; then, $|Vf|^2 \le \langle Vf, S_{\Gamma\Lambda}f\rangle$. Now, the "Moreover" part follows from the following inequality:

$$\sum_{j\in\mathbb{J}}\langle\Lambda_j f, \Theta_j S_\Lambda^{\frac{1}{2}} f\rangle - \left|\sum_{j\in\mathbb{J}}\Theta_j^*\Lambda_j f\right|^2 = \langle S_\Lambda^{\frac{1}{2}} Vf, f\rangle - |Vf|^2 \ge \langle S_\Lambda^{\frac{1}{2}} Vf, f\rangle - \langle Vf, S_{\Gamma\Lambda}f\rangle$$

$$= \langle Vf, (S_\Lambda^{\frac{1}{2}} - S_{\Gamma\Lambda})f\rangle = \sum_{j\in\mathbb{J}}\langle\Lambda_j f, \Theta_j(S_\Lambda^{\frac{1}{2}} - S_{\Gamma\Lambda})f\rangle.$$

□

Corollary 5. *Let $\Lambda = \{\Lambda_j\}_{j\in\mathbb{J}}$ be a g-frame for $\mathcal{H}$ with respect to $\{\mathcal{K}_j\}_{j\in\mathbb{J}}$ with g-frame operator S_Λ. Then, for any $\lambda \in \mathbb{R}$, for all $\mathbb{I} \subset \mathbb{J}$ and all $f \in \mathcal{H}$, we have*

$$0 \le \sum_{j\in\mathbb{I}}\langle\Lambda_j f, \Lambda_j f\rangle - \sum_{j\in\mathbb{J}}\langle\tilde{\Lambda}_j S_\mathbb{I}^\Lambda f, \tilde{\Lambda}_j S_\mathbb{I}^\Lambda f\rangle$$

$$\le (1-\lambda)\sum_{j\in\mathbb{I}^c}\langle\Lambda_j f, \Lambda_j f\rangle + \frac{\lambda^2}{4}\sum_{j\in\mathbb{J}}\langle\Lambda_j f, \Lambda_j f\rangle.$$

Proof. For each $j \in \mathbb{J}$, let Γ_j and Θ_j be the same as in the proof of Corollary 4. By Theorem 3, we have

$$\sum_{j\in\mathbb{I}}\langle\Lambda_j f, \Lambda_j f\rangle - \sum_{j\in\mathbb{J}}\langle\tilde{\Lambda}_j S_\mathbb{I}^\Lambda f, \tilde{\Lambda}_j S_\mathbb{I}^\Lambda f\rangle = \sum_{j\in\mathbb{J}}\langle\Lambda_j f, \Theta_j S_\Lambda^{\frac{1}{2}} f\rangle - \left|\sum_{j\in\mathbb{J}}\Theta_j^*\Lambda_j f\right|^2$$

$$\le -\frac{\lambda}{2}\sum_{j\in\mathbb{I}^c}\langle\Lambda_j f, \Lambda_j f\rangle + (1-\frac{\lambda}{2})\sum_{j\in\mathbb{I}^c}\langle\Lambda_j f, \Lambda_j f\rangle + \frac{\lambda^2}{4}\sum_{j\in\mathbb{J}}\langle\Lambda_j f, \Lambda_j f\rangle$$

$$= (1-\lambda)\sum_{j\in\mathbb{I}^c}\langle\Lambda_j f, \Lambda_j f\rangle + \frac{\lambda^2}{4}\sum_{j\in\mathbb{J}}\langle\Lambda_j f, \Lambda_j f\rangle.$$

By Theorem 3 again,

$$\sum_{j\in\mathbb{I}}\langle\Lambda_j f, \Lambda_j f\rangle - \sum_{j\in\mathbb{J}}\langle\tilde{\Lambda}_j S_\mathbb{I}^\Lambda f, \tilde{\Lambda}_j S_\mathbb{I}^\Lambda f\rangle = \sum_{j\in\mathbb{J}}\langle\Lambda_j f, \Theta_j S_\Lambda^{\frac{1}{2}} f\rangle - \left|\sum_{j\in\mathbb{J}}\Theta_j^*\Lambda_j f\right|^2$$

$$\ge \sum_{j\in\mathbb{J}}\langle\Lambda_j f, \Theta_j(S_\Lambda^{\frac{1}{2}} - S_{\Gamma\Lambda})f\rangle = 0,$$

and the proof is finished. □

Remark 2. *Taking $\lambda = 1$ in Corollaries 4 and 5, we can obtain Theorem 2.4 in [29].*

Funding: This research was funded by the National Natural Science Foundation of China under grant numbers 11761057 and 11561057.

Conflicts of Interest: The author declares no conflict of interest.

References

1. Duffin, R.J.; Schaeffer, A.C. A class of nonharmonic Fourier series. *Trans. Am. Math. Soc.* **1952**, *72*, 341–366. [CrossRef]
2. Bemrose, T.; Casazza, P.G.; Gröchenig, K.; Lammers, M.C.; Lynch, R.G. Weaving frames. *Oper. Matrices* **2016**, *10*, 1093–1116. [CrossRef]
3. Benedetto, J.; Powell, A.; Yilmaz, O. Sigma-Delta (ΣΔ) quantization and finite frames. *IEEE Trans. Inf. Theory* **2006**, *52*, 1990–2005. [CrossRef]

4. Casazza, P.G. The art of frame theory. *Taiwan J. Math.* **2000**, *4*, 129–201. [CrossRef]
5. Christensen, O. *An Introduction to Frames and Riesz Bases*; Birkhäuser: Boston, MA, USA, 2000.
6. Christensen, O.; Hasannasab, M. Operator representations of frames: Boundedness, duality, and stability. *Integral Equ. Oper. Theory* **2017**, *88*, 483–499. [CrossRef]
7. Christensen, O.; Hasannasab, M.; Rashidi, E. Dynamical sampling and frame representations with bounded operators. *J. Math. Anal. Appl.* **2018**, *463*, 634–644. [CrossRef]
8. Daubechies, I.; Grossmann, A.; Meyer, Y. Painless nonorthogonal expansions. *J. Math. Phys.* **1986**, *27*, 1271–1283. [CrossRef]
9. Han, D.; Sun, W. Reconstruction of signals from frame coefficients with erasures at unknown locations. *IEEE Trans. Inf. Theory* **2014**, *60*, 4013–4025. [CrossRef]
10. Strohmer, T.; Heath, R. Grassmannian frames with applications to coding and communication. *Appl. Comput. Harmon. Anal.* **2003**, *14*, 257–275. [CrossRef]
11. Sun, W. Asymptotic properties of Gabor frame operators as sampling density tends to infinity. *J. Funct. Anal.* **2010**, *258*, 913–932. [CrossRef]
12. Sun, W. G-frames and g-Riesz bases. *J. Math. Anal. Appl.* **2006**, *322*, 437–452. [CrossRef]
13. Sun, W. Stability of g-frames. *J. Math. Anal. Appl.* **2007**, *326*, 858–868. [CrossRef]
14. Li, J.Z.; Zhu, Y.C. Exact g-frames in Hilbert spaces. *J. Math. Anal. Appl.* **2011**, *374*, 201–209. [CrossRef]
15. Frank, M.; Larson, D.R. Frames in Hilbert C^*-modules and C^*-algebras. *J. Oper. Theory* **2002**, *48*, 273–314.
16. Arambašić, L. On frames for countably generated Hilbert C^*-modules. *Proc. Am. Math. Soc.* **2007**, *135*, 469–478. [CrossRef]
17. Han, D.; Jing, W.; Larson, D.R.; Li, P.T.; Mohapatra, R.N. Dilation of dual frame pairs in Hilbert C^*-modules. *Results Math.* **2013**, *63*, 241–250. [CrossRef]
18. Arambašić, L.; Bakić, D. Frames and outer frames for Hilbert C^*-modules. *Linear Multilinear Algebra* **2017**, *65*, 381–431. [CrossRef]
19. Khosravi, A.; Khosravi, B. Fusion frames and g-frames in Hilbert C^*-modules. *Int. J. Wavel. Multiresolut. Inf. Process.* **2008**, *6*, 433–446. [CrossRef]
20. Khosravi, A.; Mirzaee Azandaryani, M. Bessel multipliers in Hilbert C^*-modoles. *Banach J. Math. Anal.* **2015**, *9*, 153–163. [CrossRef]
21. Han, D.; Jing, W.; Larson, D.R.; Mohapatra, R.N. Riesz bases and their dual modular frames in Hilbert C^*-modules. *J. Math. Anal. Appl.* **2008**, *343*, 246–256. [CrossRef]
22. Alijani, A.; Dehghan, M.A. G-frames and their duals for Hilbert C^*-modules. *Bull. Iran. Math. Soc.* **2012**, *38*, 567–580.
23. Alijani, A. Generalized frames with C^*-valued bounds and their operator duals. *Filomat* **2015**, *29*, 1469–1479. [CrossRef]
24. Balan, R.; Casazza, P.G.; Edidin, D.; Kutyniok, G. A new identity for Parseval frames. *Proc. Am. Math. Soc.* **2007**, *135*, 1007–1015. [CrossRef]
25. Găvruţa, P. On some identities and inequalities for frames in Hilbert spaces. *J. Math. Anal. Appl.* **2006**, *321*, 469–478. [CrossRef]
26. Li, D.W.; Leng, J.S. On some new inequalities for fusion frames in Hilbert spaces. *Math. Inequal. Appl.* **2017**, *20*, 889–900. [CrossRef]
27. Li, D.W.; Leng, J.S. On some new inequalities for continuous fusion frames in Hilbert spaces. *Mediterr. J. Math.* **2018**, *15*, 173. [CrossRef]
28. Poria, A. Some identities and inequalities for Hilbert-Schmidt frames. *Mediterr. J. Math.* **2017**, *14*, 59. [CrossRef]
29. Xiang, Z.Q. New inequalities for g-frames in Hilbert C^*-modules. *J. Math. Inequal.* **2016**, *10*, 889–897. [CrossRef]

30. Xiang, Z.Q. New double inequalities for g-frames in Hilbert C^*-modules. *SpringerPlus* **2016**, *5*, 1025. [CrossRef]
31. Xiao, X.C.; Zeng, X.M. Some properties of g-frames in Hilbert C^*-modules. *J. Math. Anal. Appl.* **2010**, *363*, 399–408. [CrossRef]

mathematics

MDPI

Article

More on Inequalities for Weaving Frames in Hilbert Spaces

Zhong-Qi Xiang

College of Mathematics and Computer Science, Shangrao Normal University, Shangrao 334001, China; lxsy20110927@163.com; Tel.: +86-793-815-9108

Received: 14 January 2019; Accepted: 30 January 2019; Published: 2 February 2019

Abstract: In this paper, we present several new inequalities for weaving frames in Hilbert spaces from the point of view of operator theory, which are related to a linear bounded operator induced by three Bessel sequences and a scalar in the set of real numbers. It is indicated that our results are more general and cover the corresponding results recently obtained by Li and Leng. We also give a triangle inequality for weaving frames in Hilbert spaces, which is structurally different from previous ones.

Keywords: frame; weaving frame; weaving frame operator; alternate dual frame; Hilbert space

MSC: 42C15; 47B40

---

1. Introduction

Throughout this paper, $\mathbb{H}$ is a separable Hilbert space, and $\mathrm{Id}_{\mathbb{H}}$ is the identity operator on $\mathbb{H}$. The notations $\mathbb{J}$, $\mathbb{R}$, and $B(\mathbb{H})$ denote, respectively, an index set which is finite or countable, the real number set, and the family of all linear bounded operators on $\mathbb{H}$.

A sequence $\mathcal{F} = \{f_j\}_{j\in\mathbb{J}}$ of vectors in $\mathbb{H}$ is a frame (classical frame) if there are constants $A, B > 0$ such that

$$A\|x\|^2 \leq \sum_{j\in\mathbb{J}} |\langle x, f_j\rangle|^2 \leq B\|x\|^2, \quad \forall x \in \mathbb{H}. \tag{1}$$

The frame $\mathcal{F} = \{f_j\}_{j\in\mathbb{J}}$ is said to be Parseval if $A = B = 1$. If $\mathcal{F} = \{f_j\}_{j\in\mathbb{J}}$ satisfies the inequality to the right in Equation (1) we say that $\mathcal{F} = \{f_j\}_{j\in\mathbb{J}}$ is a Bessel sequence.

The appearance of frames can be tracked back to the early 1950s when they were used in the work on nonharmonic Fourier series owing to Duffin and Schaeffer [1]. We refer to [2–16] for more information on general frame theory. It should be pointed out that frames have played an important role such as in signal processing [17,18], sigma-delta quantization [19], quantum information [20], coding theory [21], and sampling theory [22], due to their nice properties.

Motivated by a problem deriving from distributed signal processing, Bemrose et al. [23] put forward the notion of (discrete) weaving frames for Hilbert spaces. The theory may be applied to deal with wireless sensor networks that require distributed processing under different frames, which could also be used in the pre-processing of signals by means of Gabor frames. Recently, weaving frames have attracted many scholars' attention, please refer to [24–30] for more information.

Balan et al. [31] discovered an interesting inequality when further discussing the remarkable Parseval frames identity arising in their work on effective algorithms for computing the reconstructions of signals, which was then extended to general frames and alternate dual frames [32], and based on the work in [31,32], some inequalities for generalized frames associated with a scalar are also established (see [33–35]). Borrowing the ideas from [34,35], Li and Leng [36] have generalized the inequalities for frames to weaving frames with a more general form. In this paper, we present several new inequalities for weaving frames and we show that our results can lead to the corresponding results in [36]. We also obtain a triangle inequality for weaving frames, which differs from previous ones in the structure.

One calls two frames $\mathcal{F} = \{f_j\}_{j\in\mathbb{J}}$ and $\mathcal{G} = \{g_j\}_{j\in\mathbb{J}}$ in $\mathbb{H}$ woven, if there exist universal constants C and D such that for each partition $\sigma \subset \mathbb{J}$, the family $\{f_j\}_{j\in\sigma} \cup \{g_j\}_{j\in\sigma^c}$ is a frame for $\mathbb{H}$ with frame bounds C and D and, in this case, we say that $\{f_j\}_{j\in\sigma} \cup \{g_j\}_{j\in\sigma^c}$ is a weaving frame.

Suppose that $\mathcal{F} = \{f_j\}_{j\in\mathbb{J}}$ and $\mathcal{G} = \{g_j\}_{j\in\mathbb{J}}$ are woven, then associated with every weaving frame $\{f_j\}_{j\in\sigma} \cup \{g_j\}_{j\in\sigma^c}$ there is a positive, self-adjoint and invertible operator, called the weaving frame operator, given below

$$S_W : \mathbb{H} \to \mathbb{H}, \quad S_W x = \sum_{j\in\sigma}\langle x, f_j\rangle f_j + \sum_{j\in\sigma^c}\langle x, g_j\rangle g_j.$$

We recall that a frame $\mathcal{H} = \{h_j\}_{j\in\mathbb{J}}$ is said to be an alternate dual frame of $\{f_j\}_{j\in\sigma} \cup \{g_j\}_{j\in\sigma^c}$ if

$$x = \sum_{j\in\sigma}\langle x, f_j\rangle h_j + \sum_{j\in\sigma^c}\langle x, g_j\rangle h_j \tag{2}$$

is valid for every $x \in \mathbb{H}$.

For each $\sigma \subset \mathbb{J}$, let $S_{\mathcal{F}}^{\sigma}$ be the positive and self-adjoint operator induced by σ and a given frame $\mathcal{F} = \{f_j\}_{j\in\mathbb{J}}$ of $\mathbb{H}$, defined by

$$S_{\mathcal{F}}^{\sigma} : \mathbb{H} \to \mathbb{H}, \quad S_{\mathcal{F}}^{\sigma} x = \sum_{j\in\sigma}\langle x, f_j\rangle f_j.$$

Let $\mathcal{F} = \{f_j\}_{j\in\mathbb{J}}$, $\mathcal{G} = \{g_j\}_{j\in\mathbb{J}}$, and $\mathcal{H} = \{h_j\}_{j\in\mathbb{J}}$ be Bessel sequences for $\mathbb{H}$, then it is easy to check that the operators

$$S_{\mathcal{FGH}} : \mathbb{H} \to \mathbb{H}, \quad S_{\mathcal{FGH}} x = \sum_{j\in\sigma}\langle x, f_j\rangle h_j + \sum_{j\in\sigma^c}\langle x, g_j\rangle h_j \tag{3}$$

and

$$S_{\mathcal{HFG}} : \mathbb{H} \to \mathbb{H}, \quad S_{\mathcal{HFG}} x = \sum_{j\in\sigma}\langle x, h_j\rangle f_j + \sum_{j\in\sigma^c}\langle x, h_j\rangle g_j \tag{4}$$

are well-defined and, further, $S_{\mathcal{FGH}}, S_{\mathcal{HFG}} \in B(\mathbb{H})$.

2. Main Results and Their Proofs

We start with the following result on operators, which will be used to prove Theorem 1.

Lemma 1. *If $P, Q, L \in B(\mathbb{H})$ satisfy $P + Q = L$, then for any $\lambda \in \mathbb{R}$,*

$$P^*P + \frac{\lambda}{2}(Q^*L + L^*Q) = Q^*Q + (1 - \frac{\lambda}{2})(P^*L + L^*P) + (\lambda - 1)L^*L \geq (\lambda - \frac{\lambda^2}{4})L^*L.$$

Proof. We have

$$P^*P + \frac{\lambda}{2}(Q^*L + L^*Q) = P^*P - \frac{\lambda}{2}(P^*L + L^*P) + \lambda L^*L,$$

and

$$\begin{aligned} Q^*Q + (1 - \frac{\lambda}{2})(P^*L + L^*P) + (\lambda - 1)L^*L &= P^*P - \frac{\lambda}{2}(P^*L + L^*P) + \lambda L^*L \\ &= (P - \frac{\lambda}{2}L)^*(P - \frac{\lambda}{2}L) + (\lambda - \frac{\lambda^2}{4})L^*L \geq (\lambda - \frac{\lambda^2}{4})L^*L. \end{aligned}$$

Thus the result holds. □

Taking 2λ instead of λ in Lemma 1 yields an immediate consequence as follows.

Corollary 1. *If $P, Q, L \in B(\mathbb{H})$ satisfy $P + Q = L$, then for any $\lambda \in \mathbb{R}$,*

$$P^*P + \lambda(Q^*L + L^*Q) = Q^*Q + (1-\lambda)(P^*L + L^*P) + (2\lambda - 1)L^*L \geq (2\lambda - \lambda^2)L^*L.$$

Theorem 1. *Suppose that two frames $\mathcal{F} = \{f_j\}_{j\in\mathbb{J}}$ and $\mathcal{G} = \{g_j\}_{j\in\mathbb{J}}$ in $\mathbb{H}$ are woven, and that $\mathcal{H} = \{h_j\}_{j\in\mathbb{J}}$ is a Bessel sequences for $\mathbb{H}$. Then for any $\sigma \subset \mathbb{J}$, for all $\lambda \in \mathbb{R}$ and all $x \in \mathbb{H}$, we have*

$$\begin{aligned} \left\|\sum_{j\in\sigma}\langle x, f_j\rangle h_j\right\|^2 + \mathrm{Re}\sum_{j\in\sigma^c}\langle x, g_j\rangle\langle h_j, S_{\mathcal{FGH}}x\rangle &= \left\|\sum_{j\in\sigma^c}\langle x, g_j\rangle h_j\right\|^2 + \mathrm{Re}\sum_{j\in\sigma}\langle x, f_j\rangle\langle h_j, S_{\mathcal{FGH}}x\rangle \\ &\geq (\lambda - \tfrac{\lambda^2}{4})\mathrm{Re}\sum_{j\in\sigma}\langle x, f_j\rangle\langle h_j, S_{\mathcal{FGH}}x\rangle + (1 - \tfrac{\lambda^2}{4})\mathrm{Re}\sum_{j\in\sigma^c}\langle x, g_j\rangle\langle h_j, S_{\mathcal{FGH}}x\rangle \end{aligned} \tag{5}$$

and

$$\begin{aligned} \left\|\sum_{j\in\sigma}\langle x, h_j\rangle f_j\right\|^2 + \mathrm{Re}\sum_{j\in\sigma^c}\langle x, h_j\rangle\langle g_j, S_{\mathcal{HFG}}x\rangle &= \left\|\sum_{j\in\sigma^c}\langle x, h_j\rangle g_j\right\|^2 + \mathrm{Re}\sum_{j\in\sigma}\langle x, h_j\rangle\langle f_j, S_{\mathcal{HFG}}x\rangle \\ &\geq (2\lambda - \lambda^2)\mathrm{Re}\sum_{j\in\sigma}\langle x, h_j\rangle\langle f_j, S_{\mathcal{HFG}}x\rangle + (1 - \lambda^2)\mathrm{Re}\sum_{j\in\sigma^c}\langle x, h_j\rangle\langle g_j, S_{\mathcal{HFG}}x\rangle, \end{aligned} \tag{6}$$

where $S_{\mathcal{FGH}}$ and $S_{\mathcal{HFG}}$ are defined respectively in Equations (3) and (4).

Proof. For any $\sigma \subset \mathbb{J}$, we define

$$Px = \sum_{j\in\sigma}\langle x, f_j\rangle h_j \quad \text{and} \quad Qx = \sum_{j\in\sigma^c}\langle x, g_j\rangle h_j, \quad \forall x \in \mathbb{H}. \tag{7}$$

Then $P, Q \in B(\mathbb{H})$, and a simple calculation gives

$$Px + Qx = \sum_{j\in\sigma}\langle x, f_j\rangle h_j + \sum_{j\in\sigma^c}\langle x, g_j\rangle h_j = S_{\mathcal{FGH}}x.$$

By Lemma 1 we obtain

$$\|Px\|^2 + \lambda\mathrm{Re}\langle S^*_{\mathcal{FGH}}Qx, x\rangle = \|Qx\|^2 + 2(1 - \frac{\lambda}{2})\mathrm{Re}\langle S^*_{\mathcal{FGH}}Px, x\rangle + (\lambda - 1)\|S_{\mathcal{FGH}}x\|^2.$$

Therefore,

$$\begin{aligned} \|Px\|^2 &= \|Qx\|^2 + 2(1 - \frac{\lambda}{2})\mathrm{Re}\langle S^*_{\mathcal{FGH}}Px, x\rangle + (\lambda - 1)\mathrm{Re}\langle S_{\mathcal{FGH}}x, S_{\mathcal{FGH}}x\rangle - \lambda\mathrm{Re}\langle S^*_{\mathcal{FGH}}Qx, x\rangle \\ &= \|Qx\|^2 + 2\mathrm{Re}\langle S^*_{\mathcal{FGH}}Px, x\rangle - \lambda\mathrm{Re}\langle (P+Q)x, S_{\mathcal{FGH}}x\rangle + (\lambda - 1)\mathrm{Re}\langle S_{\mathcal{FGH}}x, S_{\mathcal{FGH}}x\rangle \\ &= \|Qx\|^2 + 2\mathrm{Re}\langle S^*_{\mathcal{FGH}}Px, x\rangle - \mathrm{Re}\langle S_{\mathcal{FGH}}x, S_{\mathcal{FGH}}x\rangle \\ &= \|Qx\|^2 + 2\mathrm{Re}\langle Px, S_{\mathcal{FGH}}x\rangle - \mathrm{Re}\langle Px, S_{\mathcal{FGH}}x\rangle - \mathrm{Re}\langle Qx, S_{\mathcal{FGH}}x\rangle \\ &= \|Qx\|^2 + \mathrm{Re}\langle Px, S_{\mathcal{FGH}}x\rangle - \mathrm{Re}\langle Qx, S_{\mathcal{FGH}}x\rangle, \end{aligned}$$

from which we conclude that

$$\begin{aligned} &\left\|\sum_{j\in\sigma}\langle x, f_j\rangle h_j\right\|^2 + \mathrm{Re}\sum_{j\in\sigma^c}\langle x, g_j\rangle\langle h_j, S_{\mathcal{FGH}}x\rangle \\ &= \|Px\|^2 + \mathrm{Re}\langle Qx, S_{\mathcal{FGH}}x\rangle = \|Qx\|^2 + \mathrm{Re}\langle Px, S_{\mathcal{FGH}}x\rangle \\ &= \left\|\sum_{j\in\sigma^c}\langle x, g_j\rangle h_j\right\|^2 + \mathrm{Re}\sum_{j\in\sigma}\langle x, f_j\rangle\langle h_j, S_{\mathcal{FGH}}x\rangle. \end{aligned} \tag{8}$$

For the inequality in Equation (5), we apply Lemma 1 again,

$$\|Px\|^2 + \lambda \mathrm{Re}\langle S^*_{\mathcal{FGH}}Qx, x\rangle \geq (\lambda - \frac{\lambda^2}{4})\langle S^*_{\mathcal{FGH}}S_{\mathcal{FGH}}x, x\rangle$$

for any $x \in \mathbb{H}$. Hence

$$\begin{aligned} \|Px\|^2 &\geq (\lambda - \frac{\lambda^2}{4})\langle S^*_{\mathcal{FGH}}S_{\mathcal{FGH}}x, x\rangle - \lambda \mathrm{Re}\langle Qx, S_{\mathcal{FGH}}x\rangle \\ &= (\lambda - \frac{\lambda^2}{4} - \lambda)\mathrm{Re}\langle Qx, S_{\mathcal{FGH}}x\rangle + (\lambda - \frac{\lambda^2}{4})\mathrm{Re}\langle Px, S_{\mathcal{FGH}}x\rangle \\ &= (\lambda - \frac{\lambda^2}{4})\mathrm{Re}\langle Px, S_{\mathcal{FGH}}x\rangle - \frac{\lambda^2}{4}\mathrm{Re}\langle Qx, S_{\mathcal{FGH}}x\rangle, \end{aligned} \tag{9}$$

and consequently,

$$\begin{aligned} &\left\|\sum_{j\in\sigma}\langle x, f_j\rangle h_j\right\|^2 + \mathrm{Re}\sum_{j\in\sigma^c}\langle x, g_j\rangle\langle h_j, S_{\mathcal{FGH}}x\rangle = \|Px\|^2 + \mathrm{Re}\langle Qx, S_{\mathcal{FGH}}x\rangle \\ &\geq (\lambda - \frac{\lambda^2}{4})\mathrm{Re}\langle Px, S_{\mathcal{FGH}}x\rangle + (1 - \frac{\lambda^2}{4})\mathrm{Re}\langle Qx, S_{\mathcal{FGH}}x\rangle \\ &= (\lambda - \frac{\lambda^2}{4})\mathrm{Re}\sum_{j\in\sigma}\langle x, f_j\rangle\langle h_j, S_{\mathcal{FGH}}x\rangle + (1 - \frac{\lambda^2}{4})\mathrm{Re}\sum_{j\in\sigma^c}\langle x, g_j\rangle\langle h_j, S_{\mathcal{FGH}}x\rangle. \end{aligned}$$

Similar arguments hold for Equation (6), by using Corollary 1. □

Corollary 2. *Let two frames $\mathcal{F} = \{f_j\}_{j\in\mathbb{J}}$ and $\mathcal{G} = \{g_j\}_{j\in\mathbb{J}}$ in $\mathbb{H}$ be woven. Then for any $\sigma \subset \mathbb{J}$, for all $\lambda \in \mathbb{R}$ and all $x \in \mathbb{H}$, we have*

$$\begin{aligned} &\sum_{j\in\sigma}|\langle S_W^{-1}S^\sigma_{\mathcal{F}}x, f_j\rangle|^2 + \sum_{j\in\sigma^c}|\langle S_W^{-1}S^\sigma_{\mathcal{F}}x, g_j\rangle|^2 + \sum_{j\in\sigma^c}|\langle x, g_j\rangle|^2 \\ &\quad= \sum_{j\in\sigma}|\langle S_W^{-1}S^{\sigma^c}_{\mathcal{G}}x, f_j\rangle|^2 + \sum_{j\in\sigma^c}|\langle S_W^{-1}S^{\sigma^c}_{\mathcal{G}}x, g_j\rangle|^2 + \sum_{j\in\sigma}|\langle x, f_j\rangle|^2 \\ &\quad\geq (\lambda - \frac{\lambda^2}{4})\sum_{j\in\sigma}|\langle x, f_j\rangle|^2 + (1 - \frac{\lambda^2}{4})\sum_{j\in\sigma^c}|\langle x, g_j\rangle|^2. \end{aligned}$$

Proof. For each $j \in \mathbb{J}$, taking

$$h_j = \begin{cases} S_W^{-\frac{1}{2}}f_j, & j \in \sigma, \\ S_W^{-\frac{1}{2}}g_j, & j \in \sigma^c. \end{cases}$$

Then, clearly, $\mathcal{H} = \{h_j\}_{j\in\mathbb{J}}$ is a Bessel sequence for $\mathbb{H}$. Since for any $x \in \mathbb{H}$, $S_{\mathcal{FGH}}x = \sum_{j\in\sigma}\langle x, f_j\rangle S_W^{-\frac{1}{2}}f_j + \sum_{j\in\sigma^c}\langle x, g_j\rangle S_W^{-\frac{1}{2}}g_j = S_W^{-\frac{1}{2}}S_W x = S_W^{\frac{1}{2}}x$, we have $S_{\mathcal{FGH}} = S_W^{\frac{1}{2}}$. Now

$$\begin{aligned} \left\|\sum_{j\in\sigma}\langle x, f_j\rangle h_j\right\|^2 &= \left\|\sum_{j\in\sigma}\langle x, f_j\rangle S_W^{-\frac{1}{2}}f_j\right\|^2 = \left\|S_W^{-\frac{1}{2}}\sum_{j\in\sigma}\langle x, f_j\rangle f_j\right\|^2 \\ &= \|S_W^{-\frac{1}{2}}S^\sigma_{\mathcal{F}}x\|^2 = \langle S_W^{-\frac{1}{2}}S^\sigma_{\mathcal{F}}x, S_W^{-\frac{1}{2}}S^\sigma_{\mathcal{F}}x\rangle \\ &= \sum_{j\in\sigma}\langle S_W^{-1}S^\sigma_{\mathcal{F}}x, f_j\rangle\langle f_j, S_W^{-1}S^\sigma_{\mathcal{F}}x\rangle + \sum_{j\in\sigma^c}\langle S_W^{-1}S^\sigma_{\mathcal{F}}x, g_j\rangle\langle g_j, S_W^{-1}S^\sigma_{\mathcal{F}}x\rangle \\ &= \sum_{j\in\sigma}|\langle S_W^{-1}S^\sigma_{\mathcal{F}}x, f_j\rangle|^2 + \sum_{j\in\sigma^c}|\langle S_W^{-1}S^\sigma_{\mathcal{F}}x, g_j\rangle|^2. \end{aligned} \tag{10}$$

A similar discussion leads to

$$\left\|\sum_{j\in\sigma^c}\langle x,g_j\rangle h_j\right\|^2=\sum_{j\in\sigma}|\langle S_W^{-1}S_G^{\sigma^c}x,f_j\rangle|^2+\sum_{j\in\sigma^c}|\langle S_W^{-1}S_G^{\sigma^c}x,g_j\rangle|^2. \tag{11}$$

We also get

$$\mathrm{Re}\sum_{j\in\sigma}\langle x,f_j\rangle\langle h_j,S_{\mathcal{FGH}}x\rangle=\mathrm{Re}\sum_{j\in\sigma}\langle x,f_j\rangle\langle S_W^{-\frac{1}{2}}f_j,S_W^{\frac{1}{2}}x\rangle=\sum_{j\in\sigma}|\langle x,f_j\rangle|^2, \tag{12}$$

and

$$\mathrm{Re}\sum_{j\in\sigma^c}\langle x,g_j\rangle\langle h_j,S_{\mathcal{FGH}}x\rangle=\mathrm{Re}\sum_{j\in\sigma^c}\langle x,g_j\rangle\langle S_W^{-\frac{1}{2}}g_j,S_W^{\frac{1}{2}}x\rangle=\sum_{j\in\sigma^c}|\langle x,g_j\rangle|^2. \tag{13}$$

Thus the result follows from Theorem 1. □

Corollary 3. *Suppose that two frames $\mathcal{F}=\{f_j\}_{j\in\mathbb{J}}$ and $\mathcal{G}=\{g_j\}_{j\in\mathbb{J}}$ in $\mathbb{H}$ are woven. Then for any $\sigma\subset\mathbb{J}$, for all $\lambda\in\mathbb{R}$ and all $x\in\mathbb{H}$,*

$$\begin{aligned}\mathrm{Re}\Big(\sum_{j\in\sigma}\langle x,h_j\rangle\langle f_j,x\rangle\Big)+\Big\|\sum_{j\in\sigma^c}\langle x,h_j\rangle g_j\Big\|^2&=\mathrm{Re}\Big(\sum_{j\in\sigma^c}\langle x,h_j\rangle\langle g_j,x\rangle\Big)+\Big\|\sum_{j\in\sigma}\langle x,h_j\rangle f_j\Big\|^2\\&\geq(2\lambda-\lambda^2)\mathrm{Re}\Big(\sum_{j\in\sigma}\langle x,h_j\rangle\langle f_j,x\rangle\Big)+(1-\lambda^2)\mathrm{Re}\Big(\sum_{j\in\sigma^c}\langle x,h_j\rangle\langle g_j,x\rangle\Big),\end{aligned}$$

where $\mathcal{H}=\{h_j\}_{j\in\mathbb{J}}$ is an alternate dual frame of the weaving frame $\{f_j\}_{j\in\sigma}\cup\{g_j\}_{j\in\sigma^c}$.

Proof. For any $\sigma\subset\mathbb{J}$, since $\mathcal{H}=\{h_j\}_{j\in\mathbb{J}}$ is an alternate dual frame of the weaving frame $\{f_j\}_{j\in\sigma}\cup\{g_j\}_{j\in\sigma^c}$, Equation (2) gives

$$x=\sum_{j\in\sigma}\langle x,h_j\rangle f_j+\sum_{j\in\sigma^c}\langle x,h_j\rangle g_j$$

for any $x\in\mathbb{H}$ and thus, $S_{\mathcal{HFG}}=\mathrm{Id}_{\mathbb{H}}$. By Theorem 1 we obtain the relation shown in the corollary. □

Remark 1. *Corollaries 2 and 3 are respectively Theorems 7 and 9 in [36].*

Theorem 2. *Suppose that two frames $\mathcal{F}=\{f_j\}_{j\in\mathbb{J}}$ and $\mathcal{G}=\{g_j\}_{j\in\mathbb{J}}$ in $\mathbb{H}$ are woven, and that $\mathcal{H}=\{h_j\}_{j\in\mathbb{J}}$ is a Bessel sequences for $\mathbb{H}$. Then for any $\sigma\subset\mathbb{J}$, for all $\lambda\in\mathbb{R}$ and all $x\in\mathbb{H}$, we have*

$$\begin{aligned}&\mathrm{Re}\sum_{j\in\sigma}\langle x,f_j\rangle\langle h_j,S_{\mathcal{FGH}}x\rangle-\Big\|\sum_{j\in\sigma}\langle x,f_j\rangle h_j\Big\|^2\\&\quad\leq\frac{\lambda^2}{4}\mathrm{Re}\sum_{j\in\sigma^c}\langle x,g_j\rangle\langle h_j,S_{\mathcal{FGH}}x\rangle+(1-\frac{\lambda}{2})^2\mathrm{Re}\sum_{j\in\sigma}\langle x,f_j\rangle\langle h_j,S_{\mathcal{FGH}}x\rangle,\end{aligned} \tag{14}$$

and

$$\begin{aligned}&\Big\|\sum_{j\subset\sigma}\langle x,f_j\rangle h_j\Big\|^2+\Big\|\sum_{j\in\sigma^c}\langle x,g_j\rangle h_j\Big\|^2\\&\quad\geq(2\lambda-\frac{\lambda^2}{2}-1)\mathrm{Re}\sum_{j\in\sigma}\langle x,f_j\rangle\langle h_j,S_{\mathcal{FGH}}x\rangle+(1-\frac{\lambda^2}{2})\mathrm{Re}\sum_{j\in\sigma^c}\langle x,g_j\rangle\langle h_j,S_{\mathcal{FGH}}x\rangle,\end{aligned} \tag{15}$$

where $S_{\mathcal{FGH}}$ is defined in Equation (3).

Moreover, if the operators P and Q given in Equation (7) satisfy the condition that P^*Q is positive, then

$$0 \le \mathrm{Re}\sum_{j\in\sigma}\langle x, f_j\rangle\langle h_j, S_{\mathcal{FGH}}x\rangle - \Big\|\sum_{j\in\sigma}\langle x, f_j\rangle h_j\Big\|^2,$$

and

$$\Big\|\sum_{j\in\sigma}\langle x, f_j\rangle h_j\Big\|^2 + \Big\|\sum_{j\in\sigma^c}\langle x, g_j\rangle h_j\Big\|^2 \le \|S_{\mathcal{FGH}}x\|^2.$$

Proof. For any $\sigma \subset \mathbb{J}$, let P and Q be defined in Equation (7). Then all $\lambda \in \mathbb{R}$ and all $x \in \mathbb{H}$, we see from Equation (9) that

$$\begin{aligned}
&\mathrm{Re}\sum_{j\in\sigma}\langle x, f_j\rangle\langle h_j, S_{\mathcal{FGH}}x\rangle - \Big\|\sum_{j\in\sigma}\langle x, f_j\rangle h_j\Big\|^2 = \mathrm{Re}\langle Px, S_{\mathcal{FGH}}x\rangle - \|Px\|^2\\
&\le \mathrm{Re}\langle Px, S_{\mathcal{FGH}}x\rangle + \frac{\lambda^2}{4}\mathrm{Re}\langle Qx, S_{\mathcal{FGH}}x\rangle - (\lambda - \frac{\lambda^2}{4})\mathrm{Re}\langle Px, S_{\mathcal{FGH}}x\rangle\\
&= \frac{\lambda^2}{4}\mathrm{Re}\langle Qx, S_{\mathcal{FGH}}x\rangle + (1 - \lambda + \frac{\lambda^2}{4})\mathrm{Re}\langle Px, S_{\mathcal{FGH}}x\rangle\\
&= \frac{\lambda^2}{4}\mathrm{Re}\langle Qx, S_{\mathcal{FGH}}x\rangle + (1 - \frac{\lambda}{2})^2\mathrm{Re}\langle Px, S_{\mathcal{FGH}}x\rangle\\
&= \frac{\lambda^2}{4}\mathrm{Re}\sum_{j\in\sigma^c}\langle x, g_j\rangle\langle h_j, S_{\mathcal{FGH}}x\rangle + (1 - \frac{\lambda}{2})^2\mathrm{Re}\sum_{j\in\sigma}\langle x, f_j\rangle\langle h_j, S_{\mathcal{FGH}}x\rangle.
\end{aligned}$$

We next prove Equation (15). By combining Equation (8) with Equation (9) we conclude that

$$\begin{aligned}
&\Big\|\sum_{j\in\sigma}\langle x, f_j\rangle h_j\Big\|^2 + \Big\|\sum_{j\in\sigma^c}\langle x, g_j\rangle h_j\Big\|^2\\
&= \|Px\|^2 + \|Qx\|^2 = 2\|Px\|^2 + \mathrm{Re}\langle Qx, S_{\mathcal{FGH}}x\rangle - \mathrm{Re}\langle Px, S_{\mathcal{FGH}}x\rangle\\
&\ge (2\lambda - \frac{\lambda^2}{2})\mathrm{Re}\langle Px, S_{\mathcal{FGH}}x\rangle - \frac{\lambda^2}{2}\mathrm{Re}\langle Qx, S_{\mathcal{FGH}}x\rangle + \mathrm{Re}\langle Qx, S_{\mathcal{FGH}}x\rangle - \mathrm{Re}\langle Px, S_{\mathcal{FGH}}x\rangle\\
&= (2\lambda - \frac{\lambda^2}{2} - 1)\mathrm{Re}\langle Px, S_{\mathcal{FGH}}x\rangle + (1 - \frac{\lambda^2}{2})\mathrm{Re}\langle Qx, S_{\mathcal{FGH}}x\rangle\\
&= (2\lambda - \frac{\lambda^2}{2} - 1)\mathrm{Re}\sum_{j\in\sigma}\langle x, f_j\rangle\langle h_j, S_{\mathcal{FGH}}x\rangle + (1 - \frac{\lambda^2}{2})\mathrm{Re}\sum_{j\in\sigma^c}\langle x, g_j\rangle\langle h_j, S_{\mathcal{FGH}}x\rangle, \quad \forall x \in \mathbb{H}.
\end{aligned}$$

Suppose now that P^*Q is positive, then for any $x \in \mathbb{H}$,

$$\begin{aligned}
\mathrm{Re}\sum_{j\in\sigma}\langle x, f_j\rangle\langle h_j, S_{\mathcal{FGH}}x\rangle - \Big\|\sum_{j\in\sigma}\langle x, f_j\rangle h_j\Big\|^2 &= \mathrm{Re}\langle Px, S_{\mathcal{FGH}}x\rangle - \mathrm{Re}\langle Px, Px\rangle\\
&= \mathrm{Re}\langle Px, Qx\rangle = \mathrm{Re}\langle P^*Qx, x\rangle \ge 0.
\end{aligned}$$

Noting that

$$\begin{aligned}
\|Px\|^2 &= \|Qx\|^2 - \mathrm{Re}\langle Qx, S_{\mathcal{FGH}}x\rangle + \mathrm{Re}\langle Px, S_{\mathcal{FGH}}x\rangle\\
&= \mathrm{Re}\langle Qx, Qx\rangle - \mathrm{Re}\langle Qx, S_{\mathcal{FGH}}x\rangle + \mathrm{Re}\langle Px, S_{\mathcal{FGH}}x\rangle\\
&= -(\mathrm{Re}\langle Qx, S_{\mathcal{FGH}}x\rangle - \mathrm{Re}\langle Qx, Qx\rangle) + \mathrm{Re}\langle Px, S_{\mathcal{FGH}}x\rangle\\
&= -\mathrm{Re}\langle Qx, Px\rangle + \mathrm{Re}\langle Px, S_{\mathcal{FGH}}x\rangle \le \mathrm{Re}\langle Px, S_{\mathcal{FGH}}x\rangle,
\end{aligned}$$

and similarly,

$$\|Qx\|^2 \leq \mathrm{Re}\langle Qx, S_{\mathcal{FGH}}x\rangle,$$

we obtain

$$\begin{aligned}\Big\|\sum_{j\in\sigma}\langle x,f_j\rangle h_j\Big\|^2+\Big\|\sum_{j\in\sigma^c}\langle x,g_j\rangle h_j\Big\|^2 &= \|Px\|^2+\|Qx\|^2\\ &\leq \mathrm{Re}\langle Px, S_{\mathcal{FGH}}x\rangle+\mathrm{Re}\langle Qx, S_{\mathcal{FGH}}x\rangle\\ &= \mathrm{Re}\langle Px+Qx, S_{\mathcal{FGH}}x\rangle = \|S_{\mathcal{FGH}}x\|^2,\end{aligned}$$

and the proof is completed. □

Remark 2. *Suppose that the weaving frame $\{f_j\}_{j\in\sigma}\cup\{g_j\}_{j\in\sigma^c}$ is Parseval for each $\sigma\subset\mathbb{J}$, and letting $h_j=f_j$ if $j\in\sigma$ and $h_j=g_j$ if $j\in\sigma^c$, then it is easy to check that the operator P^*Q is positive.*

Corollary 4. *Suppose that two frames $\mathcal{F}=\{f_j\}_{j\in\mathbb{J}}$ and $\mathcal{G}=\{g_j\}_{j\in\mathbb{J}}$ in $\mathbb{H}$ are woven. Then for any $\sigma\subset\mathbb{J}$, for all $\lambda\in\mathbb{R}$ and all $x\in\mathbb{H}$, we have*

$$\begin{aligned}0 &\leq \sum_{j\in\sigma}|\langle x,f_j\rangle|^2-\sum_{j\in\sigma}|\langle S_W^{-1}S_{\mathcal{F}}^{\sigma}x,f_j\rangle|^2-\sum_{j\in\sigma^c}|\langle S_W^{-1}S_{\mathcal{F}}^{\sigma}x,g_j\rangle|^2\\ &\leq \frac{\lambda^2}{4}\sum_{j\in\sigma^c}|\langle x,g_j\rangle|^2+(1-\frac{\lambda}{2})^2\sum_{j\in\sigma}|\langle x,f_j\rangle|^2.\end{aligned}\tag{16}$$

$$\begin{aligned}&(2\lambda-\frac{\lambda^2}{2}-1)\sum_{j\in\sigma}|\langle x,f_j\rangle|^2+(1-\frac{\lambda^2}{2})\sum_{j\in\sigma^c}|\langle x,g_j\rangle|^2\\ &\quad\leq \sum_{j\in\sigma}|\langle S_W^{-1}S_{\mathcal{F}}^{\sigma}x,f_j\rangle|^2+\sum_{j\in\sigma^c}|\langle S_W^{-1}S_{\mathcal{F}}^{\sigma}x,g_j\rangle|^2\\ &\qquad+\sum_{j\in\sigma}|\langle S_W^{-1}S_{\mathcal{G}}^{\sigma^c}x,f_j\rangle|^2+\sum_{j\in\sigma^c}|\langle S_W^{-1}S_{\mathcal{G}}^{\sigma^c}x,g_j\rangle|^2\\ &\quad\leq \sum_{j\in\sigma}|\langle x,f_j\rangle|^2+\sum_{j\in\sigma^c}|\langle x,g_j\rangle|^2.\end{aligned}\tag{17}$$

Proof. Let $\mathcal{H}=\{h_j\}_{j\in\mathbb{J}}$ be the same as in the proof of Corollary 2. By combining Equations (10) and (12), and Theorem 2 we arrive at

$$\begin{aligned}&\sum_{j\in\sigma}|\langle x,f_j\rangle|^2-\sum_{j\in\sigma}|\langle S_W^{-1}S_{\mathcal{F}}^{\sigma}x,f_j\rangle|^2-\sum_{j\in\sigma^c}|\langle S_W^{-1}S_{\mathcal{F}}^{\sigma}x,g_j\rangle|^2\\ &\quad=\mathrm{Re}\sum_{j\in\sigma}\langle x,f_j\rangle\langle h_j,S_{\mathcal{FGH}}x\rangle-\Big\|\sum_{j\in\sigma}\langle x,f_j\rangle h_j\Big\|^2\\ &\quad\leq\frac{\lambda^2}{4}\mathrm{Re}\sum_{j\in\sigma^c}\langle x,g_j\rangle\langle h_j,S_{\mathcal{FGH}}x\rangle+(1-\frac{\lambda}{2})^2\mathrm{Re}\sum_{j\in\sigma}\langle x,f_j\rangle\langle h_j,S_{\mathcal{FGH}}x\rangle\\ &\quad=\frac{\lambda^2}{4}\sum_{j\in\sigma^c}|\langle x,g_j\rangle|^2+(1-\frac{\lambda}{2})^2\sum_{j\in\sigma}|\langle x,f_j\rangle|^2\end{aligned}$$

for each $x\in\mathbb{H}$. Let P and Q be given in Equation (7). Then a direct calculation shows that $P=S_W^{-\frac{1}{2}}S_{\mathcal{F}}^{\sigma}$ and $Q=S_W^{-\frac{1}{2}}S_{\mathcal{G}}^{\sigma^c}$ and, $P^*Q=S_{\mathcal{F}}^{\sigma}S_W^{-1}S_{\mathcal{G}}^{\sigma^c}$ as a consequence. Since $S_W^{-\frac{1}{2}}S_{\mathcal{F}}^{\sigma}S_W^{-\frac{1}{2}}$ and $S_W^{-\frac{1}{2}}S_{\mathcal{G}}^{\sigma^c}S_W^{-\frac{1}{2}}$ are positive and commutative,

$$0\leq S_W^{-\frac{1}{2}}S_{\mathcal{F}}^{\sigma}S_W^{-\frac{1}{2}}S_W^{-\frac{1}{2}}S_{\mathcal{G}}^{\sigma^c}S_W^{-\frac{1}{2}}=S_W^{-\frac{1}{2}}S_{\mathcal{F}}^{\sigma}S_W^{-1}S_{\mathcal{G}}^{\sigma^c}S_W^{-\frac{1}{2}},$$

implying that $S^{\sigma}_{\mathcal{F}} S^{-1}_{W} S^{\sigma^c}_{\mathcal{G}} = P^* Q \geq 0$. Again by Theorem 2,

$$\begin{aligned} 0 &\leq \operatorname{Re} \sum_{j\in\sigma} \langle x, f_j \rangle \langle h_j, S_{\mathcal{F}\mathcal{G}\mathcal{H}} x \rangle - \left\| \sum_{j\in\sigma} \langle x, f_j \rangle h_j \right\|^2 \\ &= \sum_{j\in\sigma} |\langle x, f_j \rangle|^2 - \sum_{j\in\sigma} |\langle S^{-1}_W S^{\sigma}_{\mathcal{F}} x, f_j \rangle|^2 - \sum_{j\in\sigma^c} |\langle S^{-1}_W S^{\sigma}_{\mathcal{F}} x, g_j \rangle|^2. \end{aligned}$$

We are now in a position to prove Equation (17). By Equations (10) and (11) we have

$$\begin{aligned} &\left\| \sum_{j\in\sigma} \langle x, f_j \rangle h_j \right\|^2 + \left\| \sum_{j\in\sigma^c} \langle x, g_j \rangle h_j \right\|^2 \\ &= \sum_{j\in\sigma} |\langle S^{-1}_W S^{\sigma}_{\mathcal{F}} x, f_j \rangle|^2 + \sum_{j\in\sigma^c} |\langle S^{-1}_W S^{\sigma}_{\mathcal{F}} x, g_j \rangle|^2 + \sum_{j\in\sigma} |\langle S^{-1}_W S^{\sigma^c}_{\mathcal{G}} x, f_j \rangle|^2 + \sum_{j\in\sigma^c} |\langle S^{-1}_W S^{\sigma^c}_{\mathcal{G}} x, g_j \rangle|^2 \end{aligned} \tag{18}$$

for any $x \in \mathbb{H}$. We also have

$$\|S_{\mathcal{F}\mathcal{G}\mathcal{H}} x\|^2 = \|S^{\frac{1}{2}}_W x\|^2 = \langle S_W x, x \rangle = \sum_{j\in\sigma} |\langle x, f_j \rangle|^2 + \sum_{j\in\sigma^c} |\langle x, g_j \rangle|^2.$$

This together with Equations (12), (13) and (18), and Theorem 2 gives Equation (17). □

Remark 3. *Inequalities (16) and (17) in Corollary 4 are respectively inequalities in Theorems 14 and 15 shown in [36].*

Suppose that $\mathcal{F} = \{f_j\}_{j\in\mathbb{J}}$, $\mathcal{G} = \{g_j\}_{j\in\mathbb{J}}$, and $\mathcal{H} = \{h_j\}_{j\in\mathbb{J}}$ are Bessel sequences for $\mathbb{H}$, and that $\{\alpha_j\}_{j\in\mathbb{J}}$ is a bounded sequence of complex numbers. For any $\sigma \subset \mathbb{J}$ and any $x \in \mathbb{H}$, we define linear bounded operators E^{σ}, E^{σ^c}, F^{σ} and F^{σ^c} respectively by

$$E^{\sigma} x = \sum_{j\in\sigma} (1-\alpha_j) \langle x, h_j \rangle f_j, \quad E^{\sigma^c} x = \sum_{j\in\sigma^c} (1-\alpha_j) \langle x, h_j \rangle g_j,$$

and

$$F^{\sigma} x = \sum_{j\in\sigma} \alpha_j \langle x, h_j \rangle f_j, \quad F^{\sigma^c} x = \sum_{j\in\sigma^c} \alpha_j \langle x, h_j \rangle g_j.$$

We are now ready to present a new triangle inequality for weaving frames.

Theorem 3. *Suppose that two frames $\mathcal{F} = \{f_j\}_{j\in\mathbb{J}}$ and $\mathcal{G} = \{g_j\}_{j\in\mathbb{J}}$ in $\mathbb{H}$ are woven. Then for any bounded sequence $\{\alpha_j\}_{j\in\mathbb{J}}$, for all $\sigma \subset \mathbb{J}$ and all $x \in \mathbb{H}$, we have*

$$\begin{aligned} \frac{3}{4}\|x\|^2 &\leq \left\| \sum_{j\in\sigma^c} \alpha_j \langle x, h_j \rangle g_j + \sum_{j\in\sigma} \alpha_j \langle x, h_j \rangle f_j \right\|^2 + \operatorname{Re}\Big(\sum_{j\in\sigma} (1-\alpha_j) \langle x, h_j \rangle \langle f_j, x \rangle + \sum_{j\in\sigma^c} (1-\alpha_j) \langle x, h_j \rangle \langle g_j, x \rangle \Big) \\ &\leq \frac{3 + \|(E^{\sigma} + E^{\sigma^c}) - (F^{\sigma} + F^{\sigma^c})\|^2}{4} \|x\|^2, \end{aligned} \tag{19}$$

where $\mathcal{H} = \{h_j\}_{j\in\mathbb{J}}$ is an alternate dual frame of the weaving frame $\{f_j\}_{j\in\sigma} \cup \{g_j\}_{j\in\sigma^c}$.

Proof. For any $\sigma \subset \mathbb{J}$, since $\mathcal{H} = \{h_j\}_{j\in\mathbb{J}}$ is an alternate dual frame of the weaving frame $\{f_j\}_{j\in\sigma} \cup \{g_j\}_{j\in\sigma^c}$, $E^{\sigma} + E^{\sigma^c} + F^{\sigma} + F^{\sigma^c} = \mathrm{Id}_{\mathbb{H}}$. For any $x \in \mathbb{H}$ we obtain

$$
\begin{aligned}
&\left\|\sum_{j\in\sigma^c} \alpha_j\langle x,h_j\rangle g_j + \sum_{j\in\sigma} \alpha_j\langle x,h_j\rangle f_j\right\|^2 + \mathrm{Re}\Big(\sum_{j\in\sigma}(1-\alpha_j)\langle x,h_j\rangle\langle f_j,x\rangle + \sum_{j\in\sigma^c}(1-\alpha_j)\langle x,h_j\rangle\langle g_j,x\rangle\Big)\\
&= \langle (F^{\sigma}+F^{\sigma^c})^*(F^{\sigma}+F^{\sigma^c})x,x\rangle + \mathrm{Re}(\langle E^{\sigma}x,x\rangle + \langle E^{\sigma^c}x,x\rangle)\\
&= \frac{1}{2}\langle (E^{\sigma}+E^{\sigma^c}+(E^{\sigma})^*+(E^{\sigma^c})^*)x,x\rangle + \langle (\mathrm{Id}_{\mathbb{H}}-(E^{\sigma}+E^{\sigma^c}))^*(\mathrm{Id}_{\mathbb{H}}-(E^{\sigma}+E^{\sigma^c}))x,x\rangle\\
&= \left\langle \left(\mathrm{Id}_{\mathbb{H}} - \frac{1}{2}(E^{\sigma}+E^{\sigma^c}+(E^{\sigma})^*+(E^{\sigma^c})^*) + (E^{\sigma}+E^{\sigma^c})^*(E^{\sigma}+E^{\sigma^c})\right)x,x\right\rangle \qquad (20)\\
&= \left\langle \left(\left((E^{\sigma}+E^{\sigma^c}) - \frac{1}{2}\mathrm{Id}_{\mathbb{H}}\right)^*\left((E^{\sigma}+E^{\sigma^c}) - \frac{1}{2}\mathrm{Id}_{\mathbb{H}}\right) + \frac{3}{4}\mathrm{Id}_{\mathbb{H}}\right)x,x\right\rangle\\
&= \left\|\left((E^{\sigma}+E^{\sigma^c}) - \frac{1}{2}\mathrm{Id}_{\mathbb{H}}\right)x\right\|^2 + \frac{3}{4}\|x\|^2\\
&\geq \frac{3}{4}\|x\|^2.
\end{aligned}
$$

On the other hand we get

$$
\begin{aligned}
&\left\|\sum_{j\in\sigma^c} \alpha_j\langle x,h_j\rangle g_j + \sum_{j\in\sigma} \alpha_j\langle x,h_j\rangle f_j\right\|^2 + \mathrm{Re}\Big(\sum_{j\in\sigma}(1-\alpha_j)\langle x,h_j\rangle\langle f_j,x\rangle + \sum_{j\in\sigma^c}(1-\alpha_j)\langle x,h_j\rangle\langle g_j,x\rangle\Big)\\
&= \langle (F^{\sigma}+F^{\sigma^c})x,(F^{\sigma}+F^{\sigma^c})x\rangle + \mathrm{Re}\langle (E^{\sigma}+E^{\sigma^c})x,x\rangle\\
&= \langle (F^{\sigma}+F^{\sigma^c})x,(F^{\sigma}+F^{\sigma^c})x\rangle + \mathrm{Re}(\langle x,x\rangle - \langle (F^{\sigma}+F^{\sigma^c})x,x\rangle)\\
&= \langle x,x\rangle - \mathrm{Re}\langle (F^{\sigma}+F^{\sigma^c})x,x\rangle + \langle (F^{\sigma}+F^{\sigma^c})x,(F^{\sigma}+F^{\sigma^c})x\rangle\\
&= \langle x,x\rangle - \mathrm{Re}\langle (F^{\sigma}+F^{\sigma^c})x,(E^{\sigma}+E^{\sigma^c})x\rangle\\
&= \langle x,x\rangle - \frac{1}{2}\langle (F^{\sigma}+F^{\sigma^c})x,(E^{\sigma}+E^{\sigma^c})x\rangle - \frac{1}{2}\langle (E^{\sigma}+E^{\sigma^c})x,(F^{\sigma}+F^{\sigma^c})x\rangle \qquad (21)\\
&= \frac{3}{4}\|x\|^2 + \frac{1}{4}\langle ((E^{\sigma}+E^{\sigma^c})+(F^{\sigma}+F^{\sigma^c}))x,((E^{\sigma}+E^{\sigma^c})+(F^{\sigma}+F^{\sigma^c}))x\rangle\\
&\quad - \frac{1}{2}\langle (F^{\sigma}+F^{\sigma^c})x,(E^{\sigma}+E^{\sigma^c})x\rangle - \frac{1}{2}\langle (E^{\sigma}+E^{\sigma^c})x,(F^{\sigma}+F^{\sigma^c})x\rangle\\
&= \frac{3}{4}\|x\|^2 + \frac{1}{4}\langle ((E^{\sigma}+E^{\sigma^c})-(F^{\sigma}+F^{\sigma^c}))x,((E^{\sigma}+E^{\sigma^c})-(F^{\sigma}+F^{\sigma^c}))x\rangle\\
&\leq \frac{3}{4}\|x\|^2 + \frac{1}{4}\|(E^{\sigma}+E^{\sigma^c})-(F^{\sigma}+F^{\sigma^c})\|^2\|x\|^2\\
&= \frac{3+\|(E^{\sigma}+E^{\sigma^c})-(F^{\sigma}+F^{\sigma^c})\|^2}{4}\|x\|^2.
\end{aligned}
$$

This along with Equation (20) yields Equation (19). □

Corollary 5. *Suppose that two frames $\mathcal{F} = \{f_j\}_{j\in\mathbb{J}}$ and $\mathcal{G} = \{g_j\}_{j\in\mathbb{J}}$ in $\mathbb{H}$ are woven. Then for all $\sigma \subset \mathbb{J}$ and all $x \in \mathbb{H}$, we have*

$$
\frac{3}{4}\|x\|^2 \leq \left\|\sum_{j\in\sigma}\langle x,h_j\rangle f_j\right\|^2 + \mathrm{Re}\sum_{j\in\sigma^c}\langle x,h_j\rangle\langle g_j,x\rangle \leq \frac{3+\|S^{\sigma^c}_{\mathcal{HG}} - S^{\sigma}_{\mathcal{HF}}\|^2}{4}\|x\|^2,
$$

where $S^{\sigma^c}_{\mathcal{HG}}, S^{\sigma}_{\mathcal{HF}} \in B(\mathbb{H})$ are defined respectively by

$$
S^{\sigma^c}_{\mathcal{HG}}x = \sum_{j\in\sigma^c}\langle x,h_j\rangle g_j \quad \text{and} \quad S^{\sigma}_{\mathcal{HF}}x = \sum_{j\in\sigma}\langle x,h_j\rangle f_j,
$$

and $\mathcal{H} = \{h_j\}_{j\in\mathbb{J}}$ *is an alternate dual frame of the weaving frame* $\{f_j\}_{j\in\sigma} \cup \{g_j\}_{j\in\sigma^c}$.

Proof. The conclusion follows by Theorem 3 if we take

$$\alpha_j = \begin{cases} 1, & j \in \sigma, \\ 0, & j \in \sigma^c. \end{cases}$$

□

Funding: This research was funded by the National Natural Science Foundation of China under grant numbers 11761057 and 11561057.

Conflicts of Interest: The author declares no conflict of interest.

References

1. Duffin, R.J.; Schaeffer, A.C. A class of nonharmonic Fourier series. *Trans. Am. Math. Soc.* **1952**, *72*, 341–366. [CrossRef]
2. Balan, R.; Wang, Y. Invertibility and robustness of phaseless reconstruction. *Appl. Comput. Harmonic Anal.* **2015**, *38*, 469–488. [CrossRef]
3. Botelho-Andrade, S.; Casazza, P.G.; Van Nguyen, H.; Tremain, J.C. Phase retrieval versus phaseless reconstruction. *J. Math. Anal. Appl.* **2016**, *436*, 131–137. [CrossRef]
4. Casazza, P.G. The art of frame theory. *Taiwan. J. Math.* **2000**, *4*, 129–201. [CrossRef]
5. Casazza, P.G.; Kutyniok, G. *Finite Frames: Theory and Applications*; Birkhäuser: Basel, Switzerland, 2013.
6. Casazza, P.G.; Ghoreishi, D.; Jose, S.; Tremain, J.C. Norm retrieval and phase retrieval by projections. *Axioms* **2017**, *6*, 6. [CrossRef]
7. Christensen, O. *An Introduction to Frames and Riesz Bases*; Birkhäuser: Boston, MA, USA, 2000.
8. Christensen, O.; Hasannasab, M. Operator representations of frames: Boundedness, duality, and stability. *Integral Equ. Oper. Theory* **2017**, *88*, 483–499. [CrossRef]
9. Christensen, O.; Hasannasab, M.; Rashidi, E. Dynamical sampling and frame representations with bounded operators. *J. Math. Anal. Appl.* **2018**, *463*, 634–644. [CrossRef]
10. Daubechies, I.; Grossmann, A.; Meyer, Y. Painless nonorthogonal expansions. *J. Math. Phys.* **1986**, *27*, 1271–1283. [CrossRef]
11. Găvruţa, P. On the Feichtinger conjecture. *Electron. J. Linear Algebra* **2013**, *26*, 546–552. [CrossRef]
12. Hasankhani Fard, M.A. Norm retrievable frames in $\mathbb{R}^n$. *Electron. J. Linear Algebra* **2016**, *31*, 425–432. [CrossRef]
13. Pehlivan, S.; Han, D.; Mohapatra, R.N. Spectrally two-uniform frames for erasures. *Oper. Matrices* **2015**, *9*, 383–399. [CrossRef]
14. Rahimi, A.; Seddighi, N. Finite equal norm Parseval wavelet frames over prime fields. *Int. J. Wavel. Multiresolut. Inf. Process.* **2017**, *15*, 1750040. [CrossRef]
15. Sahu, N.K.; Mohapatra, R.N. Frames in semi-inner product spaces. In *Mathematical Analysis and its Applicationsl*; Agrawal, P., Mohapatra, R., Singh, U., Srivastava, H., Eds.; Springer: New Delhi, India, 2015; Volume 143, pp. 149–158, ISBN 978-81-322-2484-6.
16. Xiao, X.C.; Zhou, G.R.; Zhu, Y.C. Uniform excess frames in Hilbert spaces. *Results Math.* **2018**, *73*, 108. [CrossRef]
17. Balan, R.; Casazza, P.G.; Edidin, D. On signal reconstruction without phase. *Appl. Comput. Harmonic Anal.* **2006**, *20*, 345–356. [CrossRef]
18. Han, D.; Sun, W. Reconstruction of signals from frame coefficients with erasures at unknown locations. *IEEE Trans. Inf. Theory* **2014**, *60*, 4013–4025. [CrossRef]
19. Benedetto, J.; Powell, A.; Yilmaz, O. Sigma-Delta ($\Sigma\Delta$) quantization and finite frames. *IEEE Trans. Inf. Theory* **2006**, *52*, 1990–2005. [CrossRef]
20. Jivulescu, M.A.; Găvruţa, P. Indices of sharpness for Parseval frames, quantum effects and observables. *Sci. Bull. Politeh. Univ. Timiş. Trans. Math. Phys.* **2015**, *60*, 17–29.
21. Strohmer, T.; Heath, R. Grassmannian frames with applications to coding and communication. *Appl. Comput. Harmonic Anal.* **2003**, *14*, 257–275. [CrossRef]

22. Sun, W. Asymptotic properties of Gabor frame operators as sampling density tends to infinity. *J. Funct. Anal.* **2010**, *258*, 913–932. [CrossRef]
23. Bemrose, T.; Casazza, P.G.; Gröchenig, K.; Lammers, M.C.; Lynch, R.G. Weaving frames. *Oper. Matrices* **2016**, *10*, 1093–1116. [CrossRef]
24. Casazza, P.G.; Freeman, D.; Lynch, R.G. Weaving Schauder frames. *J. Approx. Theory* **2016**, *211*, 42–60. [CrossRef]
25. Deepshikha; Vashisht, L.K. On weaving frames. *Houston J. Math.* **2018**, *44*, 887–915.
26. Deepshikha; Vashisht, L.K. Weaving K-frames in Hilbert spaces. *Results Math.* **2018**, *73*, 81. [CrossRef]
27. Khosravi, A.; Banyarani, J.S. Weaving g-frames and weaving fusion frames. *Bull. Malays. Math. Sci. Soc.* **2018**. [CrossRef]
28. Rahimi, A.; Samadzadeh, Z.; Daraby, B. Frame related operators for woven frames. *Int. J. Wavel. Multiresolut. Inf. Process.* **2018**. [CrossRef]
29. Vashisht, L.K.; Garg, S.; Deepshikha; Das, P.K. On generalized weaving frames in Hilbert spaces. *Rocky Mt. J. Math.* **2018**, *48*, 661–685. [CrossRef]
30. Vashisht, L.K.; Deepshikha. Weaving properties of generalized continuous frames generated by an iterated function system. *J. Geom. Phys.* **2016**, *110*, 282–295. [CrossRef]
31. Balan, R.; Casazza, P.G.; Edidin, D.; Kutyniok, G. A new identity for Parseval frames. *Proc. Am. Math. Soc.* **2007**, *135*, 1007–1015. [CrossRef]
32. Găvruţa, P. On some identities and inequalities for frames in Hilbert spaces. *J. Math. Anal. Appl.* **2006**, *321*, 469–478. [CrossRef]
33. Li, D.W.; Leng, J.S. On some new inequalities for fusion frames in Hilbert spaces. *Math. Inequal. Appl.* **2017**, *20*, 889–900. [CrossRef]
34. Li, D.W.; Leng, J.S. On some new inequalities for continuous fusion frames in Hilbert spaces. *Mediterr. J. Math.* **2018**, *15*, 173. [CrossRef]
35. Poria, A. Some identities and inequalities for Hilbert-Schmidt frames. *Mediterr. J. Math.* **2017**, *14*, 59. [CrossRef]
36. Li, D.W.; Leng, J.S. New inequalities for weaving frames in Hilbert spaces. *arXiv* **2018**, arXiv:1809.00863.

mathematics

Article

New Inequalities of Weaving K-Frames in Subspaces

Zhong-Qi Xiang

College of Mathematics and Computer Science, Shangrao Normal University, Shangrao 334001, China; lxsy20110927@163.com; Tel.: +86-793-815-9108

Received: 12 August 2019; Accepted: 16 September 2019; Published: 18 September 2019

Abstract: In the present paper, we obtain some new inequalities for weaving K-frames in subspaces based on the operator methods. The inequalities are associated with a sequence of bounded complex numbers and a parameter $\lambda \in \mathbb{R}$. We also give a double inequality for weaving K-frames with the help of two bounded linear operators induced by K-dual. Facts prove that our results cover those recently obtained on weaving frames due to Li and Leng, and Xiang.

Keywords: weaving frame; weaving K-frame; K-dual; pseudo-inverse

MSC: 42C15; 47B40

1. Introduction

This paper adopts the following notations: $\mathbb{J}$ is a countable index set, $\mathbb{H}$ and $\mathbb{K}$ are complex Hilbert spaces, and $\mathrm{Id}_{\mathbb{H}}$ and $\mathbb{R}$ are used to denote respectively the identical operator on $\mathbb{H}$ and the set of real numbers. As usual, we denote by $B(\mathbb{H},\mathbb{K})$ the set of all bounded linear operators on $\mathbb{H}$ and, if $\mathbb{H}=\mathbb{K}$, then $B(\mathbb{H},\mathbb{K})$ is abbreviated to $B(\mathbb{H})$.

Frames were introduced by Duffin and Schaeffer [1] in their study of nonharmonic Fourier series, which have now been used widely not only in theoretical work [2,3], but also in many application areas such as quantum mechanics [4], sampling theory [5–7], acoustics [8], and signal processing [9]. As a generalization of frames, the notion of K-frames (also known as frames for operators) was proposed by L. Găvruţa [10] when dealing with atomic decompositions for a bounded linear operator K. Please check the papers [11–17] for further information of K-frames.

Recall that a family $\{\psi_j\}_{j\in\mathbb{J}} \subset \mathbb{H}$ is called a K-frame for $\mathbb{H}$, if there exist two positive numbers A and B satisfying

$$A\|K^*f\|^2 \le \sum_{j\in\mathbb{J}} |\langle f,\psi_j\rangle|^2 \le B\|f\|^2, \quad \forall f\in\mathbb{H}.$$

The constants A and B are called K-frame bounds. If $K=\mathrm{Id}_{\mathbb{H}}$, then a K-frame turns to be a frame. In addition, if only the right-hand inequality holds, then we call $\{\psi_j\}_{j\in\mathbb{J}}$ a Bessel sequence.

Inspired by a question arising in distributed signal processing, Bemrose et al. [18] introduced the concept of weaving frames, which have interested many scholars because of their potential applications such as in wireless sensor networks and pre-processing of signals; see [19–24]. Later on, Deepshikha and Vashisht [25] applied the idea of L. Găvruţa to the case of weaving frames and thus providing us the notion of weaving K-frames.

Balan et al. [26] obtained an interesting inequality when they further examined the remarkable identity for Parseval frames deriving from their work on signal reconstruction [27]. The inequality was then extended to alternate dual frames and general frames by P. Găvruţa [28], the results in which have already been applied in quantum information theory [29]. Recently, those inequalities have been extended to some generalized versions of frames such as continuous g-frames [30], fusion frames and continuous fusion frames [31,32], Hilbert–Schmidt frames [33], and weaving frames [34,35].

Motivated by the above-mentioned works, in this paper, we establish several new inequalities for weaving K-frames in subspaces from the operator-theoretic point of view, and we show that our results can naturally lead to some corresponding results in [34,35].

One says that two frames $\Psi_1 = \{\psi_{1j}\}_{j\in\mathbb{J}}$ and $\Psi_2 = \{\psi_{2j}\}_{j\in\mathbb{J}}$ in $\mathbb{H}$ are woven, if there are universal constants C_Ψ and D_Ψ such that, for any $\sigma \subset \mathbb{J}$, $\{\psi_{1j}\}_{j\in\sigma} \cup \{\psi_{2j}\}_{j\in\sigma^c}$ is a frame for $\mathbb{H}$ with bounds C_Ψ and D_Ψ. If $C_\Psi = D_\Psi = 1$, then we call Ψ_1 and Ψ_2 1-woven. Each family $\{\psi_{1j}\}_{j\in\sigma} \cup \{\psi_{2j}\}_{j\in\sigma^c}$ is said to be a waving frame, related to which there is an invertible operator $S_{\Psi_1\Psi_2} : \mathbb{H} \to \mathbb{H}$, called the frame operator, given by

$$S_{\Psi_1\Psi_2}f = \sum_{j\in\sigma}\langle f, \psi_{1j}\rangle\psi_{1j} + \sum_{j\in\sigma^c}\langle f, \psi_{2j}\rangle\psi_{2j}.$$

Recall also that a frame $\Psi_3 = \{\psi_{3j}\}_{j\in\mathbb{J}}$ is called an alternate dual frame of $\{\psi_{1j}\}_{j\in\sigma} \cup \{\psi_{2j}\}_{j\in\sigma^c}$, if for each $f \in \mathbb{H}$ we have

$$f = \sum_{j\in\sigma}\langle f, \psi_{1j}\rangle\psi_{3j} + \sum_{j\in\sigma^c}\langle f, \psi_{2j}\rangle\psi_{3j}, \quad \forall f \in \mathbb{H}.$$

Lemma 1. *Suppose that P, Q, and K are bounded linear operators on $\mathbb{H}$ and $P + Q = K$. Then, for each $f \in \mathbb{H}$,*

$$\|Pf\|^2 + \mathrm{Re}\langle Qf, Kf\rangle \geq \frac{3}{4}\|Kf\|^2.$$

Proof. We have

$$\begin{aligned}\|Pf\|^2 + \mathrm{Re}\langle Qf, Kf\rangle &= \langle (K-Q)f, (K-Q)f\rangle + \frac{1}{2}(\langle Qf, Kf\rangle + \langle Kf, Qf\rangle)\\ &= \langle (Q^*Q - (K^*Q + Q^*K) + \frac{1}{2}(K^*Q + Q^*K))f, f\rangle + \langle K^*Kf, f\rangle\\ &= \langle (Q - \frac{1}{2}K)^*(Q - \frac{1}{2}K)f, f\rangle + \frac{3}{4}\langle K^*Kf, f\rangle \geq \frac{3}{4}\|Kf\|^2\end{aligned}$$

for any $f \in \mathbb{H}$. □

The next two lemmas are collected from the papers [36] and [32], respectively.

Lemma 2. *If $\Phi \in B(\mathbb{H}, \mathbb{K})$ has a closed range, then there is the pseudo-inverse $\Phi^\dagger \in B(\mathbb{K}, \mathbb{H})$ of Φ such that*

$$\Phi\Phi^\dagger\Phi = \Phi, \ \Phi^\dagger\Phi\Phi^\dagger = \Phi^\dagger, \ (\Phi\Phi^\dagger)^* = \Phi\Phi^\dagger, \ (\Phi^\dagger\Phi)^* = \Phi^\dagger\Phi.$$

Lemma 3. *If P and Q in $B(\mathbb{H})$ satisfy $P + Q = \mathrm{Id}_\mathbb{H}$, then, for any $\lambda \in \mathbb{R}$, we have*

$$P^*P + \lambda(Q^* + Q) = Q^*Q + (1-\lambda)(P^* + P) + (2\lambda - 1)\mathrm{Id}_\mathbb{H} \geq (2\lambda - \lambda^2)\mathrm{Id}_\mathbb{H}.$$

2. Main Results

We start with the definition on weaving K-frames due to Deepshikha and Vashisht [25].

Definition 1. *Two K-frames $\Psi_1 = \{\psi_{1j}\}_{j\in\mathbb{J}}$ and $\Psi_2 = \{\psi_{2j}\}_{j\in\mathbb{J}}$ in $\mathbb{H}$ are said to be K-woven, if there are universal constants C_Ψ and D_Ψ such that, for any $\sigma \subset \mathbb{J}$, the family $\{\psi_{1j}\}_{j\in\sigma} \cup \{\psi_{2j}\}_{j\in\sigma^c}$ is a K-frame for $\mathbb{H}$ with K-frame bounds C_Ψ and D_Ψ. In this case, the family $\{\psi_{1j}\}_{j\in\sigma} \cup \{\psi_{2j}\}_{j\in\sigma^c}$ is called a weaving K frame.*

Given a weaving K-frame $\{\psi_{1j}\}_{j\in\sigma} \cup \{\psi_{2j}\}_{j\in\sigma^c}$ for $\mathbb{H}$, recall that a Bessel sequence $\Phi = \{\phi_j\}_{j\in\mathbb{J}}$ for $\mathbb{H}$ is said to be a K-dual of $\{\psi_{1j}\}_{j\in\sigma} \cup \{\psi_{2j}\}_{j\in\sigma^c}$, if

$$Kf = \sum_{j\in\sigma}\langle f, \psi_{1j}\rangle\phi_j + \sum_{j\in\sigma^c}\langle f, \psi_{2j}\rangle\phi_j, \quad \forall f \in \mathbb{H}.$$

Let $\Psi_1 = \{\psi_{1j}\}_{j\in\mathbb{J}}$ be a given K-frame for $\mathbb{H}$. For any $\sigma \subset \mathbb{J}$, we can define a positive operator $S^{\sigma}_{\Psi_1}$ in the following way:

$$S^{\sigma}_{\Psi_1} : \mathbb{H} \to \mathbb{H}, \quad S^{\sigma}_{\Psi_1} f = \sum_{j\in\sigma} \langle f, \psi_{1j}\rangle \psi_{1j}.$$

In the following, we show that, for given two K-woven frames, we can get some inequalities under the condition that K has a closed range, which are related to a sequence of bounded complex numbers, the corresponding K-dual and a parameter $\lambda \in \mathbb{R}$.

Theorem 1. *Suppose that $K \in B(\mathbb{H})$ has a closed range and K-frames $\Psi_1 = \{\psi_{1j}\}_{j\in\mathbb{J}}$ and $\Psi_2 = \{\psi_{2j}\}_{j\in\mathbb{J}}$ in $\mathbb{H}$ are K-woven. Then,*

(i) for any $f \in Range(K)$, for all $\sigma \subset \mathbb{J}$, $\{a_j\}_{j\in\mathbb{J}} \in \ell^\infty(\mathbb{J})$, and $\lambda \in \mathbb{R}$,

$$\begin{aligned}
&\left\|\sum_{j\in\sigma} a_j \langle K^{\dagger} f, \psi_{1j}\rangle \phi_j + \sum_{j\in\sigma^c} a_j \langle K^{\dagger} f, \psi_{2j}\rangle \phi_j\right\|^2 \\
&\quad + \mathrm{Re}\left(\sum_{j\in\sigma} (1-a_j)\langle K^{\dagger} f, \psi_{1j}\rangle \langle \phi_j, f\rangle + \sum_{j\in\sigma^c} (1-a_j)\langle K^{\dagger} f, \psi_{2j}\rangle \langle \phi_j, f\rangle\right) \\
&= \left\|\sum_{j\in\sigma} (1-a_j)\langle K^{\dagger} f, \psi_{1j}\rangle \phi_j + \sum_{j\in\sigma^c} (1-a_j)\langle K^{\dagger} f, \psi_{2j}\rangle \phi_j\right\|^2 \\
&\quad + \mathrm{Re}\left(\sum_{j\in\sigma} a_j \langle K^{\dagger} f, \psi_{1j}\rangle \langle \phi_j, f\rangle + \sum_{j\in\sigma^c} a_j \langle K^{\dagger} f, \psi_{2j}\rangle \langle \phi_j, f\rangle\right) \\
&\geq (\lambda - \frac{\lambda^2}{4})\mathrm{Re}\left(\sum_{j\in\sigma} a_j \langle K^{\dagger} f, \psi_{1j}\rangle \langle \phi_j, f\rangle + \sum_{j\in\sigma^c} a_j \langle K^{\dagger} f, \psi_{2j}\rangle \langle \phi_j, f\rangle\right) \\
&\quad + (1 - \frac{\lambda^2}{4})\mathrm{Re}\left(\sum_{j\in\sigma} (1-a_j)\langle K^{\dagger} f, \psi_{1j}\rangle \langle \phi_j, f\rangle + \sum_{j\in\sigma^c} (1-a_j)\langle K^{\dagger} f, \psi_{2j}\rangle \langle \phi_j, f\rangle\right),
\end{aligned} \tag{1}$$

where $\Phi = \{\phi_j\}_{j\in\mathbb{J}}$ is a K-dual of $\{\psi_{1j}\}_{j\in\sigma} \cup \{\psi_{2j}\}_{j\in\sigma^c}$.

(ii) for any $f \in Range(K^*)$, for all $\sigma \subset \mathbb{J}$, $\{a_j\}_{j\in\mathbb{J}} \in \ell^\infty(\mathbb{J})$, and $\lambda \in \mathbb{R}$,

$$\begin{aligned}
&\left\|\sum_{j\in\sigma} a_j \langle (K^*)^{\dagger} f, \phi_j\rangle \psi_{1j} + \sum_{j\in\sigma^c} a_j \langle (K^*)^{\dagger} f, \phi_j\rangle \psi_{2j}\right\|^2 \\
&\quad + \mathrm{Re}\left(\sum_{j\in\sigma} (1-a_j)\langle (K^*)^{\dagger} f, \phi_j\rangle \langle \psi_{1j}, f\rangle + \sum_{j\in\sigma^c} (1-a_j)\langle (K^*)^{\dagger} f, \phi_j\rangle \langle \psi_{2j}, f\rangle\right) \\
&= \left\|\sum_{j\in\sigma} (1-a_j)\langle (K^*)^{\dagger} f, \phi_j\rangle \psi_{1j} + \sum_{j\in\sigma^c} (1-a_j)\langle (K^*)^{\dagger} f, \phi_j\rangle \psi_{2j}\right\|^2 \\
&\quad + \mathrm{Re}\left(\sum_{j\in\sigma} a_j \langle (K^*)^{\dagger} f, \phi_j\rangle \langle \psi_{1j}, f\rangle + \sum_{j\in\sigma^c} a_j \langle (K^*)^{\dagger} f, \phi_j\rangle \langle \psi_{2j}, f\rangle\right) \\
&\geq (2\lambda - \lambda^2)\mathrm{Re}\left(\sum_{j\in\sigma} a_j \langle (K^*)^{\dagger} f, \phi_j\rangle \langle \psi_{1j}, f\rangle + \sum_{j\in\sigma^c} a_j \langle (K^*)^{\dagger} f, \phi_j\rangle \langle \psi_{2j}, f\rangle\right) \\
&\quad + (1 - \lambda^2)\mathrm{Re}\left(\sum_{j\in\sigma} (1-a_j)\langle (K^*)^{\dagger} f, \phi_j\rangle \langle \psi_{1j}, f\rangle + \sum_{j\in\sigma^c} (1-a_j)\langle (K^*)^{\dagger} f, \phi_j\rangle \langle \psi_{2j}, f\rangle\right),
\end{aligned}$$

where $\Phi = \{\phi_j\}_{j\in\mathbb{J}}$ is a K-dual of $\{\psi_{1j}\}_{j\in\sigma} \cup \{\psi_{2j}\}_{j\in\sigma^c}$.

Proof. We define two bounded linear operators P_1 and P_2 on $\mathbb{H}$ as follows:

$$\begin{aligned} P_1 f &= \sum_{j\in\sigma} a_j \langle f, \psi_{1j}\rangle \phi_j + \sum_{j\in\sigma^c} a_j \langle f, \psi_{2j}\rangle \phi_j, \\ P_2 f &= \sum_{j\in\sigma} (1-a_j) \langle f, \psi_{1j}\rangle \phi_j + \sum_{j\in\sigma^c} (1-a_j) \langle f, \psi_{2j}\rangle \phi_j. \end{aligned} \tag{2}$$

Then, clearly, $P_1 f + P_2 f = Kf$ for each $f \in \mathbb{H}$ and thus $P_1 + P_2 = K$. Since K has a closed range, by Lemma 2, we have

$$P_1 K^\dagger + P_2 K^\dagger = KK^\dagger = P_{Range(K)},$$

where $P_{Range(K)}$ is the orthogonal projection onto $Range(K)$. Thus,

$$P_1 K^\dagger \mid_{Range(K)} + P_2 K^\dagger \mid_{Range(K)} = \mathrm{Id}_{Range(K)}.$$

By Lemma 3 (taking $\frac{\lambda}{2}$ instead of λ), we get

$$\|P_1 K^\dagger f\|^2 + \lambda \mathrm{Re}\langle P_2 K^\dagger f, f\rangle = \|P_2 K^\dagger f\|^2 + (2-\lambda)\mathrm{Re}\langle P_1 K^\dagger f, f\rangle + (\lambda - 1)\|f\|^2,$$

for any $f \in Range(K)$. Hence,

$$\begin{aligned} \|P_1 K^\dagger f\|^2 &= \|P_2 K^\dagger f\|^2 + 2\mathrm{Re}\langle P_1 K^\dagger f, f\rangle - \lambda(\mathrm{Re}\langle P_1 K^\dagger f, f\rangle + \mathrm{Re}\langle P_2 K^\dagger f, f\rangle) + (\lambda-1)\|f\|^2 \\ &= \|P_2 K^\dagger f\|^2 + 2\mathrm{Re}\langle P_1 K^\dagger f, f\rangle - \lambda\|f\|^2 + (\lambda-1)\|f\|^2 \\ &= \|P_2 K^\dagger f\|^2 + 2\mathrm{Re}\langle P_1 K^\dagger f, f\rangle - \mathrm{Re}\langle P_1 K^\dagger f, f\rangle - \mathrm{Re}\langle P_2 K^\dagger f, f\rangle. \end{aligned}$$

It follows that

$$\|P_1 K^\dagger f\|^2 + \mathrm{Re}\langle P_2 K^\dagger f, f\rangle = \|P_2 K^\dagger f\|^2 + \mathrm{Re}\langle P_1 K^\dagger f, f\rangle, \tag{3}$$

from which we arrive at

$$\begin{aligned} &\left\| \sum_{j\in\sigma} a_j \langle K^\dagger f, \psi_{1j}\rangle \phi_j + \sum_{j\in\sigma^c} a_j \langle K^\dagger f, \psi_{2j}\rangle \phi_j \right\|^2 \\ &\quad + \mathrm{Re}\left(\sum_{j\in\sigma} (1-a_j) \langle K^\dagger f, \psi_{1j}\rangle \langle \phi_j, f\rangle + \sum_{j\in\sigma^c} (1-a_j) \langle K^\dagger f, \psi_{2j}\rangle \langle \phi_j, f\rangle \right) \\ &= \left\| \sum_{j\in\sigma} (1-a_j) \langle K^\dagger f, \psi_{1j}\rangle \phi_j + \sum_{j\in\sigma^c} (1-a_j) \langle K^\dagger f, \psi_{2j}\rangle \phi_j \right\|^2 \\ &\quad + \mathrm{Re}\left(\sum_{j\in\sigma} a_j \langle K^\dagger f, \psi_{1j}\rangle \langle \phi_j, f\rangle + \sum_{j\in\sigma^c} a_j \langle K^\dagger f, \psi_{2j}\rangle \langle \phi_j, f\rangle \right). \end{aligned}$$

For the inequality in Equation (1), we apply Lemma 3 again,

$$\begin{aligned} \|P_1 K^\dagger f\|^2 &\geq (\lambda - \frac{\lambda^2}{4})\|f\|^2 - \lambda \mathrm{Re}\langle P_2 K^\dagger f, f\rangle \\ &= (\lambda - \frac{\lambda^2}{4})\mathrm{Re}\langle P_1 K^\dagger f + P_2 K^\dagger f, f\rangle - \lambda \mathrm{Re}\langle P_2 K^\dagger f, f\rangle \\ &= (\lambda - \frac{\lambda^2}{4})\mathrm{Re}\langle P_1 K^\dagger f, f\rangle - \frac{\lambda^2}{4}\mathrm{Re}\langle P_2 K^\dagger f, f\rangle. \end{aligned} \tag{4}$$

Thus, for any $f \in Range(K)$,

$$\begin{aligned}
&\left\|\sum_{j\in\sigma} a_j\langle K^{\dagger}f,\psi_{1j}\rangle\phi_j + \sum_{j\in\sigma^c} a_j\langle K^{\dagger}f,\psi_{2j}\rangle\phi_j\right\|^2 \\
&\qquad + \mathrm{Re}\Big(\sum_{j\in\sigma}(1-a_j)\langle K^{\dagger}f,\psi_{1j}\rangle\langle\phi_j,f\rangle + \sum_{j\in\sigma^c}(1-a_j)\langle K^{\dagger}f,\psi_{2j}\rangle\langle\phi_j,f\rangle\Big) \\
&\geq (\lambda-\frac{\lambda^2}{4})\mathrm{Re}\langle P_1K^{\dagger}f,f\rangle + (1-\frac{\lambda^2}{4})\mathrm{Re}\langle P_2K^{\dagger}f,f\rangle \\
&= (\lambda-\frac{\lambda^2}{4})\mathrm{Re}\Big(\sum_{j\in\sigma} a_j\langle K^{\dagger}f,\psi_{1j}\rangle\langle\phi_j,f\rangle + \sum_{j\in\sigma^c} a_j\langle K^{\dagger}f,\psi_{2j}\rangle\langle\phi_j,f\rangle\Big) \\
&\qquad + (1-\frac{\lambda^2}{4})\mathrm{Re}\Big(\sum_{j\in\sigma}(1-a_j)\langle K^{\dagger}f,\psi_{1j}\rangle\langle\phi_j,f\rangle + \sum_{j\in\sigma^c}(1-a_j)\langle K^{\dagger}f,\psi_{2j}\rangle\langle\phi_j,f\rangle\Big).
\end{aligned}$$

(ii) The proof is similar to (i), so we omit the details. □

Corollary 1. *Suppose that two frames $\Psi_1 = \{\psi_{1j}\}_{j\in\mathbb{J}}$ and $\Psi_2 = \{\psi_{2j}\}_{j\in\mathbb{J}}$ in $\mathbb{H}$ are woven. Then, for any $f \in \mathbb{H}$, for all $\sigma \subset \mathbb{J}$ and all $\lambda \in \mathbb{R}$, we have*

$$\begin{aligned}
&\sum_{j\in\sigma^c}|\langle f,\psi_{2j}\rangle|^2 + \sum_{j\in\sigma}|\langle S^{\sigma}_{\Psi_1}f, S^{-1}_{\Psi_1\Psi_2}\psi_{1j}\rangle|^2 + \sum_{j\in\sigma^c}|\langle S^{\sigma}_{\Psi_1}f, S^{-1}_{\Psi_1\Psi_2}\psi_{2j}\rangle|^2 \\
&\quad = \sum_{j\in\sigma}|\langle f,\psi_{1j}\rangle|^2 + \sum_{j\in\sigma}|\langle S^{\sigma^c}_{\Psi_2}f, S^{-1}_{\Psi_1\Psi_2}\psi_{1j}\rangle|^2 + \sum_{j\in\sigma^c}|\langle S^{\sigma^c}_{\Psi_2}f, S^{-1}_{\Psi_1\Psi_2}\psi_{2j}\rangle|^2 \\
&\quad \geq (\lambda-\frac{\lambda^2}{4})\sum_{j\in\sigma}|\langle f,\psi_{1j}\rangle|^2 + (1-\frac{\lambda^2}{4})\sum_{j\in\sigma^c}|\langle f,\psi_{2j}\rangle|^2.
\end{aligned}$$

Proof. Letting $K^{\dagger} = \mathrm{Id}_{\mathbb{H}}$ and

$$\phi_j = \begin{cases} S^{-1/2}_{\Psi_1\Psi_2}\psi_{1j}, & j\in\sigma, \\ S^{-1/2}_{\Psi_1\Psi_2}\psi_{2j}, & j\in\sigma^c. \end{cases}$$

In addition, taking $S^{-1/2}_{\Psi_1\Psi_2}\psi_{1j}$, $S^{-1/2}_{\Psi_1\Psi_2}\psi_{2j}$ and $S^{1/2}_{\Psi_1\Psi_2}f$ instead of ψ_{1j}, ψ_{2j} and f respectively in (i) of Theorem 1 leads to

$$\begin{aligned}
&\left\|\sum_{j\in\sigma} a_j\langle f,\psi_{1j}\rangle S^{-1/2}_{\Psi_1\Psi_2}\psi_{1j} + \sum_{j\in\sigma^c} a_j\langle f,\psi_{2j}\rangle S^{-1/2}_{\Psi_1\Psi_2}\psi_{2j}\right\|^2 \\
&\qquad + \mathrm{Re}\Big(\sum_{j\in\sigma}(1-a_j)\langle f,\psi_{1j}\rangle\langle\psi_{1j},f\rangle + \sum_{j\in\sigma^c}(1-a_j)\langle f,\psi_{2j}\rangle\langle\psi_{2j},f\rangle\Big) \\
&= \left\|\sum_{j\in\sigma}(1-a_j)\langle f,\psi_{1j}\rangle S^{-1/2}_{\Psi_1\Psi_2}\psi_{1j} + \sum_{j\in\sigma^c}(1-a_j)\langle f,\psi_{2j}\rangle S^{-1/2}_{\Psi_1\Psi_2}\psi_{2j}\right\|^2 \\
&\qquad + \mathrm{Re}\Big(\sum_{j\in\sigma} a_j\langle f,\psi_{1j}\rangle\langle\psi_{1j},f\rangle + \sum_{j\in\sigma^c} a_j\langle f,\psi_{2j}\rangle\langle\psi_{2j},f\rangle\Big) \\
&\geq (\lambda-\frac{\lambda^2}{4})\mathrm{Re}\Big(\sum_{j\in\sigma} a_j\langle f,\psi_{1j}\rangle\langle\psi_{1j},f\rangle + \sum_{j\in\sigma^c} a_j\langle f,\psi_{2j}\rangle\langle\psi_{2j},f\rangle\Big) \\
&\qquad + (1-\frac{\lambda^2}{4})\mathrm{Re}\Big(\sum_{j\in\sigma}(1-a_j)\langle f,\psi_{1j}\rangle\langle\psi_{1j},f\rangle + \sum_{j\in\sigma^c}(1-a_j)\langle f,\psi_{2j}\rangle\langle\psi_{2j},f\rangle\Big). \qquad (5)
\end{aligned}$$

A direction calculation shows that

$$\begin{aligned}\left\|\sum_{j\in\sigma}\langle f,\psi_{1j}\rangle S_{\Psi_1\Psi_2}^{-1/2}\psi_{1j}\right\|^2 &= \left\|S_{\Psi_1\Psi_2}^{-1/2}\sum_{j\in\sigma}\langle f,\psi_{1j}\rangle\psi_{1j}\right\|^2 = \|S_{\Psi_1\Psi_2}^{-1/2}S_{\Psi_1}^{\sigma}f\|^2 \\ &= \langle S_{\Psi_1\Psi_2}^{-1/2}S_{\Psi_1}^{\sigma}f, S_{\Psi_1\Psi_2}^{-1/2}S_{\Psi_1}^{\sigma}f\rangle = \langle S_{\Psi_1\Psi_2}S_{\Psi_1\Psi_2}^{-1}S_{\Psi_1}^{\sigma}f, S_{\Psi_1\Psi_2}^{-1}S_{\Psi_1}^{\sigma}f\rangle \\ &= \sum_{j\in\sigma}\langle S_{\Psi_1\Psi_2}^{-1}S_{\Psi_1}^{\sigma}f,\psi_{1j}\rangle\langle\psi_{1j},S_{\Psi_1\Psi_2}^{-1}S_{\Psi_1}^{\sigma}f\rangle + \sum_{j\in\sigma^c}\langle S_{\Psi_1\Psi_2}^{-1}S_{\Psi_1}^{\sigma}f,\psi_{2j}\rangle\langle\psi_{2j},S_{\Psi_1\Psi_2}^{-1}S_{\Psi_1}^{\sigma}f\rangle \\ &= \sum_{j\in\sigma}|\langle S_{\Psi_1}^{\sigma}f,S_{\Psi_1\Psi_2}^{-1}\psi_{1j}\rangle|^2 + \sum_{j\in\sigma^c}|\langle S_{\Psi_1}^{\sigma}f,S_{\Psi_1\Psi_2}^{-1}\psi_{2j}\rangle|^2, \qquad (6)\end{aligned}$$

and, similarly,

$$\left\|\sum_{j\in\sigma^c}\langle f,\psi_{2j}\rangle S_{\Psi_1\Psi_2}^{-1/2}\psi_{2j}\right\|^2 = \sum_{j\in\sigma}|\langle S_{\Psi_2}^{\sigma^c}f,S_{\Psi_1\Psi_2}^{-1}\psi_{1j}\rangle|^2 + \sum_{j\in\sigma^c}|\langle S_{\Psi_2}^{\sigma^c}f,S_{\Psi_1\Psi_2}^{-1}\psi_{2j}\rangle|^2. \qquad (7)$$

Thus, the result follows if, in Equation (5), we take $a_j = \begin{cases}1, & j\in\sigma,\\ 0, & j\in\sigma^c.\end{cases}$ □

Corollary 2. *Suppose that two frames $\Psi_1 = \{\psi_{1j}\}_{j\in\mathbb{J}}$ and $\Psi_2 = \{\psi_{2j}\}_{j\in\mathbb{J}}$ in $\mathbb{H}$ are woven. Then, for any $\sigma\subset\mathbb{J}$, for all $\lambda\in\mathbb{R}$ and all $f\in\mathbb{H}$, we have*

$$\begin{aligned}&\left\|\sum_{j\in\sigma}\langle f,\phi_j\rangle\psi_{1j}\right\|^2 + \mathrm{Re}\sum_{j\in\sigma^c}\langle f,\phi_j\rangle\langle\psi_{2j},f\rangle \\ &= \left\|\sum_{j\in\sigma^c}\langle f,\phi_j\rangle\psi_{2j}\right\|^2 + \mathrm{Re}\sum_{j\in\sigma}\langle f,\phi_j\rangle\langle\psi_{1j},f\rangle \\ &\geq (2\lambda-\lambda^2)\mathrm{Re}\sum_{j\in\sigma}\langle f,\phi_j\rangle\langle\psi_{1j},f\rangle + (1-\lambda^2)\mathrm{Re}\sum_{j\in\sigma^c}\langle f,\phi_j\rangle\langle\psi_{2j},f\rangle,\end{aligned}$$

where $\Phi = \{\phi_j\}_{j\in\mathbb{J}}$ is an alternate dual of $\{\psi_{1j}\}_{j\in\sigma}\cup\{\psi_{2j}\}_{j\in\sigma^c}$.

Proof. The result follows immediately from (ii) in Theorem 1 when taking $K^{\dagger} = \mathrm{Id}_{\mathbb{H}}$ and

$$a_j = \begin{cases}1, & j\in\sigma,\\ 0, & j\in\sigma^c.\end{cases}$$

□

Suppose that two frames $\Psi_1 = \{\psi_{1j}\}_{j\in\mathbb{J}}$ and $\Psi_2 = \{\psi_{2j}\}_{j\in\mathbb{J}}$ in $\mathbb{H}$ are 1-woven. For any $\sigma\subset\mathbb{J}$ and any $j\in\mathbb{J}$, taking $\phi_j = \begin{cases}\psi_{1j}, & j\in\sigma,\\ \psi_{2j}, & j\in\sigma^c.\end{cases}$ Then, obviously, $\Phi = \{\phi_j\}_{j\in\mathbb{J}}$ is an alternate dual of the frame $\{\psi_{1j}\}_{j\in\sigma}\cup\{\psi_{2j}\}_{j\in\sigma^c}$. Thus, Corollary 2 provides us a direct consequence as follows.

Corollary 3. *Let the two frames $\Psi_1 = \{\psi_{1j}\}_{j\in\mathbb{J}}$ and $\Psi_2 = \{\psi_{2j}\}_{j\in\mathbb{J}}$ in $\mathbb{H}$ be 1-woven. Then, for any $\sigma\subset\mathbb{J}$, for all $\lambda\subset\mathbb{R}$ and all $f\in\mathbb{H}$, we have*

$$\begin{aligned}\left\|\sum_{j\in\sigma}\langle f,\psi_{1j}\rangle\psi_{1j}\right\|^2 + \sum_{j\in\sigma^c}|\langle f,\psi_{2j}\rangle|^2 &= \left\|\sum_{j\in\sigma^c}\langle f,\psi_{2j}\rangle\psi_{2j}\right\|^2 + \sum_{j\in\sigma}|\langle f,\psi_{1j}\rangle|^2 \\ &\geq (2\lambda-\lambda^2)\sum_{j\in\sigma}|\langle f,\psi_{1j}\rangle|^2 + (1-\lambda^2)\sum_{j\in\sigma^c}|\langle f,\psi_{2j}\rangle|^2.\end{aligned}$$

Remark 1. *Corollaries 1 and 2 are respectively Theorems 7 and 9 in [34], and Theorem 5 in [34] can be obtained if we put $\lambda = \frac{1}{2}$ in Corollary 3.*

Theorem 2. *Suppose that $K \in B(\mathbb{H})$ has a closed range and that K-frames $\Psi_1 = \{\psi_{1j}\}_{j\in\mathbb{J}}$ and $\Psi_2 = \{\psi_{2j}\}_{j\in\mathbb{J}}$ in $\mathbb{H}$ are K-woven. Then, for any $f \in Range(K)$, for all $\sigma \subset \mathbb{J}$, $\{a_j\}_{j\in\mathbb{J}} \in \ell^\infty(\mathbb{J})$, and $\lambda \in \mathbb{R}$,*

$$\begin{aligned}
&\Big\|\sum_{j\in\sigma} a_j\langle K^\dagger f, \psi_{1j}\rangle\phi_j + \sum_{j\in\sigma^c} a_j\langle K^\dagger f, \psi_{2j}\rangle\phi_j\Big\|^2 + \Big\|\sum_{j\in\sigma}(1-a_j)\langle K^\dagger f, \psi_{1j}\rangle\phi_j + \sum_{j\in\sigma^c}(1-a_j)\langle K^\dagger f, \psi_{2j}\rangle\phi_j\Big\|^2 \\
&\geq (2\lambda - \frac{\lambda^2}{2} - 1)\mathrm{Re}\Big(\sum_{j\in\sigma} a_j\langle K^\dagger f, \psi_{1j}\rangle\langle\phi_j, f\rangle + \sum_{j\in\sigma^c} a_j\langle K^\dagger f, \psi_{2j}\rangle\langle\phi_j, f\rangle\Big) \\
&\quad + (1 - \frac{\lambda^2}{2})\mathrm{Re}\Big(\sum_{j\in\sigma}(1-a_j)\langle K^\dagger f, \psi_{1j}\rangle\langle\phi_j, f\rangle + \sum_{j\in\sigma^c}(1-a_j)\langle K^\dagger f, \psi_{2j}\rangle\langle\phi_j, f\rangle\Big),
\end{aligned}$$

where $\Phi = \{\phi_j\}_{j\in\mathbb{J}}$ is a K-dual of $\{\psi_{1j}\}_{j\in\sigma} \cup \{\psi_{2j}\}_{j\in\sigma^c}$.

Moreover, if $(P_1K^\dagger)^*P_2K^\dagger$ is a positive operator, then

$$\begin{aligned}
&\Big\|\sum_{j\in\sigma} a_j\langle K^\dagger f, \psi_{1j}\rangle\phi_j + \sum_{j\in\sigma^c} a_j\langle K^\dagger f, \psi_{2j}\rangle\phi_j\Big\|^2 \\
&\quad + \Big\|\sum_{j\in\sigma}(1-a_j)\langle K^\dagger f, \psi_{1j}\rangle\phi_j + \sum_{j\in\sigma^c}(1-a_j)\langle K^\dagger f, \psi_{2j}\rangle\phi_j\Big\|^2 \leq \|f\|^2
\end{aligned}$$

for any $f \in Range(K)$, where P_1 and P_2 are given in Equation (2).

Proof. For any $f \in Range(K)$, for all $\sigma \subset \mathbb{J}$, $\{a_j\}_{j\in\mathbb{J}} \in \ell^\infty(\mathbb{J})$, and $\lambda \in \mathbb{R}$, we know, by combining Equation (3) and Lemma 3, that

$$\begin{aligned}
&\Big\|\sum_{j\in\sigma} a_j\langle K^\dagger f, \psi_{1j}\rangle\phi_j + \sum_{j\in\sigma^c} a_j\langle K^\dagger f, \psi_{2j}\rangle\phi_j\Big\|^2 + \Big\|\sum_{j\in\sigma}(1-a_j)\langle K^\dagger f, \psi_{1j}\rangle\phi_j + \sum_{j\in\sigma^c}(1-a_j)\langle K^\dagger f, \psi_{2j}\rangle\phi_j\Big\|^2 \\
&= \|P_1K^\dagger f\|^2 + \|P_2K^\dagger f\|^2 = 2\|P_2K^\dagger f\|^2 + \mathrm{Re}\langle P_1K^\dagger f, f\rangle - \mathrm{Re}\langle P_2K^\dagger f, f\rangle \\
&\geq (2 - \frac{\lambda^2}{2})\|f\|^2 - (4 - 2\lambda)\mathrm{Re}\langle P_1K^\dagger f, f\rangle + \mathrm{Re}\langle P_1K^\dagger f, f\rangle - \mathrm{Re}\langle P_2K^\dagger f, f\rangle \\
&= (2\lambda - \frac{\lambda^2}{2} - 1)\mathrm{Re}\langle P_1K^\dagger f, f\rangle + (1 - \frac{\lambda^2}{2})\mathrm{Re}\langle P_2K^\dagger f, f\rangle \\
&= (2\lambda - \frac{\lambda^2}{2} - 1)\mathrm{Re}\Big(\sum_{j\in\sigma} a_j\langle K^\dagger f, \psi_{1j}\rangle\langle\phi_j, f\rangle + \sum_{j\in\sigma^c} a_j\langle K^\dagger f, \psi_{2j}\rangle\langle\phi_j, f\rangle\Big) \\
&\quad + (1 - \frac{\lambda^2}{2})\mathrm{Re}\Big(\sum_{j\in\sigma}(1-a_j)\langle K^\dagger f, \psi_{1j}\rangle\langle\phi_j, f\rangle + \sum_{j\in\sigma^c}(1-a_j)\langle K^\dagger f, \psi_{2j}\rangle\langle\phi_j, f\rangle\Big).
\end{aligned}$$

For the "Moreover" part, we have for any $f \in Range(K)$ that

$$\begin{aligned}
\|P_1K^\dagger f\|^2 &= \|P_2K^\dagger f\|^2 - \mathrm{Re}\langle P_2K^\dagger f, f\rangle + \mathrm{Re}\langle P_1K^\dagger f, f\rangle \\
&= \mathrm{Re}\langle P_2K^\dagger f, P_2K^\dagger f\rangle - \mathrm{Re}\langle P_2K^\dagger f, f\rangle + \mathrm{Re}\langle P_1K^\dagger f, f\rangle \\
&= -(\mathrm{Re}\langle P_2K^\dagger f, P_1K^\dagger f + P_2K^\dagger f\rangle - \mathrm{Re}\langle P_2K^\dagger f, P_2K^\dagger f\rangle) + \mathrm{Re}\langle P_1K^\dagger f, f\rangle \\
&= -\mathrm{Re}\langle P_2K^\dagger f, P_1K^\dagger f\rangle + \mathrm{Re}\langle P_1K^\dagger f, f\rangle \leq \mathrm{Re}\langle P_1K^\dagger f, f\rangle.
\end{aligned}$$

With a similar discussion, we can show that $\|P_2K^{\dagger}f\|^2 \leq \mathrm{Re}\langle P_2K^{\dagger}f, f\rangle$. Thus,

$$\Big\|\sum_{j\in\sigma} a_j\langle K^{\dagger}f, \psi_{1j}\rangle\phi_j + \sum_{j\in\sigma^c} a_j\langle K^{\dagger}f, \psi_{2j}\rangle\phi_j\Big\|^2 + \Big\|\sum_{j\in\sigma}(1-a_j)\langle K^{\dagger}f, \psi_{1j}\rangle\phi_j + \sum_{j\in\sigma^c}(1-a_j)\langle K^{\dagger}f, \psi_{2j}\rangle\phi_j\Big\|^2$$
$$\leq \mathrm{Re}\langle P_1K^{\dagger}f, f\rangle + \mathrm{Re}\langle P_2K^{\dagger}f, f\rangle = \mathrm{Re}\langle P_1K^{\dagger}f + P_2K^{\dagger}f, f\rangle = \|f\|^2.$$

□

Corollary 4. *Suppose that two frames $\Psi_1 = \{\psi_{1j}\}_{j\in\mathbb{J}}$ and $\Psi_2 = \{\psi_{2j}\}_{j\in\mathbb{J}}$ in $\mathbb{H}$ are woven. Then, for any $\sigma \subset \mathbb{J}$, for all $\lambda \in \mathbb{R}$ and all $f \in \mathbb{H}$, we have*

$$\begin{aligned}
&(2\lambda - \frac{\lambda^2}{2} - 1)\sum_{j\in\sigma}|\langle f, \psi_{1j}\rangle|^2 + (1 - \frac{\lambda^2}{2})\sum_{j\in\sigma^c}|\langle f, \psi_{2j}\rangle|^2 \\
&\leq \sum_{j\in\sigma}|\langle S^{\sigma}_{\Psi_1}f, S^{-1}_{\Psi_1\Psi_2}\psi_{1j}\rangle|^2 + \sum_{j\in\sigma^c}|\langle S^{\sigma}_{\Psi_1}f, S^{-1}_{\Psi_1\Psi_2}\psi_{2j}\rangle|^2 \\
&\quad + \sum_{j\in\sigma}|\langle S^{\sigma^c}_{\Psi_2}f, S^{-1}_{\Psi_1\Psi_2}\psi_{1j}\rangle|^2 + \sum_{j\in\sigma^c}|\langle S^{\sigma^c}_{\Psi_2}f, S^{-1}_{\Psi_1\Psi_2}\psi_{2j}\rangle|^2 \\
&\leq \sum_{j\in\sigma}|\langle f, \psi_{1j}\rangle|^2 + \sum_{j\in\sigma^c}|\langle f, \psi_{2j}\rangle|^2.
\end{aligned} \tag{8}$$

Proof. Letting $K^{\dagger} = \mathrm{Id}_{\mathbb{H}}$ and for any $\sigma \subset \mathbb{J}$, taking

$$a_j = \begin{cases} 1, & j\in\sigma, \\ 0, & j\in\sigma^c, \end{cases} \qquad \phi_j = \begin{cases} S^{-1/2}_{\Psi_1\Psi_2}\psi_{1j}, & j\in\sigma, \\ S^{-1/2}_{\Psi_1\Psi_2}\psi_{2j}, & j\in\sigma^c. \end{cases}$$

If, now, we replace ψ_{1j}, ψ_{2j} and f in the left-hand inequality of Theorem 2 respectively by $S^{-1/2}_{\Psi_1\Psi_2}\psi_{1j}$, $S^{-1/2}_{\Psi_1\Psi_2}\psi_{2j}$ and $S^{1/2}_{\Psi_1\Psi_2}f$, then

$$\begin{aligned}
&\Big\|\sum_{j\in\sigma}\langle f, \psi_{1j}\rangle S^{-1/2}_{\Psi_1\Psi_2}\psi_{1j}\Big\|^2 + \Big\|\sum_{j\in\sigma^c}\langle f, \psi_{2j}\rangle S^{-1/2}_{\Psi_1\Psi_2}\psi_{2j}\Big\|^2 \\
&\geq (2\lambda - \frac{\lambda^2}{2} - 1)\mathrm{Re}\sum_{j\in\sigma}\langle f, \psi_{1j}\rangle\langle\psi_{1j}, f\rangle + (1 - \frac{\lambda^2}{2})\mathrm{Re}\sum_{j\in\sigma^c}\langle f, \psi_{2j}\rangle\langle\psi_{2j}, f\rangle \\
&= (2\lambda - \frac{\lambda^2}{2} - 1)\sum_{j\in\sigma}|\langle f, \psi_{1j}\rangle|^2 + (1 - \frac{\lambda^2}{2})\sum_{j\in\sigma^c}|\langle f, \psi_{2j}\rangle|^2.
\end{aligned}$$

This along with Equations (6) and (7) gives the left-hand inequality in Equation (8), and the proof of the right-hand inequality is similar and we omit the details. □

Theorem 3. *Suppose that $K \in B(\mathbb{H})$ has a closed range and that K-frames $\Psi_1 = \{\psi_{1j}\}_{j\in\mathbb{J}}$ and $\Psi_2 = \{\psi_{2j}\}_{j\in\mathbb{J}}$ in $\mathbb{H}$ are K-woven. Then, for all $\sigma \subset \mathbb{J}$, for any $\{a_j\}_{j\in\mathbb{J}} \in \ell^{\infty}(\mathbb{J})$, $\lambda \in \mathbb{R}$ and $f \in Range(K)$,*

$$\begin{aligned}
&\mathrm{Re}\Big(\sum_{j\in\sigma} a_j\langle K^{\dagger}f, \psi_{1j}\rangle\langle\phi_j, f\rangle + \sum_{j\in\sigma^c} a_j\langle K^{\dagger}f, \psi_{2j}\rangle\langle\phi_j, f\rangle\Big) - \Big\|\sum_{j\in\sigma} a_j\langle K^{\dagger}f, \psi_{1j}\rangle\phi_j + \sum_{j\in\sigma^c} a_j\langle K^{\dagger}f, \psi_{2j}\rangle\phi_j\Big\|^2 \\
&\leq (1 - \frac{\lambda}{2})^2\mathrm{Re}\Big(\sum_{j\in\sigma} a_j\langle K^{\dagger}f, \psi_{1j}\rangle\langle\phi_j, f\rangle + \sum_{j\in\sigma^c} a_j\langle K^{\dagger}f, \psi_{2j}\rangle\langle\phi_j, f\rangle\Big) \\
&\quad + \frac{\lambda^2}{4}\mathrm{Re}\Big(\sum_{j\in\sigma}(1-a_j)\langle K^{\dagger}f, \psi_{1j}\rangle\langle\phi_j, f\rangle + \sum_{j\in\sigma^c}(1-a_j)\langle K^{\dagger}f, \psi_{2j}\rangle\langle\phi_j, f\rangle\Big),
\end{aligned}$$

where $\Phi = \{\phi_j\}_{j\in\mathbb{J}}$ *is a K-dual of* $\{\psi_{1j}\}_{j\in\sigma} \cup \{\psi_{2j}\}_{j\in\sigma^c}$.

Moreover, if $(P_1K^\dagger)^*P_2K^\dagger \geq 0$, *then*

$$\mathrm{Re}\Big(\sum_{j\in\sigma} a_j\langle K^\dagger f, \psi_{1j}\rangle\langle\phi_j, f\rangle + \sum_{j\in\sigma^c} a_j\langle K^\dagger f, \psi_{2j}\rangle\langle\phi_j, f\rangle\Big) - \Big\|\sum_{j\in\sigma} a_j\langle K^\dagger f, \psi_{1j}\rangle\phi_j + \sum_{j\in\sigma^c} a_j\langle K^\dagger f, \psi_{2j}\rangle\phi_j\Big\|^2 \geq 0$$

for any $f \in Range(K)$, *where* P_1 *and* P_2 *are given in Equation (2).*

Proof. For all $\sigma \subset \mathbb{J}$, for any $\{a_j\}_{j\in\mathbb{J}} \in \ell^\infty(\mathbb{J})$, $\lambda \in \mathbb{R}$ and $f \in Range(K)$, we see from Equation (4) that

$$\begin{aligned}
&\mathrm{Re}\Big(\sum_{j\in\sigma} a_j\langle K^\dagger f, \psi_{1j}\rangle\langle\phi_j, f\rangle + \sum_{j\in\sigma^c} a_j\langle K^\dagger f, \psi_{2j}\rangle\langle\phi_j, f\rangle\Big) - \Big\|\sum_{j\in\sigma} a_j\langle K^\dagger f, \psi_{1j}\rangle\phi_j + \sum_{j\in\sigma^c} a_j\langle K^\dagger f, \psi_{2j}\rangle\phi_j\Big\|^2 \\
&= \mathrm{Re}\langle P_1K^\dagger f, f\rangle - \|P_1K^\dagger f\|^2 \\
&\leq \mathrm{Re}\langle P_1K^\dagger f, f\rangle - (\lambda - \frac{\lambda^2}{4})\mathrm{Re}\langle P_1K^\dagger f, f\rangle + \frac{\lambda^2}{4}\mathrm{Re}\langle P_2K^\dagger f, f\rangle \\
&= (1-\frac{\lambda}{2})^2\mathrm{Re}\Big(\sum_{j\in\sigma} a_j\langle K^\dagger f, \psi_{1j}\rangle\langle\phi_j, f\rangle + \sum_{j\in\sigma^c} a_j\langle K^\dagger f, \psi_{2j}\rangle\langle\phi_j, f\rangle\Big) \\
&\quad + \frac{\lambda^2}{4}\mathrm{Re}\Big(\sum_{j\in\sigma}(1-a_j)\langle K^\dagger f, \psi_{1j}\rangle\langle\phi_j, f\rangle + \sum_{j\in\sigma^c}(1-a_j)\langle K^\dagger f, \psi_{2j}\rangle\langle\phi_j, f\rangle\Big).
\end{aligned}$$

Suppose now that $(P_1K^\dagger)^*P_2K^\dagger$ is a positive operator. Then

$$\begin{aligned}
&\mathrm{Re}\Big(\sum_{j\in\sigma} a_j\langle K^\dagger f, \psi_{1j}\rangle\langle\phi_j, f\rangle + \sum_{j\in\sigma^c} a_j\langle K^\dagger f, \psi_{2j}\rangle\langle\phi_j, f\rangle\Big) - \Big\|\sum_{j\in\sigma} a_j\langle K^\dagger f, \psi_{1j}\rangle\phi_j + \sum_{j\in\sigma^c} a_j\langle K^\dagger f, \psi_{2j}\rangle\phi_j\Big\|^2 \\
&= \mathrm{Re}\langle P_1K^\dagger f, f\rangle - \|P_1K^\dagger f\|^2 = \mathrm{Re}\langle P_1K^\dagger f, P_1K^\dagger f + P_2K^\dagger f\rangle - \mathrm{Re}\langle P_1K^\dagger f, P_1K^\dagger f\rangle \\
&= \mathrm{Re}\langle P_1K^\dagger f, P_2K^\dagger f\rangle = \mathrm{Re}\langle f, (P_1K^\dagger)^*P_2K^\dagger f\rangle \geq 0.
\end{aligned}$$

□

Corollary 5. *Let the two frames* $\Psi_1 = \{\psi_{1j}\}_{j\in\mathbb{J}}$ *and* $\Psi_2 = \{\psi_{2j}\}_{j\in\mathbb{J}}$ *in* $\mathbb{H}$ *be woven. Then, for any* $\sigma \subset \mathbb{J}$, *for all* $\lambda \in \mathbb{R}$ *and all* $f \in \mathbb{H}$, *we have*

$$\begin{aligned}
0 &\leq \sum_{j\in\sigma}|\langle f, \psi_{1j}\rangle|^2 - \sum_{j\in\sigma}|\langle S^\sigma_{\Psi_1} f, S^{-1}_{\Psi_1\Psi_2}\psi_{1j}\rangle|^2 - \sum_{j\in\sigma^c}|\langle S^\sigma_{\Psi_1} f, S^{-1}_{\Psi_1\Psi_2}\psi_{2j}\rangle|^2 \\
&\leq (1-\frac{\lambda}{2})^2\sum_{j\in\sigma}|\langle f, \psi_{1j}\rangle|^2 + \frac{\lambda^2}{4}\sum_{j\in\sigma^c}|\langle f, \psi_{2j}\rangle|^2.
\end{aligned}$$

Proof. The proof is similar to Corollary 4 by using Theorem 3, so we omit it. □

Remark 2. *Corollaries 4 and 5 are respectively Theorems 15 and 14 in [34].*

We conclude the paper with a double inequality for K-weaving frames stated as follows.

Theorem 4. *Suppose that K-frames $\Psi_1 = \{\psi_{1j}\}_{j\in\mathbb{J}}$ and $\Psi_2 = \{\psi_{2j}\}_{j\in\mathbb{J}}$ in $\mathbb{H}$ are K-woven. Then, for any $\sigma \subset \mathbb{J}$, for all $\{a_j\}_{j\in\mathbb{J}} \in \ell^\infty(\mathbb{J})$ and all $f \in \mathbb{H}$, we have*

$$\begin{aligned}\frac{3}{4}\|Kf\|^2 &\le \left\|\sum_{j\in\sigma} a_j\langle f,\psi_{1j}\rangle\phi_j + \sum_{j\in\sigma^c} a_j\langle f,\psi_{2j}\rangle\phi_j\right\|^2 \\ &\quad + \mathrm{Re}\left(\sum_{j\in\sigma}(1-a_j)\langle f,\psi_{1j}\rangle\langle\phi_j,Kf\rangle + \sum_{j\in\sigma^c}(1-a_j)\langle f,\psi_{2j}\rangle\langle\phi_j,Kf\rangle\right) \\ &\le \frac{3\|K\|^2 + \|P_1-P_2\|^2}{4}\|f\|^2,\end{aligned}$$

where P_1 and P_2 are given in Equation (2), and $\Phi = \{\phi_j\}_{j\in\mathbb{J}}$ is a K-dual of $\{\psi_{1j}\}_{j\in\sigma} \cup \{\psi_{2j}\}_{j\in\sigma^c}$.

Proof. For any $\sigma \subset \mathbb{J}$, for all $\{a_j\}_{j\in\mathbb{J}} \in \ell^\infty(\mathbb{J})$ and all $f \in \mathbb{H}$, it is easy to check that $P_1 + P_2 = K$. By Lemma 1, we get

$$\begin{aligned}&\left\|\sum_{j\in\sigma} a_j\langle f,\psi_{1j}\rangle\phi_j + \sum_{j\in\sigma^c} a_j\langle f,\psi_{2j}\rangle\phi_j\right\|^2 + \mathrm{Re}\left(\sum_{j\in\sigma}(1-a_j)\langle f,\psi_{1j}\rangle\langle\phi_j,Kf\rangle + \sum_{j\in\sigma^c}(1-a_j)\langle f,\psi_{2j}\rangle\langle\phi_j,Kf\rangle\right) \\ &\quad = \|P_1f\|^2 + \mathrm{Re}\langle P_2f,Kf\rangle \ge \frac{3}{4}\|Kf\|^2.\end{aligned}$$

We also have

$$\begin{aligned}&\left\|\sum_{j\in\sigma} a_j\langle f,\psi_{1j}\rangle\phi_j + \sum_{j\in\sigma^c} a_j\langle f,\psi_{2j}\rangle\phi_j\right\|^2 + \mathrm{Re}\left(\sum_{j\in\sigma}(1-a_j)\langle f,\psi_{1j}\rangle\langle\phi_j,Kf\rangle + \sum_{j\in\sigma^c}(1-a_j)\langle f,\psi_{2j}\rangle\langle\phi_j,Kf\rangle\right) \\ &\quad = \langle P_1f,P_1f\rangle + \frac{1}{2}\langle P_2f,Kf\rangle + \frac{1}{2}\langle Kf,P_2f\rangle \\ &\quad = \langle P_1f,P_1f\rangle + \frac{1}{2}\langle (K-P_1)f,Kf\rangle + \frac{1}{2}\langle Kf,(K-P_1)f\rangle \\ &\quad = \langle Kf,Kf\rangle - \frac{1}{2}[\langle P_1f,Kf\rangle - \langle P_1f,P_1f\rangle] - \frac{1}{2}[\langle Kf,P_1f\rangle - \langle P_1f,P_1f\rangle] \\ &\quad = \langle Kf,Kf\rangle - \frac{1}{2}\langle P_1f,P_2f\rangle - \frac{1}{2}\langle P_2f,P_1f\rangle \\ &\quad = \frac{3}{4}\langle Kf,Kf\rangle + \frac{1}{4}\langle P_1f+P_2f,P_1f+P_2f\rangle - \frac{1}{2}\langle P_1f,P_2f\rangle - \frac{1}{2}\langle P_2f,P_1f\rangle \\ &\quad = \frac{3}{4}\langle Kf,Kf\rangle + \frac{1}{4}\langle (P_1-P_2)f,(P_1-P_2)f\rangle \\ &\quad \le \frac{3}{4}\|K\|^2\|f\|^2 + \frac{1}{4}\|P_1-P_2\|^2\|f\|^2 = \frac{3\|K\|^2 + \|P_1-P_2\|^2}{4}\|f\|^2,\end{aligned}$$

and the proof is over. □

Remark 3. *Theorem 3 in [35] can be obtained when taking $K = \mathrm{Id}_{\mathbb{H}}$ in Theorem 4.*

Funding: This research was funded by the National Natural Science Foundation of China under Grant Nos. 11761057 and 11561057.

Conflicts of Interest: The author declares no conflict of interest.

References

1. Duffin, R.J.; Schaeffer, A.C. A class of nonharmonic Fourier series. *Trans. Am. Math. Soc.* **1952**, *72*, 341–366. [CrossRef]
2. Dai, F. Characterizations of function spaces on the sphere using frames. *Trans. Am. Math. Soc.* **2006**, *359*, 567–589. [CrossRef]

3. Pesenson, I.Z. Paley-Wiener-Schwartz nearly Parseval frames on noncompact symmetric spaces. In *Commutative and Noncommutative Harmonic Analysis and Applications*; Mayeli, A., Iosevich, A., Jorgensen, P.E.T., Ólafsson, G., Eds.; American Mathematical Society: Providence, RI, USA, 2013; Volume 603, pp. 55–71, ISBN 978-0-8218-9493-4.
4. Gazeau, J.-P. *Coherent States in Quantum Physics*; Wiley-VCH: Berlin, Germany, 2009.
5. Feichtinger, H.G.; Gröchenig, K. Theory and practice of irregular sampling. In *Wavelets: Mathematics and Applications*; Benedetto, J., Frazier, M., Eds.; CRC Press: Boca Raton, FL, USA, 1994; pp. 305–363, ISBN 978-0849382710.
6. Poon, C. A consistent and stable approach to generalized sampling. *J. Fourier Anal. Appl.* **2014**, *20*, 985–1019. [CrossRef]
7. Sun, W. Asymptotic properties of Gabor frame operators as sampling density tends to infinity. *J. Funct. Anal.* **2010**, *258*, 913–932. [CrossRef]
8. Balazs, P.; Holighaus, N.; Necciari, T.; Stoeva, D.T. Frame theory for signal processing in psychoacoustics. In *Excursions in Harmonic Analysis*; Balan, R., Benedetto, J., Czaja, W., Dellatorre, M., Okoudjou, K., Eds.; Birkhäuser: Cham, Switzerland, 2017; Volume 5, pp. 225–268, ISBN 978-3-319-54710-7.
9. Bölcskei, H.; Hlawatsch, F.; Feichtinger, H.G. Frame-theoretic analysis of oversampled filter banks. *IEEE Trans. Signal Process.* **1998**, *46*, 3256–3268. [CrossRef]
10. Găvruţa, L. Frames for operators. *Appl. Comput. Harmon. Anal.* **2012**, *32*, 139–144. [CrossRef]
11. Guo, X.X. Canonical dual *K*-Bessel sequences and dual *K*-Bessel generators for unitary systems of Hilbert spaces. *J. Math. Anal. Appl.* **2016**, *444*, 598–609. [CrossRef]
12. Jia, M.; Zhu, Y.C. Some results about the operator perturbation of a *K*-frame. *Results Math.* **2018**, *73*, 138. [CrossRef]
13. Johnson, P.S.; Ramu, G. Class of bounded operators associated with an atomic system. *Tamkang J. Math.* **2015**, *46*, 85–90. [CrossRef]
14. Poumai, K.T.; Jahan, S. Atomic systems for operators. *Int. J. Wavelets Multiresolut. Inf. Process.* **2019**, *17*, 1850066. [CrossRef]
15. Poumai, K.T.; Jahan, S. On *K*-Atomic Decompositions in Banach Spaces. *Electron. J. Math. Anal. Appl.* **2018**, *6*, 183–197.
16. Xiang, Z.Q.; Li, Y.M. Frame sequences and dual frames for operators. *ScienceAsia* **2016**, *42*, 222–230. [CrossRef]
17. Xiao, X.C.; Zhu, Y.C.; Găvruţa, L. Some properties of *K*-frames in Hilbert spaces. *Results Math.* **2013**, *63*, 1243–1255. [CrossRef]
18. Bemrose, T.; Casazza, P.G.; Gröchenig, K.; Lammers, M.C.; Lynch, R.G. Weaving frames. *Oper. Matrices* **2016**, *10*, 1093–1116. [CrossRef]
19. Casazza, P.G.; Freeman, D.; Lynch, R.G. Weaving Schauder frames. *J. Approx. Theory* **2016**, *211*, 42–60. [CrossRef]
20. Deepshikha; Vashisht, L.K. On weaving frames. *Houston J. Math.* **2018**, *44*, 887–915.
21. Khosravi, A.; Banyarani, J.S. Weaving g-frames and weaving fusion frames. *Bull. Malays. Math. Sci. Soc.* **2018**, *42*, 3111–3129. [CrossRef]
22. Rahimi, A.; Samadzadeh, Z.; Daraby, B. Frame related operators for woven frames. *Int. J. Wavelets Multiresolut. Inf. Process.* **2019**, *17*, 1950010. [CrossRef]
23. Vashisht, L.K.; Garg, S.; Deepshikha; Das, P.K. On generalized weaving frames in Hilbert spaces. *Rocky Mt. J. Math.* **2018**, *48*, 661–685. [CrossRef]
24. Vashisht, L.K.; Deepshikha. Weaving properties of generalized continuous frames generated by an iterated function system. *J. Geom. Phys.* **2016**, *110*, 282–295. [CrossRef]
25. Deepshikha; Vashisht, L.K. Weaving *K*-frames in Hilbert spaces. *Results Math.* **2018**, *73*, 81. [CrossRef]
26. Balan, R.; Casazza, P.G.; Edidin, D.; Kutyniok, G. A new identity for Parseval frames. *Proc. Am. Math. Soc.* **2007**, *135*, 1007–1015. [CrossRef]
27. Balan, R.; Casazza, P.G.; Edidin, D. On signal reconstruction without phase. *Appl. Comput. Harmon. Anal.* **2006**, *20*, 345–356. [CrossRef]
28. Găvruţa, P. On some identities and inequalities for frames in Hilbert spaces. *J. Math. Anal. Appl.* **2006**, *321*, 469–478. [CrossRef]
29. Jivulescu, M.A.; Găvruţa, P. Indices of sharpness for Parseval frames, quantum effects and observables. *Sci. Bull. Politeh. Univ. Timiş. Trans. Math. Phys.* **2015**, *60*, 17–29.

30. Fu, Y.L.; Zhang, W. Some new inequalities for dual continuous g-frames. *Mathematics* **2019**, *7*, 662. [CrossRef]
31. Li, D.W.; Leng, J.S. On some new inequalities for fusion frames in Hilbert spaces. *Math. Inequal. Appl.* **2017**, *20*, 889–900. [CrossRef]
32. Li, D.W.; Leng, J.S. On some new inequalities for continuous fusion frames in Hilbert spaces. *Mediterr. J. Math.* **2018**, *15*, 173. [CrossRef]
33. Poria, A. Some identities and inequalities for Hilbert-Schmidt frames. *Mediterr. J. Math.* **2017**, *14*, 59. [CrossRef]
34. Li, D.W.; Leng, J.S. New inequalities for weaving frames in Hilbert spaces. *arXiv* **2018**, arXiv:1809.00863
35. Xiang, Z.Q. More on inequalities for weaving frames in Hilbert spaces. *Mathematics* **2019**, *7*, 141. [CrossRef]
36. Christensen, O. *An Introduction to Frames and Riesz Bases*; Birkhäuser: Boston, MA, USA, 2000.

mathematics

MDPI

Article

A New Gronwall–Bellman Inequality in Frame of Generalized Proportional Fractional Derivative

Jehad Alzabut [1,*], Weerawat Sudsutad [2], Zeynep Kayar [3] and Hamid Baghani [4]

1 Department of Mathematics and General Sciences, Prince Sultan University, P.O. Box 66833, 11586 Riyadh, Saudi Arabia
2 Department of General Education, Faculty of Science and Health Technology, Navamindradhiraj University, Bangkok 10300, Thailand
3 Department of Mathematics, Van Yuzuncu Yil University, 65080 Van, Turkey
4 Department of Mathematics, Faculty of Mathematics, University of Sistan and Baluchestan, Zahedan, Iran
* Correspondence: jalzabut@psu.edu.sa

Received: 19 July 2019; Accepted: 12 August 2019; Published: 15 August 2019

Abstract: New versions of a Gronwall–Bellman inequality in the frame of the generalized (Riemann–Liouville and Caputo) proportional fractional derivative are provided. Before proceeding to the main results, we define the generalized Riemann–Liouville and Caputo proportional fractional derivatives and integrals and expose some of their features. We prove our main result in light of some efficient comparison analyses. The Gronwall–Bellman inequality in the case of weighted function is also obtained. By the help of the new proposed inequalities, examples of Riemann–Liouville and Caputo proportional fractional initial value problems are presented to emphasize the solution dependence on the initial data and on the right-hand side.

Keywords: Gronwall–Bellman inequality; proportional fractional derivative; Riemann–Liouville and Caputo proportional fractional initial value problem

1. Introduction

Integral inequalities have been used as fabulous instruments to explore the qualitative properties of differential equations [1]. Over the years, there have appeared many inequalities which have been established by many authors such as Ostrowski type inequality, Hardy type inequality, Olsen type inequality, Gagliardo–Nirenberg type inequality, Lyapunove type inequality, Opial type inequality and Hermite–Hadamard type inequality [2,3]. However, the most common and significant inequality is the Gronwall–Bellman inequality, which they introduced in [4,5]. The Gronwall–Bellman inequality allows one to provide an estimate for a function that is known to satisfy a certain integral inequality by the solution of the corresponding integral equation. In particular, it has been employed to provide a comparison that can be used to prove uniqueness of a solution to an initial value problem (see some recent relevant papers [6–9]).

Fractional differential equations (FDEs) is a rich area of research that has widespread applications in science and engineering. Indeed, it describes a large number of nonlinear phenomena in different fields such as physics, chemistry, biology, viscoelasticity, control hypothesis, speculation, fluid dynamics, hydrodynamics, aerodynamics, information processing system networking, notable and picture processing, control theory, etc. FDEs also provide marvellous tools for the depiction of memory and inherited properties of many materials and processes. In view of recent developments, one can consequently conclude that FDEs have emerged significant achievements in the last couple of decades [10–16]. The study of integral equations in the scope of non-integer-order equations has been in the spotlight in the recent years. Many mathematicians in the field of applied and pure mathematics have dedicated their efforts to extend, generalize and refine the integral inequalities carried over from integer order equations to the non-integer order equations. Meanwhile, different definitions of fractional derivatives have been recently introduced [17,18]. The Gronwall–Bellman inequality, which is our concern herein, has been under investigation and different versions of it have been established for different types of fractional operators [19–25].

In this paper, new versions for a Gronwall–Bellman inequality in the frame of the newly defined generalized (Riemann–Liouville and Caputo) proportional fractional derivative are provided. Before proceeding to the main results, we define the generalized Riemann–Liouville and Caputo proportional fractional derivatives and integrals and expose some of their features [26]. We prove our main result in light of some efficient comparison analysis. The Gronwall–Bellman inequality in the case of a weighted function is also obtained. By the help of the new proposed inequalities, examples of Riemann–Liouville and Caputo generalized proportional fractional initial value problems are presented to emphasize the solution dependence on the initial data and on the right-hand side. It worth mentioning that the new proposed derivative is well-behaved. Indeed, it has nonlocal character and satisfies the semigroup or the so-called index property. Besides, the resulting inequalities converge to the classical ones upon considering particular cases of the derivative. That is, our results not only extend the classical inequalities but also generalize the existing ones for non-integer-order equations.

2. The GPF Derivatives and Integrals

We assemble in this section some fundamental preliminaries that are used throughout the remaining part of the paper. For their justifications and proofs, the reader can consult the work in [26].

In control theory, a proportional derivative controller (PDC) for controller output u at time t with two tuning parameters has the algorithm

$$u(t) = \kappa_p E(t) + \kappa_d \frac{d}{dt} E(t),$$

where κ_p is the proportional gain, κ_d is the derivative gain, and E is the input deviation or the error between the state variable and the process variable. Recent investigations have shown that PDC has direct incorporation in the control of complex networks models (see [27] for more details).

For $\rho \in [0,1]$, let the functions $\kappa_0, \kappa_1 : [0,1] \times \mathbb{R} \to [0,\infty)$ be continuous such that for all $t \in \mathbb{R}$ we have

$$\lim_{\rho \to 0^+} \kappa_1(\rho,t) = 1, \quad \lim_{\rho \to 0^+} \kappa_0(\rho,t) = 0, \quad \lim_{\rho \to 1^-} \kappa_1(\rho,t) = 0, \quad \lim_{\rho \to 1^-} \kappa_0(\rho,t) = 1,$$

and $\kappa_1(\rho,t) \neq 0$, $\rho \in [0,1)$, $\kappa_0(\rho,t) \neq 0$, $\rho \in (0,1]$. Then, Anderson et al. [28] defined the proportional derivative of order ρ by

$$D^\rho \xi(t) = \kappa_1(\rho,t)\xi(t) + \kappa_0(\rho,t)\xi'(t) \tag{1}$$

provided that the right-hand side exists at $t \in \mathbb{R}$ and $\xi' := \frac{d}{dt}\xi$. For the operator given in Equation (1), κ_1 is a type of proportional gain κ_p, κ_0 is a type of derivative gain κ_d, ξ is the error and $u = D^\rho \xi$ is the controller output. The reader can consult the work in [29] for more details about the control theory of the proportional derivative and its component functions. We only consider here the case when $\kappa_1(\rho,t) = 1-\rho$ and $\kappa_0(\rho,t) = \rho$. Therefore, Equation (1) becomes

$$D^\rho \xi(t) = (1-\rho)\xi(t) + \rho\xi'(t). \tag{2}$$

It is easy to find that $\lim_{\rho \to 0^+} D^\rho \xi(t) = \xi(t)$ and $\lim_{\rho \to 1^-} D^\rho \xi(t) = \xi'(t)$. Thus, the derivative in Equation (2) is somehow more general than the conformable derivative, which certainly does not converge to the original functions as ρ tends to 0.

In what follows, we define the generalized proportional fractional (GPF) integral and derivative:

Definition 1 ([26]). *For $0 < \rho \leq 1$, $\alpha \in \mathbb{C}$ and $Re(\alpha) > 0$, the GPF integral of ξ of order α is*

$$({}_aI^{\alpha,\rho}\xi)(t) = \frac{1}{\rho^\alpha\Gamma(\alpha)}\int_a^t e^{\frac{\rho-1}{\rho}(t-\tau)}(t-\tau)^{\alpha-1}\xi(\tau)d\tau = \rho^{-\alpha}e^{\frac{\rho-1}{\rho}t}\left({}_aI^{\alpha}(e^{\frac{1-\rho}{\rho}t}\xi(t))\right). \tag{3}$$

Definition 2 ([26]). *For $0 < \rho \leq 1$, $\alpha \in \mathbb{C}$, $Re(\alpha) \geq 0$ and $n = [Re(\alpha)] + 1$. Then, the Riemann–Liouville type GPF derivative of f of order α is*

$$({}_aD^{\alpha,\rho}\xi)(t) = D^{n,\rho}\,{}_aI^{n-\alpha,\rho}\xi(t) = \frac{D_t^{n,\rho}}{\rho^{n-\alpha}\Gamma(n-\alpha)}\int_a^t e^{\frac{\rho-1}{\rho}(t-\tau)}(t-\tau)^{n-\alpha-1}\xi(\tau)d\tau. \tag{4}$$

Remark 1. *If we let $\rho = 1$ in Definition 2, then one can obtain the left Riemann–Liouville fractional derivative [12,14,15]. Moreover, it is obvious that*

$$\lim_{\alpha\to 0}(D^{\alpha,\rho}\xi)(t) = \xi(t) \quad and \quad \lim_{\alpha\to 1}(D^{\alpha,\rho}\xi)(t) = (D^{\rho}\xi)(t).$$

Proposition 1 ([26]). *Let $\alpha, \beta \in \mathbb{C}$ be such that $Re(\alpha) \geq 0$ and $Re(\beta) > 0$. Then, for any $0 < \rho \leq 1$, we have*

(1) $\left({}_aI^{\alpha,\rho}e^{\frac{\rho-1}{\rho}t}(t-a)^{\beta-1}\right)(x) = \frac{\Gamma(\beta)}{\Gamma(\beta+\alpha)\rho^\alpha}e^{\frac{\rho-1}{\rho}x}(x-a)^{\alpha+\beta-1}, \quad Re(\alpha) > 0.$

(2) $\left({}_aD^{\alpha,\rho}e^{\frac{\rho-1}{\rho}t}(t-a)^{\beta-1}\right)(x) = \frac{\rho^\alpha\Gamma(\beta)}{\Gamma(\beta-\alpha)}e^{\frac{\rho-1}{\rho}x}(x-a)^{\beta-1-\alpha}, \quad Re(\alpha) \geq 0.$

In the following lemmas, we expose some features of Riemann–Liouville type GPF operator. The first result concerns with the index property of GPF which is of great significance.

Lemma 1 ([26]). *If $0 < \rho \leq 1$, $Re(\alpha) > 0$ and $Re(\beta) > 0$. For a continuous function ξ defined on $[a, \infty)$, we have*

$${}_aI^{\alpha,\rho}({}_aI^{\beta,\rho}\xi)(t) = {}_aI^{\beta,\rho}({}_aI^{\alpha,\rho}\xi)(t) = ({}_aI^{\alpha+\beta,\rho}\xi)(t). \tag{5}$$

The action of the operator ${}_aD^{\alpha,\rho}$ on the integral operator is demonstrated in the following results.

Lemma 2 ([26]). *Let $0 < \rho \leq 1$, $0 \leq m < [Re(\alpha)] + 1$ and ξ be integrable in each interval $[a, t], t > a$. Then,*

$${}_aD^{m,\rho}({}_aI^{\alpha,\rho}\xi)(t) = ({}_aI^{\alpha-m,\rho}\xi)(t). \tag{6}$$

Corollary 1 ([26]). *Let $0 < \rho \leq 1$, $0 < Re(\beta) < Re(\alpha)$ and $m - 1 < Re(\beta) \leq m$. Then, we have*

$${}_aD^{\beta,\rho}({}_aI^{\alpha,\rho}\xi)(t) = ({}_aI^{\alpha-\beta,\rho}\xi)(t).$$

Lemma 3 ([26]). *Let f be integrable on $t \geq a$ and $Re[\alpha] > 0$, $0 < \rho \leq 1$, $n = [Re(\alpha)] + 1$. Then, we have*

$${}_aD^{\alpha,\rho}({}_aI^{\alpha,\rho}\xi)(t) = \xi(t).$$

Lemma 4 ([26]). *Let $0 < \rho \leq 1$, $Re(\alpha) > 0$, $n = [Re(\alpha)] + 1$, $\xi \in L_1(a, b)$ and $({}_aI^{\alpha,\rho}\xi)(t) \in AC^n[a, b]$. Then,*

$${}_aI^{\alpha,\rho}({}_aD^{\alpha,\rho}\xi)(t) = \xi(t) - e^{\frac{\rho-1}{\rho}(t-a)}\sum_{j=1}^{n}({}_aI^{j-\alpha,\rho}\xi)(a^+)\frac{(t-a)^{\alpha-j}}{\rho^{\alpha-j}\Gamma(\alpha+1-j)}. \tag{7}$$

The GPF derivative of Caputo type is defined as follows:

Definition 3 ([26]). *For $0 < \rho \le 1$, $\alpha \in \mathbb{C}$, $Re(\alpha) \ge 0$ and $n = [Re(\alpha)] + 1$. Then, the GPF derivative of Caputo type of ξ of order α is*

$$({}_a^C D^{\alpha,\rho} \xi)(t) = {}_a I^{n-\alpha,\rho} (D^{n,\rho} \xi)(t) = \frac{1}{\rho^{n-\alpha}\Gamma(n-\alpha)} \int_a^t e^{\frac{\rho-1}{\rho}(t-\tau)} (t-\tau)^{n-\alpha-1} (D^{n,\rho}\xi)(\tau) d\tau. \tag{8}$$

Proposition 2 ([26]). *Let $\alpha, \beta \in \mathbb{C}$ be such that $Re(\alpha) > 0$ and $Re(\beta) > 0$. Then, for any $0 < \rho \le 1$ and $n = [Re(\alpha)] + 1$, we have*

$$({}_a^C D^{\alpha,\rho} e^{\frac{\rho-1}{\rho}t} (t-a)^{\beta-1})(x) = \frac{\rho^\alpha \Gamma(\beta)}{\Gamma(\beta-\alpha)} e^{\frac{\rho-1}{\rho}x} (x-a)^{\beta-1-\alpha}, \quad Re(\beta) > n.$$

For $k = 0, 1, \ldots, n-1$, we have $({}_a^C D^{\alpha,\rho} e^{\frac{\rho-1}{\rho}t} (t-a)^k)(x) = 0$.

Lemma 5 ([26]). *For $\rho \in (0,1]$, $Re(\alpha) > 0$ and $n = [Re(\alpha)] + 1$. Then, we have*

$$ {}_a I^{\alpha,\rho} ({}_a^C D^{\alpha,\rho} \xi)(t) = \xi(t) - e^{\frac{\rho-1}{\rho}(t-a)} \sum_{k=0}^{n-1} \frac{({}_a D^{k,\rho}\xi)(a)}{\rho^k k!} (t-a)^k. \tag{9}$$

3. Main Results

This section is devoted to provide our main results of this paper. We formulate new versions of the Gronwall–Bellman inequality within GPF operators in Riemann–Liouville and Caputo settings.

3.1. *Gronwall–Bellman Inequality via the GPF Derivative of Riemann–Liouville Type*

Consider the following generalized proportional Riemann–Liouville fractional initial value problem

$$\begin{cases} ({}_a D^{\alpha,\rho} y)(t) = f(t, y(t)), \quad 0 < \alpha \le 1, \quad t \in [a,b], \\ \lim\limits_{t\to a^+} ({}_a I^{1-\alpha,\rho} y)(t) = y(a) = y_a. \end{cases} \tag{10}$$

Applying the operator ${}_a I^{\alpha,\rho}$ to both sides of Equation (10), we obtain

$$y(t) = e^{\frac{\rho-1}{\rho}(t-a)} (t-a)^{\alpha-1} y(a) + {}_a I^{\alpha,\rho} f(t, y(t)), \tag{11}$$

In the following, we present a comparison result for the GPF integral operator.

Theorem 1. *Let η and ζ be nonnegative continuous functions defined on $[a,b]$ and satisfying*

$$\eta(t) \ge e^{\frac{\rho-1}{\rho}(t-a)} (t-a)^{\alpha-1} \eta(a) + {}_a I^{\alpha,\rho} f(t, \eta(t)), \tag{12}$$

and

$$\zeta(t) \le e^{\frac{\rho-1}{\rho}(t-a)} (t-a)^{\alpha-1} \zeta(a) + {}_a I^{\alpha,\rho} f(t, \zeta(t)), \tag{13}$$

respectively. Suppose further that f satisfies a one-sided Lipschitz condition of the form

$$f(t,x) - f(t,y) \le \frac{L}{e^{\frac{\rho-1}{\rho}(a-t)} [e^{\frac{\rho-1}{\rho}(t-a)} (t-a)^{\alpha-1} + \frac{(t-a)^\alpha}{\alpha} + 1]} (x-y), \ \text{for} \ x \ge y, \ L > 0, \tag{14}$$

and $f(t,y)$ is nondecreasing in y. Then, $\eta(a) \ge \zeta(a)$ and $L < (1 + \frac{\alpha}{(t-a)^\alpha}) \Gamma(\alpha) \rho^\alpha e^{-\frac{\rho-1}{\rho}(t-a)}$ imply that $\eta(t) \ge \zeta(t)$ for all $t \in [a,b]$.

Proof. We start by setting

$$\eta_\varepsilon(t) = \eta(t) + \varepsilon\big[e^{\frac{\rho-1}{\rho}(t-a)}(t-a)^{\alpha-1} + \frac{(t-a)^\alpha}{\alpha} + 1\big], \text{ for small } \varepsilon > 0, \tag{15}$$

so that we have

$$\eta_\varepsilon(a) = \eta(a) + \varepsilon > \eta(a) \text{ and } \eta_\varepsilon(t) > \eta(t), \ t \in [a,b]. \tag{16}$$

It follows that

$$\begin{aligned}\eta_\varepsilon(t) \geq\ & e^{\frac{\rho-1}{\rho}(t-a)}(t-a)^{\alpha-1}\eta(a) + \frac{1}{\Gamma(\alpha)\rho^\alpha}\int_a^t e^{\frac{\rho-1}{\rho}(t-s)}(t-s)^{\alpha-1}f(s,\eta(s))ds \\ +\ & \varepsilon\Big[e^{\frac{\rho-1}{\rho}(t-a)}(t-a)^{\alpha-1} + \frac{(t-a)^\alpha}{\alpha} + 1\Big]\end{aligned}$$

or

$$\begin{aligned}\eta_\varepsilon(t) \geq\ & e^{\frac{\rho-1}{\rho}(t-a)}(t-a)^{\alpha-1}\eta(a) + \frac{1}{\Gamma(\alpha)\rho^\alpha}\int_a^t e^{\frac{\rho-1}{\rho}(t-s)}(t-s)^{\alpha-1}f(s,\eta(s))ds \\ -\ & \frac{1}{\Gamma(\alpha)\rho^\alpha}\int_a^t e^{\frac{\rho-1}{\rho}(t-s)}(t-s)^{\alpha-1}f(s,\eta_\varepsilon(s))ds \\ +\ & \frac{1}{\Gamma(\alpha)\rho^\alpha}\int_a^t e^{\frac{\rho-1}{\rho}(t-s)}(t-s)^{\alpha-1}f(s,\eta_\varepsilon(s))ds \\ +\ & \varepsilon\Big[e^{\frac{\rho-1}{\rho}(t-a)}(t-a)^{\alpha-1} + \frac{(t-a)^\alpha}{\alpha} + 1\Big].\end{aligned}$$

Using the Lipschitz condition in Equation (14) and the relations in Equations (15) and (16), we obtain

$$\begin{aligned}\eta_\varepsilon(t) \geq\ & e^{\frac{\rho-1}{\rho}(t-a)}(t-a)^{\alpha-1}\eta_\varepsilon(a) - \frac{\varepsilon L}{\Gamma(\alpha)\rho^\alpha}\int_a^t (t-s)^{\alpha-1}ds \\ +\ & \frac{1}{\Gamma(\alpha)\rho^\alpha}\int_a^t e^{\frac{\rho-1}{\rho}(t-s)}(t-s)^{\alpha-1}f(s,\eta_\varepsilon(s))ds + \varepsilon\Big[\frac{(t-a)^\alpha}{\alpha} + 1\Big]\end{aligned}$$

Since $\int_a^t (t-s)^{\alpha-1}ds = \frac{(t-a)^\alpha}{\alpha}$ and $L < (1 + \frac{\alpha}{(t-a)^\alpha})\Gamma(\alpha)\rho^\alpha e^{-\frac{\rho-1}{\rho}(t+a)}$, we arrive at

$$\eta_\varepsilon(t) > e^{\frac{\rho-1}{\rho}(t-a)}(t-a)^{\alpha-1}\eta_\varepsilon(a) + \frac{1}{\Gamma(\alpha)\rho^\alpha}\int_a^t e^{\frac{\rho-1}{\rho}(t-s)}(t-s)^{\alpha-1}f(s,\eta_\varepsilon(s))ds.$$

The remaining part of the proof can be completed by adopting the same steps followed in the proof of Theorem 2.1 in [30,31] to get $\eta_\varepsilon(t) \geq \zeta(t), t \in [a,b]$. However, and since ε is arbitrary, we conclude that $\eta(t) \geq \zeta(t), t \in [a,b]$ holds true.

□

Remark 2. *The Lipschitz condition in Equation (14) can be relaxed by relaxing the upper bound for the constant L.*

For our purpose, we replace $f(t,y(t))$ in Equation (11) by $x(t)y(t)$ where $|x(t)| < 1, \ t \in [a,b]$. Define the following operator

$$\Omega_x\phi = {}_aI^{\alpha,\rho}x(t)\phi(t). \tag{17}$$

The following results are important in the proof of the main theorem. We only state these lemmas as their proofs are straightforward.

Lemma 6. *For any constant* λ*, one has*

$$\left|\Omega_{\lambda} e^{\frac{\rho-1}{\rho}(t-a)}(t-a)^{\alpha-1}\right| \leq \Omega_{|\lambda|} e^{\frac{\rho-1}{\rho}(t-a)}(t-a)^{\alpha-1}. \tag{18}$$

Lemma 7. *For any constant* λ*, one has*

$$\left|\Omega_{\lambda}^{n} e^{\frac{\rho-1}{\rho}(t-a)}(t-a)^{\alpha-1}\right| = \frac{|\lambda|^{n}(t-a)^{(n+1)\alpha-1}\Gamma(\alpha)}{\rho^{n\alpha}\Gamma((n+1)\alpha)} e^{\frac{\rho-1}{\rho}(t-a)}, \quad n = 0, 1, 2, \cdots. \tag{19}$$

Lemma 8. *Let* $\lambda > 0$ *be such that* $|y(t)| \leq \lambda$ *for* $t \in [a, b]$*. Then,*

$$\left|\Omega_{y}^{n} e^{\frac{\rho-1}{\rho}(t-a)}(t-a)^{\alpha-1}\right| \leq \Omega_{\lambda}^{n} e^{\frac{\rho-1}{\rho}(t-a)}(t-a)^{\alpha-1}, \quad n = 0, 1, 2, \cdots. \tag{20}$$

Theorems 1 and 2 together give us the desired proportional Riemann–Liouville fractional Gronwall–Bellman-type inequality.

Theorem 2. *Let* y *be a nonnegative function on* $[a, b]$*. Then, the GPF integral equation*

$$y(t) = e^{\frac{\rho-1}{\rho}(t-a)}(t-a)^{\alpha-1}y(a) + {}_{a}I^{\alpha,\rho}x(t)y(t), \quad t \in [a, b], \tag{21}$$

has a solution

$$y(t) = y(a)\sum_{k=0}^{\infty}\Omega_{x}^{k} e^{\frac{\rho-1}{\rho}(t-a)}(t-a)^{\alpha-1}. \tag{22}$$

Proof. The proof is accomplished by applying the successive approximation method. Set

$$y_0(t) = e^{\frac{\rho-1}{\rho}(t-a)}(t-a)^{\alpha-1}y(a)$$

and

$$y_n(t) = e^{\frac{\rho-1}{\rho}(t-a)}(t-a)^{\alpha-1}y(a) + {}_{a}I^{\alpha,\rho}x(t)y_{n-1}(t), \quad n \geq 1.$$

We observe that

$$\begin{aligned} y_1(t) &= e^{\frac{\rho-1}{\rho}(t-a)}(t-a)^{\alpha-1}y(a) + {}_{a}I^{\alpha,\rho}x(t)y_0(t) \\ &= y(a)\Omega_{x}^{0} e^{\frac{\rho-1}{\rho}(t-a)}(t-a)^{\alpha-1} + y(a)\Omega_{x}^{1} e^{\frac{\rho-1}{\rho}(t-a)}(t-a)^{\alpha-1}, \end{aligned}$$

and

$$\begin{aligned} y_2(t) &= e^{\frac{\rho-1}{\rho}(t-a)}(t-a)^{\alpha-1}y(a) + {}_{a}I^{\alpha,\rho}x(t)y_1(t) \\ &= y(a)\Omega_{x}^{0} e^{\frac{\rho-1}{\rho}(t-a)}(t-a)^{\alpha-1} + \Omega_{x}^{1}\left[y(a)\Omega_{x}^{0} e^{\frac{\rho-1}{\rho}(t-a)}(t-a)^{\alpha-1} + y(a)\Omega_{x}^{1} e^{\frac{\rho-1}{\rho}(t-a)}(t-a)^{\alpha-1}\right] \\ &= y(a)\Omega_{x}^{0} e^{\frac{\rho-1}{\rho}(t-a)}(t-a)^{\alpha-1} + y(a)\Omega_{x}^{1} e^{\frac{\rho-1}{\rho}(t-a)}(t-a)^{\alpha-1} + y(a)\Omega_{x}^{2} e^{\frac{\rho-1}{\rho}(t-a)}(t-a)^{\alpha-1}. \end{aligned}$$

It follows inductively that

$$y_n(t) = y(a)\sum_{k=0}^{n}\Omega_{x}^{k} e^{\frac{\rho-1}{\rho}(t-a)}(t-a)^{\alpha-1}, \quad n \geq 0. \tag{23}$$

Formally, taking the limit as $n \to \infty$ to obtain

$$y(t) = y(a) \sum_{k=0}^{\infty} \Omega_x^k e^{\frac{\rho-1}{\rho}(t-a)} (t-a)^{\alpha-1}. \tag{24}$$

We use Lemmas 6–8, the comparison test and the d'Alembert ratio test to show the absolute convergence of the series in Equation (24). Indeed, the infinite series

$$\sum_{n=0}^{\infty} \frac{\lambda^n (t-a)^{(n+1)\alpha-1} \Gamma(\alpha)}{\rho^{n\alpha} \Gamma((n+1)\alpha)} e^{\frac{\rho-1}{\rho}(t-a)},$$

is convergent for all $t \in [a,b]$ and for all $0 < \lambda, \rho \leq 1$. Let a_n be defined as

$$a_n = \frac{\lambda^n (t-a)^{(n+1)\alpha-1} \Gamma(\alpha)}{\rho^{n\alpha} \Gamma((n+1)\alpha)} e^{\frac{\rho-1}{\rho}(t-a)}. \tag{25}$$

Then, we have

$$\lim_{n\to\infty} \left| \frac{a_{n+1}}{a_n} \right| = \frac{\lambda(t-a)^\alpha}{\rho^\alpha} \lim_{n\to\infty} \left| \frac{\Gamma((n+1)\alpha)}{\Gamma((n+2)\alpha)} \right|.$$

Next, we use Stirling approximation formula for the Gamma function $x\Gamma(x) \sim \sqrt{2\pi x} \left(\frac{x}{e}\right)^x$, where x is large enough. It is a straightforward computation using this formula to show that

$$\lim_{x\to\infty} \frac{x\Gamma(x)}{\sqrt{2\pi x}\left(\frac{x}{e}\right)^x} = 1 \quad \text{and} \quad \lim_{x\to\infty} \left(\frac{x}{x+1}\right)^x = \frac{1}{e},$$

which are all we need. Hence, we have

$$\lim_{n\to\infty} \frac{(n+1)\alpha\Gamma((n+1)\alpha)}{\sqrt{2\pi(n+1)\alpha}\left(\frac{(n+1)\alpha}{e}\right)^{(n+1)\alpha}} = 1 \quad \text{and} \quad \lim_{n\to\infty} \frac{(n+2)\alpha\Gamma((n+2)\alpha)}{\sqrt{2\pi(n+2)\alpha}\left(\frac{(n+2)\alpha}{e}\right)^{(n+2)\alpha}} = 1.$$

Thus,

$$\begin{aligned}
\lim_{n\to\infty} \left| \frac{a_{n+1}}{a_n} \right| &= \frac{\lambda(t-a)^\alpha}{\rho^\alpha} \lim_{n\to\infty} \left| \frac{\Gamma((n+1)\alpha)}{\Gamma((n+2)\alpha)} \right| \\
&= \frac{\lambda(t-a)^\alpha}{\rho^\alpha} \lim_{n\to\infty} \left[\sqrt{\frac{n+2}{n+1}} \left(\frac{n+1}{n+2}\right)^\alpha \left(\frac{e}{\alpha}\right)^\alpha \left(\frac{n+1}{n+2}\right)^{n\alpha} \left(\frac{1}{n+2}\right)^\alpha \right] \\
&= \frac{\lambda(t-a)^\alpha}{\rho^\alpha} \left[\left(\frac{e}{\alpha}\right)^\alpha \left(\frac{1}{e}\right)^\alpha 0 \right] \\
&= 0 < 1.
\end{aligned}$$

Hence, convergence is guaranteed. Besides, one can easily show that Equation (22) solves Equation (21). □

Remark 3. *Note that Equation (22) solves the inequality*

$$\zeta(t) \leq e^{\frac{\rho-1}{\rho}(t-a)} (t-a)^{\alpha-1} \zeta(a) + {}_a I^{\alpha,\rho} \zeta(t) y(t), \quad t \in [a,b], \tag{26}$$

where ζ and y are nonnegative real valued functions such that $0 \leq y(t) < \lambda < 1$.

Now, we are in a position to state the main theorem, which is a new version of the Gronwall–Bellman inequality within the generalized proportional fractional Riemann–Liouville settings.

Corollary 2. *Let ζ and y be nonnegative real valued functions such that $0 \le y(t) < \lambda < 1$ and*

$$\zeta(t) \le e^{\frac{\rho-1}{\rho}(t-a)}(t-a)^{\alpha-1}\zeta(a) + {}_aI^{\alpha,\rho}\zeta(t)y(t), \quad t \in [a,b]. \tag{27}$$

Then,

$$\zeta(t) \le \zeta(a)\sum_{k=0}^{\infty}\Omega_y^k e^{\frac{\rho-1}{\rho}(t-a)}(t-a)^{\alpha-1}. \tag{28}$$

The proof of the corollary is a straightforward implementation of Theorems 1 and 2. Indeed, it is immediately obtained by setting $\eta(t) = \zeta(a)\sum_{k=0}^{\infty}\Omega_y^k e^{\frac{\rho-1}{\rho}(t-a)}(t-a)^{\alpha-1}$.

3.2. Gronwall–Bellman Inequality via the GPF Derivative of Caputo Type

Consider the following generalized proportional Caputo fractional initial value problem

$$\begin{cases} ({}_a^CD^{\alpha,\rho}y)(t) = f(t,y(t)), \quad 0 < \alpha \le 1, \quad t \in [a,b], \\ y(a) = y_a. \end{cases} \tag{29}$$

Applying the operator ${}_aI^{\alpha,\rho}$ to both sides of Equation (29), we obtain

$$y(t) = e^{\frac{\rho-1}{\rho}(t-a)}y(a) + {}_aI^{\alpha,\rho}f(t,y(t)), \tag{30}$$

The results of this subsection resemble the ones proved in Section 3.1. To avoid redundancy, therefore, we skip some steps of the proofs. We start by the following comparison result for the generalized proportional Caputo fractional integral operator.

Theorem 3. *Let η and ζ be nonnegative continuous functions defined on $[a,b]$ and satisfy*

$$\eta(t) \ge e^{\frac{\rho-1}{\rho}(t-a)}\eta(a) + {}_aI^{\alpha,\rho}f(t,\eta(t)), \tag{31}$$

and

$$\zeta(t) \le e^{\frac{\rho-1}{\rho}(t-a)}\zeta(a) + {}_aI^{\alpha,\rho}f(t,\zeta(t)), \tag{32}$$

respectively. Suppose further that f satisfies one-sided Lipschitz condition of the form

$$f(t,x) - f(t,y) \le \frac{L}{e^{\frac{\rho-1}{\rho}(a-t)}\left[e^{\frac{\rho-1}{\rho}(t-a)} + \frac{(t-a)^\alpha}{\alpha}\right]}(x-y), \ \text{for} \ x \ge y, \ L > 0, \tag{33}$$

and $f(t,y)$ is nondecreasing in y. Then, $\eta(a) \ge \zeta(a)$ and $L < \Gamma(\alpha)\rho^\alpha e^{-\frac{\rho-1}{\rho}(t-a)}$ imply that $\eta(t) \ge \zeta(t)$ for all $t \in [a,b]$.

The proof of the above theorem can be completed by setting $\eta_\varepsilon(t) = \eta(t) + \varepsilon\left[e^{\frac{\rho-1}{\rho}(t-a)} + \frac{(t-a)^\alpha}{\alpha}\right]$, for small $\varepsilon > 0$, and following similar steps as the proof of Theorem 1.

In the sequel, we replace $f(t,y(t))$ in Equation (30) by $x(t)y(t)$, where $|x(t)| < 1$, $t \in [a,b]$. Define the following operator

$$\Phi_x\phi = {}_aI^{\alpha,\rho}x(t)\phi(t). \tag{34}$$

In similar manner, the following lemmas are formulated for Caputo type operator.

Lemma 9. *For any constant λ, one has*

$$|\Phi_{\lambda} e^{\frac{\rho-1}{\rho}(t-a)}| \leq \Phi_{|\lambda|} e^{\frac{\rho-1}{\rho}(t-a)}. \tag{35}$$

Lemma 10. *For any constant λ, one has*

$$|\Phi_{\lambda}^{n} e^{\frac{\rho-1}{\rho}(t-a)}| = \frac{|\lambda|^{n}(t-a)^{n\alpha}}{\rho^{n\alpha}\Gamma(n\alpha+1)} e^{\frac{\rho-1}{\rho}(t-a)}, \quad n = 0, 1, 2, \cdots. \tag{36}$$

Lemma 11. *Let $\lambda > 0$ be such that $|y(t)| \leq \lambda$ for $t \in [a, b]$. Then,*

$$|\Phi_{y}^{n} e^{\frac{\rho-1}{\rho}(t-a)}| = \Phi_{\lambda}^{n} e^{\frac{\rho-1}{\rho}(t-a)}, \qquad n = 0, 1, 2, \cdots. \tag{37}$$

Theorem 4. *Let y be a nonnegative function on $[a, b]$. Then, the generalized proportional fractional integral equation*

$$y(t) = e^{\frac{\rho-1}{\rho}(t-a)} y(a) + {}_{a}I^{\alpha,\rho} x(t) y(t), \quad t \in [a, b], \tag{38}$$

has a solution

$$y(t) = y(a) \sum_{k=0}^{\infty} \Phi_{x}^{k} e^{\frac{\rho-1}{\rho}(t-a)}. \tag{39}$$

Proof. We employ the successive approximation method to complete the proof. Set

$$\begin{aligned} y_0(t) &= e^{\frac{\rho-1}{\rho}(t-a)} y(a) \\ y_n(t) &= e^{\frac{\rho-1}{\rho}(t-a)} y(a) + {}_{a}I^{\alpha,\rho} x(t) y_{n-1}(t), \quad n \geq 1. \end{aligned}$$

We observe that

$$y_1(t) = y(a)\Phi_{x}^{0} e^{\frac{\rho-1}{\rho}(t-a)} + y(a)\Phi_{x}^{1} e^{\frac{\rho-1}{\rho}(t-a)}$$

and

$$\begin{aligned} y_2(t) &= e^{\frac{\rho-1}{\rho}(t-a)} y(a) + {}_{a}I^{\alpha,\rho} x(t) y_1(t) \\ &= y(a)\Phi_{x}^{0} e^{\frac{\rho-1}{\rho}(t-a)} + y(a)\Phi_{x}^{1} e^{\frac{\rho-1}{\rho}(t-a)} + y(a)\Phi_{x}^{2} e^{\frac{\rho-1}{\rho}(t-a)}. \end{aligned}$$

It follows inductively that $y_n(t) = y(a) \sum_{k=0}^{n} \Phi_{x}^{k} e^{\frac{\rho-1}{\rho}(t-a)}$. Taking the limit as $n \to \infty$ to obtain

$$y(t) = y(a) \sum_{k=0}^{\infty} \Phi_{x}^{k} e^{\frac{\rho-1}{\rho}(t-a)}. \tag{40}$$

Following the same arguments as in the proof of Theorem 2, we use Lemmas 9–11, the comparison test and the d'Alembert ratio test to show the absolute convergence of the series in Equation (40). Moreover, it is clear to verify that Equation (39) solves Equation (38). The proof is finished. □

Remark 4. *Note that Equation (39) solves the inequality*

$$\zeta(t) \leq e^{\frac{\rho-1}{\rho}(t-a)} \zeta(a) + {}_{a}I^{\alpha,\rho} \zeta(t) y(t), \quad t \in [a, b], \tag{41}$$

where ζ and y are nonnegative functions on $[a, b]$ such that $0 \leq y(t) < \lambda < 1$.

The Gronwall–Bellman inequality in generalized proportional Caputo fractional is stated as follows.

Corollary 3. *Let ζ and y be nonnegative real valued functions such that $0 \leq y(t) < \lambda < 1$ and*

$$\zeta(t) \leq e^{\frac{\rho-1}{\rho}(t-a)}\zeta(a) + {}_aI^{\alpha,\rho}\zeta(t)y(t), \qquad t \in [a,b]. \tag{42}$$

Then,

$$\zeta(t) \leq \zeta(a)\sum_{k=0}^{\infty} \Phi_y^k e^{\frac{\rho-1}{\rho}(t-a)}. \tag{43}$$

To prove Equation (43), we set $\eta(t) = \zeta(a)\sum_{k=0}^{\infty} \Phi_y^k e^{\frac{\rho-1}{\rho}(t-a)}$ and the rest follows as a direct application of Theorems 3 and 4.

4. Gronwall–Bellman Inequality via Weighted Function

In this section, we extend the results obtained in Section 3 to the case of weighted function. The analysis can be carried out for the Riemann–Liouville and Caputo operators. However, we only present the results for the case of Riemann–Liouville proportional fractional operator. Unlike previous relevant results in the literature [32], the weighted function w in the following first two theorems requires no monotonic restriction.

Theorem 5. *Let η, ζ, w be nonnegative continuous functions on $[a,b]$ where η and ζ satisfy*

$$\eta(t) \geq e^{\frac{\rho-1}{\rho}(t-a)}(t-a)^{\alpha-1}\eta(a) + w(t)_aI^{\alpha,\rho}f(t,\eta(t)), \tag{44}$$

and

$$\zeta(t) \leq e^{\frac{\rho-1}{\rho}(t-a)}(t-a)^{\alpha-1}\zeta(a) + w(t)_aI^{\alpha,\rho}f(t,\zeta(t)), \tag{45}$$

respectively. Suppose further that f satisfies one-sided Lipschitz condition of the form

$$f(t,x) - f(t,y) \leq \frac{L}{e^{\frac{\rho-1}{\rho}(a-t)}\left[e^{\frac{\rho-1}{\rho}(t-a)}(t-a)^{\alpha-1} + w(t)\frac{(t-a)^{\alpha}}{\alpha} + 1\right]}(x-y), \ \textit{for} \ x \geq y, \ L > 0, \tag{46}$$

and $f(t,y)$ is nondecreasing in y. Then, $\eta(a) \geq \zeta(a)$ and $L < \left(1 + \frac{\alpha}{w(t)(t-a)^{\alpha}}\right)\Gamma(\alpha)\rho^{\alpha}e^{-\frac{\rho-1}{\rho}(t-a)}$ imply that $\eta(t) \geq \zeta(t)$ for all $t \in [a,b]$.

To prove the above theorem, we set $\eta_\varepsilon(t) = \eta(t) + \varepsilon\left[e^{\frac{\rho-1}{\rho}(t-a)} + w(t)\frac{(t-a)^{\alpha}}{\alpha} + 1\right]$, for small $\varepsilon > 0$, and follow similar steps as the proof of Theorem 1.

Remark 5. *The Lipschitz condition in Equation (46) can be relaxed by relaxing the upper bound for the constant L.*

Theorem 6. *Let x, y be nonnegative functions on $[a,b]$ and w be a nonnegative continuous function defined on $[a,b]$. Further, assume that $|x(t)| < 1$ for $t \in [a,b]$ and $\max_{t\in[a,b]} w(t) = M$. Then, the generalized proportional fractional integral equation*

$$y(t) = e^{\frac{\rho-1}{\rho}(t-a)}(t-a)^{\alpha-1}y(a) + w(t)_aI^{\alpha,\rho}x(t)y(t), \quad t \in [a,b], \tag{47}$$

has a solution

$$y(t) = y(a)\Omega_x^0 e^{\frac{\rho-1}{\rho}(t-a)}(t-a)^{\alpha-1} + y(a)w(t)\sum_{k=1}^{\infty} M^{k-1}\Omega_x^k e^{\frac{\rho-1}{\rho}(t-a)}(t-a)^{\alpha-1}. \tag{48}$$

Remark 6. *Note that Equation (48) solves the inequality*

$$\zeta(t) \leq e^{\frac{\rho-1}{\rho}(t-a)}(t-a)^{\alpha-1}\zeta(a) + w(t)_a I^{\alpha,\rho}\zeta(t)y(t), \quad t \in [a,b], \tag{49}$$

where ζ, y *are nonnegative functions on* $[a,b]$ *and* w *is a nonnegative continuous function defined on* $[a,b]$ *and* $0 \leq y(t) < \lambda < 1$ *and* $\max_{t\in[a,b]} w(t) = M$.

The Gronwall–Bellman inequality in case of weighted function w is stated as follows.

Theorem 7. *Let* ζ, y *be nonnegative functions on* $[a,b]$ *and* w *be a nonnegative continuous function defined on* $[a,b]$. *Further, assume that* $0 \leq y(t) < \lambda < 1$ *for* $t \in [a,b]$ *and* $\max_{t\in[a,b]} w(t) = M$ *and*

$$\zeta(t) \leq e^{\frac{\rho-1}{\rho}(t-a)}(t-a)^{\alpha-1}\zeta(a) + w(t)_a I^{\alpha,\rho}\zeta(t)y(t), \quad t \in [a,b]. \tag{50}$$

Then,

$$\zeta(t) \leq \zeta(a)\Omega_y^0 e^{\frac{\rho-1}{\rho}(t-a)}(t-a)^{\alpha-1} + \zeta(a)w(t)\sum_{k=1}^{\infty} M^{k-1}\Omega_y^k e^{\frac{\rho-1}{\rho}(t-a)}(t-a)^{\alpha-1}. \tag{51}$$

If the weighted function w possesses a monotonic behavior, then Theorem 6 and Theorem 7 can be reformulated, respectively, in the following forms.

Theorem 8. *Let* y, x *be nonnegative functions on* $[a,b]$ *and* w *be a nonnegative continuous function defined on* $[a,b]$. *Further, assume that* $|x(t)| < 1$ *for* $t \in [a,b]$ *and* w *is a nondecreasing function. Then, the generalized proportional fractional integral equation*

$$y(t) = e^{\frac{\rho-1}{\rho}(t-a)}(t-a)^{\alpha-1}y(a) + w(t)_a I^{\alpha,\rho}x(t)y(t), \quad t \in [a,b], \tag{52}$$

has a solution

$$y(t) = y(a)\sum_{k=0}^{\infty} w^k(t)\Omega_x^k e^{\frac{\rho-1}{\rho}(t-a)}(t-a)^{\alpha-1}. \tag{53}$$

Theorem 9. *Let* ζ, y *be nonnegative functions on* $[a,b]$ *and* w *be a nonnegative continuous function defined on* $[a,b]$. *Assume that* $0 \leq y(t) < \lambda < 1$ *for* $t \in [a,b]$ *and* w *is a nondecreasing function and*

$$\zeta(t) \leq e^{\frac{\rho-1}{\rho}(t-a)}(t-a)^{\alpha-1}\zeta(a) + w(t)_a I^{\alpha,\rho}\zeta(t)y(t), \quad t \in [a,b]. \tag{54}$$

Then,

$$\zeta(t) \leq \zeta(a)\sum_{k=0}^{\infty} w^k(t)\Omega_y^k e^{\frac{\rho-1}{\rho}(t-a)}(t-a)^{\alpha-1}. \tag{55}$$

5. Applications

In this section, two examples of Riemann–Liouville and Caputo generalized proportional fractional initial value problems are presented. By the help of the new proposed Gronwall–Bellman inequalities in Theorems 2 and 3, we show that the solution of the initial value problems depend on the initial data and on the right-hand side.

Consider the proportional Riemann–Liouville fractional initial value problem in Equation (10). In the remaining part of this section, we assume that the nonlinearity function $f(t,y)$ satisfies a Lipschitz condition with a constant $L \in [0,1)$ for all (t,y).

Example 1. *Consider the following Riemann–Liouville proportional fractional initial value problems of the form*

$$({}_aD^{\alpha,\rho}\beta)(t) = f(t,\beta(t)), \quad \lim_{t\to a^+}({}_aI^{1-\alpha,\rho}\beta)(t) = \beta(a) = \beta_0, \quad 0<\alpha\le 1,\, t\in[a,b], \tag{56}$$

and

$$({}_aD^{\alpha,\rho}\gamma)(t) = f(t,\gamma(t)), \quad \lim_{t\to a^+}({}_aI^{1-\alpha,\rho}\gamma)(t) = \gamma(a) = \gamma_0, \quad 0<\alpha\le 1,\, t\in[a,b]. \tag{57}$$

We claim that a small change in the initial condition implies a small change in the solution.

Proof. Applying the generalized proportional fractional integral operator in Equations (56) and (57), we have

$$\beta(t) = e^{\frac{\rho-1}{\rho}(t-a)}(t-a)^{\alpha-1}\beta_0 + {}_aI^{\alpha,\rho}f(t,\beta(t)),$$

and

$$\gamma(t) = e^{\frac{\rho-1}{\rho}(t-a)}(t-a)^{\alpha-1}\gamma_0 + {}_aI^{\alpha,\rho}f(t,\gamma(t)).$$

It follows that

$$\beta(t)-\gamma(t) = e^{\frac{\rho-1}{\rho}(t-a)}(t-a)^{\alpha-1}(\beta_0-\gamma_0) + {}_aI^{\alpha,\rho}[f(t,\beta(t))-f(t,\gamma(t))].$$

Taking the absolute value, we obtain

$$\begin{aligned}|\beta(t)-\gamma(t)| &\le e^{\frac{\rho-1}{\rho}(t-a)}(t-a)^{\alpha-1}|\beta_0-\gamma_0| + {}_aI^{\alpha,\rho}|f(t,\beta(t))-f(t,\gamma(t))| \\ &\le e^{\frac{\rho-1}{\rho}(t-a)}(t-a)^{\alpha-1}|\beta_0-\gamma_0| + L_aI^{\alpha,\rho}|\beta(t)-\gamma(t)|.\end{aligned}$$

By employing Theorem 2, we get

$$\begin{aligned}|\beta(t)-\gamma(t)| &\le |\beta_0-\gamma_0|\sum_{k=0}^{\infty}\Omega_L^k e^{\frac{\rho-1}{\rho}(t-a)}(t-a)^{\alpha-1} \\ &= |\beta_0-\gamma_0|\sum_{k=0}^{\infty}\frac{L^k(t-a)^{(k+1)\alpha-1}\Gamma(\alpha)}{\rho^{k\alpha}\Gamma((k+1)\alpha)}e^{\frac{\rho-1}{\rho}(t-a)}.\end{aligned}$$

Consider the initial value problem

$$\begin{cases}({}_aD^{\alpha,\rho}\nu)(t) = f(t,\nu(t)), \quad 0<\alpha\le 1, \quad t\in[a,b] \\ \lim_{t\to a^+}({}_aI^{1-\alpha,\rho}\nu)(t) = \nu(a) = \beta_n,\end{cases} \tag{58}$$

where $\beta_n \to \beta_0$. If the solution of Equation (58) is denoted by ν_n, then, for all $t\in[a,b]$, we have

$$|\beta(t)-\nu_n(t)| \le |\beta_0-\beta_n|\sum_{k=0}^{\infty}\frac{L^k(t-a)^{(k+1)\alpha-1}\Gamma(\alpha)}{\rho^{k\alpha}\Gamma((k+1)\alpha)}e^{\frac{\rho-1}{\rho}(t-a)}.$$

Hence, $|\beta(t)-\nu_n(t)| \to 0$ when $\beta_n \to \beta_0$ as $n\to\infty$. We conclude that a small change in the initial condition implies a small change in the solution. □

Example 2. *Consider the following Caputo generalized proportional fractional initial value problems of the form*

$$({}_a^CD^{\alpha,\rho}\beta)(t) = f(t,\beta(t)), \quad \beta(a) = \beta_0, \quad \alpha\in(0,1], \quad t\in[a,b]. \tag{59}$$

and

$$({}^{C}_{a}D^{\alpha,\rho}\sigma)(t) = f(t,\sigma(t)) + g(t,\sigma(t)), \quad \sigma(a) = \sigma_0, \quad \alpha \in (0,1], \quad t \in [a,b]. \tag{60}$$

We claim that the solution of Equation (60) depends continuously on the right-hand side of Equation (60) if $|g(t,\sigma)| \le Ke^{\frac{\rho-1}{\rho}(t-a)}$ *for all* $t \in [a,b]$ *and for a positive number K.*

Proof. If the solution of Equation (60) is denoted by σ, then, for all $t \in [a,b]$, we have

$$\begin{aligned}|\beta(t) - \sigma(t)| &\le e^{\frac{\rho-1}{\rho}(t-a)}|\beta_0 - \sigma_0| + {}_aI^{\alpha,\rho}|f(t,\beta(t)) - f(t,\sigma(t))| + {}_aI^{\alpha,\rho}|g(t,\sigma(t))| \\ &\le e^{\frac{\rho-1}{\rho}(t-a)}|\beta_0 - \sigma_0| + L\, {}_aI^{\alpha,\rho}|\beta(t) - \sigma(t)| + {}_aI^{\alpha,\rho}|g(t,\sigma(t))|.\end{aligned}$$

By the assumption, we have

$$|\beta(t) - \sigma(t)| \le e^{\frac{\rho-1}{\rho}(t-a)}|\beta_0 - \sigma_0| + L\, {}_aI^{\alpha,\rho}|\beta(t) - \sigma(t)| + {}_aI^{\alpha,\rho}Ke^{\frac{\rho-1}{\rho}(t-a)}$$

or

$$|\beta(t) - \sigma(t)| + \frac{K}{L}e^{\frac{\rho-1}{\rho}(t-a)} \le e^{\frac{\rho-1}{\rho}(t-a)}\left(|\beta_0 - \sigma_0| + \frac{K}{L}\right) + L\, {}_aI^{\alpha,\rho}\left(|\beta(t) - \sigma(t)| + \frac{K}{L}e^{\frac{\rho-1}{\rho}(t-a)}\right).$$

Let $r(t) = |\beta(t) - \sigma(t)| + \frac{K}{L}e^{\frac{\rho-1}{\rho}(t-a)}$. Then, if we apply Theorem 3, we obtain

$$r(t) \le \left(|\beta_0 - \sigma_0| + \frac{K}{L}\right)\sum_{k=0}^{\infty}\Phi_L^k e^{\frac{\rho-1}{\rho}(t-a)},$$

or

$$|\beta(t) - \sigma(t)| \le \left(|\beta_0 - \sigma_0| + \frac{K}{L}\right)\sum_{k=0}^{\infty}\frac{L^k(t-a)^{(k+1)\alpha-1}}{\rho^{k\alpha}\Gamma((k+1)\alpha)}e^{\frac{\rho-1}{\rho}(t-a)} - \frac{K}{L}e^{\frac{\rho-1}{\rho}(t-a)}.$$

For $a \le t \le b$, letting $Ke^{\frac{\rho-1}{\rho}(t-a)} < \delta$ implies that

$$\begin{aligned}|\beta(t) - \sigma(t)| &\le |\beta_0 - \sigma_0|\sum_{k=0}^{\infty}\frac{L^k(t-a)^{(k+1)\alpha-1}}{K\rho^{k\alpha}\Gamma((k+1)\alpha)}\delta + \frac{\delta}{L}\left[\sum_{k=0}^{\infty}\frac{L^k(t-a)^{(k+1)\alpha-1}}{\rho^{k\alpha}\Gamma((k+1)\alpha)} - 1\right] \\ &\le \delta\left\{|\beta_0 - \sigma_0|\sum_{k=0}^{\infty}\frac{L^k b^{(k+1)\alpha-1}}{K\rho^{k\alpha}\Gamma((k+1)\alpha)} + \frac{1}{L}\left[\sum_{k=0}^{\infty}\frac{L^k b^{(k+1)\alpha-1}}{\rho^{k\alpha}\Gamma((k+1)\alpha)} - 1\right]\right\} = \epsilon,\end{aligned}$$

which implies that a small change on the right-hand side of Equation (59) implies a small change in its solution. □

6. Conclusions

One of the most crucial issues in the theory of differential equations is to study qualitative properties for solutions of these equations. Integral inequalities are significant instruments that facilitate exploring such properties. In this paper, we accommodate a newly defined generalized proportional fractional (GPF) derivative to establish new versions for the well-known Gronwall–Bellman inequality. We prove our results in the frame of GPF operators within the Riemann–Liouville and Caputo settings. The main results are also extended to the weighted function case. One can easily figure out that the current results generalize the ones previously obtained in the literature. Indeed, the case $\rho = 1$ covers the results of classical Riemann–Liouville and Caputo fractional derivatives. As an application, we provide two efficient examples that demonstrate the solution dependence on the initial data and on the right-hand side of the initial value problems.

The results of this paper have strong potential to be used for establishing new substantial investigations in the future for equations involving the GPF operators.

Author Contributions: All authors contributed equally and significantly to this paper. All authors have read and approved the final version of the manuscript.

Funding: The first author would like to thank Prince Sultan University for funding this work through research group Nonlinear Analysis Methods in Applied Mathematics (NAMAM) group number RG-DES-2017-01-17.

Acknowledgments: The authors would like to express their sincere thanks to the handling editor and the referees for their constructive comments and suggestions. We believe that their efforts have significantly helped in improving the contents of this paper.

Conflicts of Interest: The authors declare that they have no competing interests.

References

1. Hardy, G.H.; Littlewood, J.E.; Pólya, G. *Inequalities*; Cambridge University Press: London, UK, 1952.
2. Bainov, D.D.; Simeonov, P.S. *Integral Inequalities and Applications (Mathematics and Its Applications)*; Springer: Berlin, Germany, 1992.
3. Cloud, J.M.; Drachman, C.B.; Lebedev, P.L. *Inequalities with Applications to Engineering*; Springer: Berlin, Germany, 2014.
4. Grönwall-Bellman, T.H. Note on the derivatives with respect to a parameter of the solutions of a system of differential equations. *Ann. Math.* **1919**, *20*, 292–296. [CrossRef]
5. Bellman, R. The stability of solutions of linear differential equations. *Duke Math. J.* **1943**, *10* , 643–647. [CrossRef]
6. Rasmussen, D.L. Gronwall's inequality for functions of two independent variables. *J. Math. Anal. Appl.* **1976**, *55*, 407–417. [CrossRef]
7. Dragomir, S.S. *Some Gronwall Type Inequalities and Applications*; Nova Science Pub Inc.: Hauppauge, NY, USA, 2003.
8. Lin, X. A note on Gronwall's inequality on time scales. *Abstr. Appl. Anal.* **2014**, *2014*, 623726. [CrossRef]
9. Wang, W.; Feng, Y.; Wang, Y. Nonlinear Gronwall—Bellman type inequalities and their applications. *Mathematics* **2017**, *5*, 31. [CrossRef]
10. Hilfer, R. *Applications of Fractional Calculus in Physics*; Word Scientific: Singapore, 2000.
11. Debnath, L. Recent applications of fractional calculus to science and engineering. *Int. J. Math. Math. Sci.* **2003**, *54*, 3413–3442. [CrossRef]
12. Kilbas, A.; Srivastava, H.M.; Trujillo, J.J. *Theory and Application of Fractional Differential Equations. North-Holland Mathematical Studies*; Elsevier (North-Holland) Science Publishers: Amsterdam, The Netherlands, 2006; Volume 204.
13. Magin, R.L. *Fractional Calculus in Bioengineering*; Begell House: Redding, CT, USA, 2006.
14. Samko, S.G.; Kilbas, A.A.; Marichev, O.I. *Fractional Integrals and Derivatives: Theory and Applications*; Gordon and Breach Science: Yverdon, Switzerland, 1993.
15. Podlubny, I. *Fractional Differential Equations. Mathematics in Science and Engineering 198*; Academic Press: San Diego, CA, USA, 1999.
16. He, J.H.; Ji, F.Y. Two-scale mathematics and fractional calculus for thermodynamics. *Therm. Sci.* **2019**. [CrossRef]
17. Caputo, M.; Fabrizio, M. A new definition of fractional derivative without singular kernel. *Progr. Fract. Differ. Appl.* **2015**, *1*, 73–85.
18. Atangana, A.; Baleanu, D. New fractional derivatives with non-local and non-singular kernels. *Therm. Sci.* **2016**, *20*, 757–763. [CrossRef]
19. Ye, H.; Gao, J.; Ding, Y. A generalized Gronwall inequality and its application to a fractional differential equation. *J. Math. Anal. Appl.* **2007**, *328*, 1075–1081. [CrossRef]
20. Ferreira, R.A.C. A Discrete fractional Gronwall inequality. *Proc. Am. Math. Soc.* **2012**, *5*, 1605–1612. [CrossRef]
21. Abdeljawad, T.; Alzabut, J. The q-fractional analogue for Gronwall-type inequality. *J. Funct. Spaces Appl.* **2013**. [CrossRef]

22. Abdeljawad, T.; Alzabut, J.; Baleanu, D. A generalized q-fractional Gronwall inequality and its applications to nonlinear delay q-fractional difference systems. *J. Inequal. Appl.* **2016**, *240*, 1–13. [CrossRef]
23. Zhang, Z.; Wei, Z. A generalized Gronwall inequality and its application to fractional neutral evolution inclusions. *J. Inequal. Appl.* **2016**, *45*, 1–18. [CrossRef]
24. Sarikaya, M.Z. Gronwall type inequalities for conformable fractional integrals. *Konuralp J. Math.* **2016**, *4*, 217–222.
25. Alzabut, J.; Abdeljawad, T. A generalized discrete fractional Gronwall inequality and its application on the uniqueness of solutions for nonlinear delay fractional difference system. *Appl. Anal. Discrete Math.* **2018**, *12*, 36–48. [CrossRef]
26. Jarad, F.; Abdeljawad, T.; Alzabut, J. Generalized fractional derivatives generated by a class of local proportional derivatives. *Eur. Phys. J. Spec. Top.* **2017**, *226*, 3457–3471. [CrossRef]
27. Ding, D.; Zhang, X.; Cao, J.; Wang, N.; Liang, D. Bifurcation control of complex networks model via PD controller. *Neurocomputing* **2016**, *175*, 1–9.
28. Anderson, D.R.; Ulness, D.J. Newly defined conformable derivatives. *Adv. Dyn. Syst. Appl.* **2015**, *10*, 109–137.
29. Anderson, D.R. Second-order self-adjoint differential equations using a proportional–derivative controller. *Commun. Appl. Nonlinear Anal.* **2017**, *24*, 17–48.
30. Lakshmikantham, V.; Vatsala, A.S. Basic theory of fractional differential equations. *Nonlinear Anal.* **2008**, *69*, 2677–2682. [CrossRef]
31. Denton, Z.; Vatsala, A.S. Fractional integral inequalities and applications. *Comput. Math. Appl.* **2010**, *59*, 1087–1094. [CrossRef]
32. Alzabut, J.; Abdeljawad, T.; Jarad, F.; Sudsutad, W. A Gronwall inequality via the generalized proportional fractional derivative with applications. *J. Inequal. Appl.* **2019**, *101*, 1–12. [CrossRef]

mathematics

Article

Hermite-Hadamard Type Inequalities for Interval (h_1, h_2)-Convex Functions

Yanrong An [1], Guoju Ye [2,*], Dafang Zhao [3] and Wei Liu [2]

1 School of Business, Nanjing University, Nanjing 210098, China; yrannu@163.com
2 College of Science, Hohai University, Nanjing 210098, China; liuw626@hhu.edu.cn
3 School of Mathematics and Statistics, Hubei Normal University, Huangshi 435002, China; dafangzhao@163.com
* Correspondence: ygjhhu@163.com; Tel.: +86-134-0586-3846

Received: 18 April 2019; Accepted: 14 May 2019; Published: 17 May 2019

Abstract: We introduce the concept of interval (h_1, h_2)-convex functions. Under the new concept, we establish some new interval Hermite-Hadamard type inequalities, which generalize those in the literature. Also, we give some interesting examples.

Keywords: Hermite-Hadamard inequality; interval-valued functions; (h_1, h_2)-convex

1. Introduction

Interval analysis was introduced in numerical analysis by Moore in the celebrated book [1]. Over the past 50 years, it has attracted considerable interest and has been applied in various fields, such as interval differential equations [2], aeroelasticity [3], aerodynamic load analysis [4], and so on. For more profound results and applications, see [5–9].

It is known that inequalities play an important role in almost all branches of mathematics as well as in other areas of science. Among the many types of inequalities, those carrying the names of Jensen, Hermite-Hadamard, Hardy, Ostrowski, Minkowski and Opial et al. have a deep significance and have made a great impact in substantial fields of research. Recently, some of these inequalities have been extended to interval-valued functions by Chalco-Cano et al.; see, e.g., [10–16]. Surprisingly enough, interval Hermite-Hadamard type inequalities has perhaps not received enough attention [17]. For convenience, we recall the classical Hermite-Hadamard inequality. Let f be convex, then

$$f\left(\frac{u+v}{2}\right) \leq \frac{1}{v-u}\int_u^v f(t)dt \leq \frac{f(u)+f(v)}{2}.$$

This inequality has been developed for different classes of convexity [18–26]. Especially, since the h-convex concept was proposed by Varosanec in 2007 [27], a number of authors have already studied more refined Hermite-Hadamard inequalities involving h-convex functions [28–33].

In 2018, Awan et al. introduced (h_1, h_2)-convex functions and proved the following inequality [34]:

Theorem 1. *Let $f : [u, v] \to \mathbb{R}$. If f is (h_1, h_2)-convex, and $h_1(\frac{1}{2})h_2(\frac{1}{2}) \neq 0$. Then*

$$\frac{1}{2h_1(\frac{1}{2})h_2(\frac{1}{2})}f\left(\frac{u+v}{2}\right) \leq \frac{1}{v-u}\int_u^v f(t)dt \leq [f(u)+f(v)]\int_0^1 h_1(x)h_2(1-x)dx.$$

Motivated by Awan et al., our main objective is to generalize the results above by constructing interval Hermite-Hadamard type inequalities for (h_1, h_2)-convex functions. Also, we present some examples to illustrate our theorems. Our results generalize some known inequalities presented in [17,32,34,35]. Furthermore, the present results can be considered as tools for further research in interval convex analysis, interval nonlinear programming, inequalities for fuzzy-interval-valued functions, among others.

We give preliminaries in Section 2. In Section 3, we introduce interval (h_1, h_2)-convex concept, and obtain some interval Hermite-Hadamard type inequalities. Moreover, some interesting examples are given. In Section 4, we give conclusions and future work.

2. Preliminaries

For the basic notations and definitions on interval analysis, see [17]. The family of all intervals and positive intervals of $\mathbb{R}$ are denoted by $\mathbb{R}_\mathcal{I}$ and $\mathbb{R}_\mathcal{I}^+$, respectively. For interval $[\underline{u}, \overline{u}]$ and $[\underline{v}, \overline{v}]$, the Hausdorff distance is defined by

$$d([\underline{u}, \overline{u}], [\underline{v}, \overline{v}]) = \max\left\{|\underline{u} - \underline{v}|, |\overline{u} - \overline{v}|\right\}.$$

Then, $(\mathbb{R}_\mathcal{I}, d)$ is complete.

A set of numbers $\{t_{i-1}, \xi_i, t_i\}_{i=1}^m$ is said to be a tagged partition P of $[u, v]$ if

$$u = t_0 < t_1 < \cdots < t_m = v$$

and if $t_{i-1} \leq \xi_i \leq t_i$ for all $i = 1, 2, \ldots, m$. Moreover, if we let $\Delta t_i = t_i - t_{i-1}$, then the partition is called δ-fine if $\Delta t_i < \delta$ for each i. We denote by $\mathcal{P}(\delta, [u, v])$ the family of all δ-fine partitions of $[u, v]$. Given $P \in \mathcal{P}(\delta, [u, v])$, we define a integral sum of $f : [u, v] \to \mathbb{R}_\mathcal{I}$ as follows:

$$S(f, P, \delta, [u, v]) = \sum_{i=1}^{m} f(\xi_i)(t_i - t_{i-1}).$$

Definition 1. *Let $f : [u, v] \to \mathbb{R}_\mathcal{I}$. f is called IR-integrable on $[u, v]$ with IR-integral $A = (IR) \int_u^v f(t)dt$, if there exists an $A \in \mathbb{R}_\mathcal{I}$ such that for any $\epsilon > 0$ there exists a $\delta > 0$ such that*

$$d(S(f, P, \delta, [u, v]), A) < \epsilon$$

for each $P \in \mathcal{P}(\delta, [u, v])$. Let $\mathcal{IR}_{([u,v])}$ denote the set of all IR-integrable functions on $[u, v]$.

Definition 2. *Let $h_1, h_2 : [0, 1] \subseteq J \to \mathbb{R}^+$ such that $h_1, h_2 \not\equiv 0$ (Awan et al. [34]). $f : J \to \mathbb{R}^+$ is called (h_1, h_2)-convex, or that $f \in SX((h_1, h_2), J, \mathbb{R})$, if for any $s, t \in J$ and $x \in (0, 1)$ one has*

$$f(xs + (1 - x)t) \leq h_1(x)h_2(1 - x)f(s) + h_1(1 - x)h_2(x)f(t). \quad (1)$$

Remark 1. *If $h_2 \equiv 1$, then Definition 2 reduces to h-convex in [27].*

If $h_1 = h_2 \equiv 1$, then Definition 2 reduces to P-function in [18].

If $h_1(t) = t^s$, $h_2 \equiv 1$, then Definition 2 reduces to s-convex in [36].

We end this section of preliminaries by introducing the new concept of interval (h_1, h_2)-convexity. This idea is inspired by Costa [12]. Note that for interval $[\underline{u}, \overline{u}]$ and $[\underline{v}, \overline{v}]$, the inclusion " $\subseteq$ " is defined by

$$[\underline{u}, \overline{u}] \subseteq [\underline{v}, \overline{v}] \Longleftrightarrow \underline{v} \leq \underline{u}, \ \overline{u} \leq \overline{v}.$$

Definition 3. *Let $h_1, h_2 : [0, 1] \subseteq J \to \mathbb{R}^+$ such that $h_1, h_2 \not\equiv 0$. $f : J \to \mathbb{R}_\mathcal{I}^+$ is called interval (h_1, h_2)-convex, if for all $s, t \in J$ and $x \in (0, 1)$ one has*

$$h_1(x)h_2(1 - x)f(s) + h_1(1 - x)h_2(x)f(t) \subseteq f(xs + (1 - x)t). \quad (2)$$

The set of all interval (h_1, h_2)-convex function is denoted by $SX((h_1, h_2), J, \mathbb{R}_\mathcal{I}^+)$.

3. Interval Hermite-Hadamard Type Inequality

In what follows, let $H(x,y) = h_1(x)h_2(y)$ for $x, y \in [0,1]$.

Theorem 2. *Let $f : [u,v] \to \mathbb{R}_{\mathcal{I}}^+$, $h_1, h_2 : [0,1] \to \mathbb{R}^+$ and $H(\frac{1}{2},\frac{1}{2}) \neq 0$. If $f \in SX((h_1,h_2),[u,v],\mathbb{R}_{\mathcal{I}}^+)$ and $f \in \mathcal{IR}_{([u,v])}$, then*

$$\frac{1}{2H(\frac{1}{2},\frac{1}{2})} f\Big(\frac{u+v}{2}\Big) \supseteq \frac{1}{v-u}\int_u^v f(t)dt \supseteq [f(u)+f(v)]\int_0^1 H(x,1-x)dx. \tag{3}$$

Proof. By hypothesis, we have

$$H\Big(\frac{1}{2},\frac{1}{2}\Big)f(xu+(1-x)v) + H\Big(\frac{1}{2},\frac{1}{2}\Big)f((1-x)u+xv) \subseteq f\Big(\frac{u+v}{2}\Big).$$

Then

$$\int_0^1 \underline{f}(xu+(1-x)v)dx + \int_0^1 \underline{f}((1-x)u+xv)dx \geq \frac{1}{H(\frac{1}{2},\frac{1}{2})}\int_0^1 \underline{f}\Big(\frac{u+v}{2}\Big)dx,$$
$$\int_0^1 \overline{f}(xu+(1-x)v)dx + \int_0^1 \overline{f}((1-x)u+xv)dx \leq \frac{1}{H(\frac{1}{2},\frac{1}{2})}\int_0^1 \overline{f}\Big(\frac{u+v}{2}\Big)dx.$$

It follows that

$$\frac{2}{v-u}\int_u^v \underline{f}(t)dt \geq \frac{1}{H(\frac{1}{2},\frac{1}{2})}\int_0^1 \underline{f}\Big(\frac{u+v}{2}\Big)dx = \frac{1}{H(\frac{1}{2},\frac{1}{2})}\underline{f}\Big(\frac{u+v}{2}\Big),$$
$$\frac{2}{v-u}\int_u^v \overline{f}(t)dt \leq \frac{1}{H(\frac{1}{2},\frac{1}{2})}\int_0^1 \overline{f}\Big(\frac{u+v}{2}\Big)dx = \frac{1}{H(\frac{1}{2},\frac{1}{2})}\overline{f}\Big(\frac{u+v}{2}\Big).$$

This implies

$$\frac{1}{H(\frac{1}{2},\frac{1}{2})}\Big[\underline{f}\Big(\frac{u+v}{2}\Big), \overline{f}\Big(\frac{u+v}{2}\Big)\Big] \supseteq \frac{2}{v-u}\Big[\int_u^v \underline{f}(t)dt, \int_u^v \overline{f}(t)dt\Big].$$

Thus,

$$\frac{1}{2H(\frac{1}{2},\frac{1}{2})} f\Big(\frac{u+v}{2}\Big) \supseteq \frac{1}{v-u}\int_u^v f(t)dt.$$

In the same way as above, we have

$$\frac{1}{v-u}\int_u^v f(t)dt \supseteq [f(u)+f(v)]\int_0^1 H(x,1-x)dx,$$

and the result follows. □

Remark 2. *If $H(x,y) \equiv h_1(x)$, then Theorem 2 reduces to ([17], Theorem 4.1).*

If $h_1(x) = x^s$, $h_2 \equiv 1$, then Theorem 2 reduces to ([37], Theorem 4).

If $h_1 = h_2 \equiv 1$, then inequality (3) in Theorem 2 reduces to inequality for P-function.

If $\underline{f} = \overline{f}$, then Theorem 2 reduces to ([34], Theorem 1). Furthermore, If $h_2 = 1$, then we get ([32], Theorem 6).

Example 1. *Suppose that $h_1(x) = x$, $h_2(x) \equiv 1$ for $x \in [0,1]$, $[u,v] = [-1,1]$, and $f : [u,v] \to \mathbb{R}_{\mathcal{I}}^+$ be defined by $f(t) = [t^2, 4 - e^t]$. Then*

$$\frac{1}{2H\left(\frac{1}{2},\frac{1}{2}\right)} f\left(\frac{u+v}{2}\right) = f(0) = [0,3],$$
$$\frac{1}{v-u}\int_u^v f(t)dt = \frac{1}{2}\left[\int_{-1}^{1} t^2 dt, \int_{-1}^{1}(6-e^t)dt\right] = \left[\frac{1}{3}, 4 - \frac{e-e^{-1}}{2}\right],$$
$$[f(u)+f(v)]\int_0^1 H(x,1-x)dx = \left[1, 4 - \frac{e+e^{-1}}{2}\right].$$

Then, we obtain that

$$[0,3] \supseteq \left[\frac{1}{3}, 4 - \frac{e-e^{-1}}{2}\right] \supseteq \left[\frac{1}{3}, 4 - \frac{e+e^{-1}}{2}\right].$$

Consequently, Theorem 2 is verified.

The next result generalizes Theorem 3.1 of [35] and Theorem 4.3 of [17].

Theorem 3. *Let $f : [u,v] \to \mathbb{R}_{\mathcal{I}}^+$, $h_1, h_2 : [0,1] \to \mathbb{R}^+$ and $H\left(\frac{1}{2},\frac{1}{2}\right) \neq 0$. If $f \in SX((h_1,h_2),[u,v],\mathbb{R}_{\mathcal{I}}^+)$ and $f \in \mathcal{IR}_{([u,v])}$, then*

$$\frac{1}{4H^2\left(\frac{1}{2},\frac{1}{2}\right)} f\left(\frac{u+v}{2}\right) \supseteq \Delta_1 \supseteq \frac{1}{v-u}\int_u^v f(t)dt$$
$$\supseteq \Delta_2 \supseteq [f(u)+f(v)]\left[\frac{1}{2} + H\left(\frac{1}{2},\frac{1}{2}\right)\right]\int_0^1 H(x,1-x)dx,$$

where

$$\Delta_1 = \frac{1}{4H\left(\frac{1}{2},\frac{1}{2}\right)}\left[f\left(\frac{3u+v}{4}\right) + f\left(\frac{u+3v}{4}\right)\right],$$
$$\Delta_2 = \left[\frac{f(u)+f(v)}{2} + f\left(\frac{u+v}{2}\right)\right]\int_0^1 H(x,1-x)dx.$$

Proof. For $[u, \frac{u+v}{2}]$, one has

$$H\left(\frac{1}{2},\frac{1}{2}\right) f\left(xu + (1-x)\frac{u+v}{2}\right) + H\left(\frac{1}{2},\frac{1}{2}\right) f\left((1-x)u + x\frac{u+v}{2}\right)$$
$$\subseteq f\left(\frac{xu + (1-x)\frac{u+v}{2}}{2} + \frac{(1-x)u + x\frac{u+v}{2}}{2}\right) = f\left(\frac{3u+v}{4}\right).$$

Consequently, we get

$$\frac{1}{4H\left(\frac{1}{2},\frac{1}{2}\right)} f\left(\frac{3u+v}{4}\right) \supseteq \frac{1}{v-u}\int_u^{\frac{u+v}{2}} f(t)dt.$$

In the same way as above, for $[\frac{u+v}{2}, v]$, we have

$$\frac{1}{4H\left(\frac{1}{2},\frac{1}{2}\right)} f\left(\frac{u+3v}{4}\right) \supseteq \frac{1}{v-u}\int_{\frac{u+v}{2}}^{v} f(t)dt.$$

Hence,

$$\Delta_1 = \frac{1}{4H\left(\frac{1}{2},\frac{1}{2}\right)}\left[f\left(\frac{3u+v}{4}\right)+f\left(\frac{u+3v}{4}\right)\right] \supseteq \frac{1}{v-u}\int_u^v f(t)dt.$$

Thanks to Theorem 2, one has

$$\begin{aligned}
&\frac{1}{4\left[H\left(\frac{1}{2},\frac{1}{2}\right)\right]^2} f\left(\frac{u+v}{2}\right)\\
&= \frac{1}{4\left[H\left(\frac{1}{2},\frac{1}{2}\right)\right]^2} f\left(\frac{1}{2}\cdot\frac{3u+v}{4}+\frac{1}{2}\cdot\frac{u+3v}{4}\right)\\
&\supseteq \frac{1}{4\left[H\left(\frac{1}{2},\frac{1}{2}\right)\right]^2}\left[H\left(\frac{1}{2},\frac{1}{2}\right)f\left(\frac{3u+v}{4}\right)+H\left(\frac{1}{2},\frac{1}{2}\right)f\left(\frac{u+3v}{4}\right)\right]\\
&\supseteq \Delta_1\\
&\supseteq \frac{1}{v-u}\int_u^v f(t)dt\\
&\supseteq \frac{1}{2}\left[f(u)+f(v)+2f\left(\frac{u+v}{2}\right)\right]\int_0^1 H(x,1-x)dx\\
&= \Delta_2\\
&\supseteq \left[\frac{f(u)+f(v)}{2}+H\left(\frac{1}{2},\frac{1}{2}\right)(f(u)+f(v))\right]\int_0^1 H(x,1-x)dx\\
&\supseteq [f(u)+f(v)]\left[\frac{1}{2}+H\left(\frac{1}{2},\frac{1}{2}\right)\right]\int_0^1 H(x,1-x)dx,
\end{aligned}$$

and the result follows. □

Example 2. *Furthermore, by Example 1, we have*

$$\begin{aligned}
\Delta_1 &= \frac{1}{2}\left[f\left(-\frac{1}{2}\right),f\left(\frac{1}{2}\right)\right] = \left[\frac{1}{4},4-\frac{e^{\frac{1}{2}}+e^{-\frac{1}{2}}}{2}\right],\\
\Delta_2 &= \frac{1}{2}\left(\left[1,4-\frac{e+e^{-1}}{2}\right]+[0,3]\right) = \left[\frac{1}{2},\frac{7}{2}-\frac{e+e^{-1}}{4}\right],\\
&[f(u)+f(v)]\left[\frac{1}{2}+H\left(\frac{1}{2},\frac{1}{2}\right)\right]\int_0^1 H(x,1-x)dx = \left[1,4-\frac{e+e^{-1}}{2}\right].
\end{aligned}$$

Then, we obtain that

$$[0,3] \supseteq \left[\frac{1}{4},4-\frac{e^{\frac{1}{2}}+e^{-\frac{1}{2}}}{2}\right] \supseteq \left[\frac{1}{3},4-\frac{e-e^{-1}}{2}\right] \supseteq \left[\frac{1}{2},\frac{7}{2}-\frac{e+e^{-1}}{4}\right] \supseteq \left[1,4-\frac{e+e^{-1}}{2}\right].$$

Consequently, Theorem 3 is verified.

Similarly, we get the following result, which generalizes Theorem 3 of [34] and Theorem 4.5 of [17].

Theorem 4. *Let* $f,g:[u,v]\to\mathbb{R}_{\mathcal{I}}^{+}$, $h_1,h_2:[0,1]\to\mathbb{R}^{+}$ *and* $H\left(\frac{1}{2},\frac{1}{2}\right)\neq 0$. *If* $f,g\in SX((h_1,h_2),[u,v],\mathbb{R}_{\mathcal{I}}^{+})$ *and* $fg\in\mathcal{IR}_{([u,v])}$, *then*

$$\frac{1}{v-u}\int_u^v f(t)g(t)dt \supseteq M(u,v)\int_0^1 H^2(x,1-x)dx + N(u,v)\int_0^1 H(x,x)H(1-x,1-x)dx,$$

where

$$M(u,v) = f(u)g(u) + f(v)g(v),\ N(u,v) = f(u)g(v) + f(v)g(u).$$

Example 3. *Suppose that* $h_1(x) = x$, $h_2(x) \equiv 1$, $[u,v] = [0,1]$ *and*

$$f(t) = [t^2, 4 - e^t], g(t) = [t, 3 - t^2].$$

Then

$$\begin{aligned}
&\frac{1}{v-u}\int_u^v f(t)g(t)dt = \int_0^1 [t^3, (4-e^t)(3-t^2)]dt = \left[\frac{1}{4}, \frac{35}{3} - 2e\right],\\
&M(u,v)\int_0^1 H^2(x,1-x)dx = M(0,1)\int_0^1 x^2 dx = \left[\frac{1}{3}, \frac{17}{3} - \frac{2}{3}e\right],\\
&N(u,v)\int_0^1 H(x,x)H(1-x,1-x)dx = N(0,1)\int_0^1 x^2 dx = \left[0, 3 - \frac{e}{2}\right].
\end{aligned}$$

It follows that

$$\left[\frac{1}{4}, \frac{35}{3} - 2e\right] \supseteq \left[\frac{1}{3}, \frac{17}{3} - \frac{2}{3}e\right] + \left[0, 3 - \frac{e}{2}\right] = \left[\frac{1}{3}, \frac{26}{3} - \frac{7}{6}e\right].$$

Consequently, Theorem 4 is verified.

The next result generalizes Theorem 2 of [34] and Theorem 4.6 of [17].

Theorem 5. *Let* $f, g : [u,v] \to \mathbb{R}_{\mathcal{I}}^+$, $h_1, h_2 : [0,1] \to \mathbb{R}^+$, *and* $H(\frac{1}{2}, \frac{1}{2}) \neq 0$. *If* $f, g \in SX((h_1,h_2), [u,v], \mathbb{R}_{\mathcal{I}}^+)$ *and* $fg \in \mathcal{IR}_{([u,v])}$, *then*

$$\begin{aligned}
\frac{1}{2H^2(\frac{1}{2},\frac{1}{2})} f\Big(\frac{u+v}{2}\Big) g\Big(\frac{u+v}{2}\Big) \supseteq \frac{1}{v-u}\int_u^v f(t)g(t)dt + N(u,v)\int_0^1 H^2(x,1-x)dx\\
+ M(u,v)\int_0^1 H(x,x)H(1-x,1-x)dx.
\end{aligned}$$

Proof. By hypothesis, one has

$$\begin{aligned}
&f\Big(\frac{u+v}{2}\Big) g\Big(\frac{u+v}{2}\Big)\\
&\supseteq H^2\Big(\frac{1}{2},\frac{1}{2}\Big)\Big[\underline{f}(xu+(1-x)v)\underline{g}(xu+(1-x)v), \overline{f}(xu+(1-x)v)\overline{g}(xu+(1-x)v)\Big]\\
&+ H^2\Big(\frac{1}{2},\frac{1}{2}\Big)\Big[\underline{f}(xu+(1-x)v)\underline{g}((1-x)u+xv), \overline{f}(xu+(1-x)v)\overline{g}((1-x)u+xv)\Big]\\
&+ H^2\Big(\frac{1}{2},\frac{1}{2}\Big)\Big[\underline{f}((1-x)u+xv)\underline{g}(xu+(1-x)v), \overline{f}((1-x)u+xv)\overline{g}(xu+(1-x)v)\Big]\\
&+ H^2\Big(\frac{1}{2},\frac{1}{2}\Big)\Big[\underline{f}((1-x)u+xv)\underline{g}((1-x)u+xv), \overline{f}((1-x)u+xv)\overline{g}((1-x)u+xv)\Big]\\
&\supseteq H^2\Big(\frac{1}{2},\frac{1}{2}\Big)\Big[f(xu+(1-x)v)g(xu+(1-x)v) + f((1-x)u+xv)g((1-x)u+xv)\Big]\\
&+ H^2\Big(\frac{1}{2},\frac{1}{2}\Big)\Big[\big(H(x,1-x)f(u) + H(1-x,x)f(v)\big)\big(H(1-x,x)g(u) + H(x,1-x)g(v)\big)\\
&+ \Big(H(1-x,x)f(u) + H(x,1-x)f(v)\Big)\Big(H(x,1-x)g(u) + H(1-x,x)g(v)\Big)\Big]\\
&= H^2\Big(\frac{1}{2},\frac{1}{2}\Big)\Big[f(xu+(1-x)v)g(xu+(1-x)v) + f((1-x)u+xv)g((1-x)u+xv)\Big]\\
&+ 2H^2\Big(\frac{1}{2},\frac{1}{2}\Big)\Big[H(x,x)H(1-x,1-x)M(u,v) + H^2(x,1-x)N(u,v)\Big]
\end{aligned}$$

Integrating over $[0,1]$, and the result follows. $\square$

Example 4. *Furthermore, by Example 3, we get*

$$\frac{1}{2H^2\left(\frac{1}{2},\frac{1}{2}\right)} f\left(\frac{u+v}{2}\right)g\left(\frac{u+v}{2}\right) = 2f\left(\frac{1}{2}\right)g\left(\frac{1}{2}\right) = \left[\frac{1}{4}, 22 - \frac{11}{2}\sqrt{e}\right],$$

$$N(u,v)\int_0^1 H^2(x,1-x)dx = N(0,1)\int_0^1 x^2 dx = \left[0, 6-e\right],$$

$$M(u,v)\int_0^1 H(x,x)H(1-x,1-x)dx = M(0,1)\int_0^1 (x-x^2)dx = \left[\frac{1}{6}, \frac{17}{6} - \frac{e}{3}\right].$$

It follows that

$$\left[\frac{1}{4}, 22 - \frac{11}{2}\sqrt{e}\right] \supseteq \left[0, 6-e\right] + \left[\frac{1}{6}, \frac{17}{6} - \frac{e}{3}\right] + \left[\frac{1}{4}, \frac{35}{3} - 2e\right] = \left[\frac{5}{12}, \frac{123}{6} - \frac{10}{3}e\right].$$

Consequently, Theorem 5 is verified.

4. Conclusions

We introduced interval (h_1, h_2)-convex and presented some new interval Hermite-Hadamard type inequalities. Our results generalize some known Hermite-Hadamard type inequalities and will be useful in developing the theory of interval differential (or integral) inequalities and interval convex analysis. As a future research direction, we intend to investigate inequalities for fuzzy-interval-valued functions, and some applications in interval nonlinear programming.

Author Contributions: All authors contributed equally to the writing of this paper. All authors read and approved the final manuscript.

Funding: This research is supported by the National Key Research and Development Program of China (2018YFC1508106), the Fundamental Research Funds for the Central Universities (2017B19714 and 2017B07414) and Natural Science Foundation of Jiangsu Province (BK20180500).

Conflicts of Interest: The authors declare no conflict of interest.

References

1. Moore, R.E. *Interval Analysis*; Prentice-Hall, Inc.: Englewood Cliffs, NJ, USA, 1966.
2. Gasilov, N.A.; Amrahov, Ş.E. Solving a nonhomogeneous linear system of interval differential equations. *Soft Comput.* **2018**, *22*, 3817–3828. [CrossRef]
3. Li, Y.; Wang, T.H. Interval analysis of the wing divergence. *Aerosp. Sci. Technol.* **2018**, *74*, 17–21. [CrossRef]
4. Zhu, J.J.; Qiu, Z.P. Interval analysis for uncertain aerodynamic loads with uncertain-but-bounded parameters. *J. Fluid. Struct.* **2018**, *81*, 418–436. [CrossRef]
5. Chalco-Cano, Y.; Rufián-Lizana, A.; Román-Flores, H.; Jiménez-Gamero, M.D. Calculus for interval-valued functions using generalized Hukuhara derivative and applications. *Fuzzy Sets Syst.* **2013**, *219*, 49–67. [CrossRef]
6. Chalco-Cano, Y.; Silva, G.N.; Rufián-Lizana, A. On the Newton method for solving fuzzy optimization problems. *Fuzzy Sets Syst.* **2015**, *272*, 60–69. [CrossRef]
7. Entani, T.; Inuiguchi, M. Pairwise comparison based interval analysis for group decision aiding with multiple criteria. *Fuzzy Sets Syst.* **2015**, *274*, 79–96. [CrossRef]
8. Osuna-Gómez, R.; Chalco-Cano, Y.; Hernández-Jiménez, B.; Ruiz-Garzón, G. Optimality conditions for generalized differentiable interval-valued functions. *Inf. Sci.* **2015**, *321*, 136–146. [CrossRef]
9. Moore, R.E.; Kearfott, R.B.; Cloud, M.J. *Introduction to Interval Analysis*; SIAM: Philadelphia, PA, USA, 2009.
10. Chalco-Cano, Y.; Flores-Franulič, A.; Román-Flores, H. Ostrowski type inequalities for interval-valued functions using generalized Hukuhara derivative. *Comput. Appl. Math.* **2012**, *31*, 457–472.

11. Chalco-Cano, Y.; Lodwick, W.A.; Condori-Equice, W. Ostrowski type inequalities and applications in numerical integration for interval-valued functions. *Soft Comput.* **2015**, *19*, 3293–3300. [CrossRef]
12. Costa, T.M. Jensen's inequality type integral for fuzzy-interval-valued functions. *Fuzzy Sets Syst.* **2017**, *327*, 31–47. [CrossRef]
13. Costa, T.M.; Román-Flores, H. Some integral inequalities for fuzzy-interval-valued functions. *Inf. Sci.* **2017**, *420*, 110–125. [CrossRef]
14. Costa, T.M.; Román-Flores, H.; Chalco-Cano, Y. Opial-type inequalities for interval-valued functions. *Fuzzy Sets Syst.* **2019**, *358*, 48–63. [CrossRef]
15. Flores-Franulič, A.; Chalco-Cano, Y.; Román-Flores, H. An Ostrowski type inequality for interval-valued functions. In Proceedings of the IFSA World Congress and NAFIPS Annual Meeting, Edmonton, AB, Canada, 24–28 June 2013; pp. 1459–1462.
16. Román-Flores, H.; Chalco-Cano, Y.; Lodwick, W.A. Some integral inequalities for interval-valued functions. *Comput. Appl. Math.* **2018**, *37*, 1306–1318. [CrossRef]
17. Zhao, D.F.; An, T.Q.; Ye, G.J.; Liu, W. New Jensen and Hermite-Hadamard type inequalities for h-convex interval-valued functions. *J. Inequal. Appl.* **2018**, *2018*, 302. [CrossRef]
18. Dragomir, S.S.; Pečarić, J.; Persson, L.E. Some inequalities of Hadamard type. *Soochow J. Math.* **1995**, *21*, 335–341.
19. Dragomir, S.S. Inequalities of Hermite-Hadamard type for functions of selfadjoint operators and matrices. *J. Math. Inequal.* **2017**, *11*, 241–259. [CrossRef]
20. Latif, M.A. On some new inequalities of Hermite-Hadamard type for functions whose derivatives are s-convex in the second sense in the absolute value. *Ukrainian Math. J.* **2016**, *67*, 1552–1571. [CrossRef]
21. Noor, M.A.; Cristescu, G.; Awan, M.U. Generalized Fractional Hermite-Hadamard Inequalities for twice differentiable s-convex functions. *Filomat* **2014**, *24*, 191–197. [CrossRef]
22. Noor, M.A.; Noor, K.I.; Awan, M.U.; Li, J. On Hermite-Hadamard inequalities for h-preinvex functions. *Filomat* **2014**, *24*, 1463–1474. [CrossRef]
23. Noor, M.A.; Noor, K.I.; Iftikhar, S.; Ionescu, C. Hermite-Hadamard inequalities for co-ordinated harmonic convex functions. *Politehn. Univ. Bucharest Sci. Bull. Ser. A Appl. Math. Phys.* **2017**, *79*, 25–34.
24. Noor, M.A.; Noor, K.I.; Mihai, M.V.; Awan, M.U. Fractional Hermite-Hadamard inequalities for some classes of differentiable preinvex functions. *Politehn. Univ. Bucharest Sci. Bull. Ser. A Appl. Math. Phys.* **2016**, *78*, 163–174.
25. Xi, B.-Y.; He, C.-Y.; Qi, F. Some new inequalities of the Hermite-Hadamard type for extended $((s_1, m_1) - (s_2, m_2))$-convex functions on co-ordinates. *Cogent Math.* **2016**, *3*, 1267300. [CrossRef]
26. Xi, B.-Y.; Qi, F. Inequalities of Hermite-Hadamard type for extended s-convex functions and applications to means. *J. Nonlinear Convex Anal.* **2015**, *6*, 873–890.
27. Varošanec, S. On h-convexity. *J. Math. Anal. Appl.* **2007**, *326*, 303–311. [CrossRef]
28. Bombardelli, M.; Varošanec, S. Properties of h-convex functions related to the hermite-hadamard-fejér inequalities. *Comput. Math. Appl.* **2009**, *58*, 1869–1877. [CrossRef]
29. Dragomir, S.S. Inequalities of hermite-hadamard type for h-convex functions on linear spaces. *Proyecciones J. Math.* **2015**, *32*, 323–341. [CrossRef]
30. Latif, M.A.; Alomari, M. On Hadmard-Type Inequalities for h-Convex Functions on the Co-ordinates. *Int. J. Math. Anal.* **2009**, *3*, 1645–1656.
31. Matłoka, M. On Hadamard's inequality for h-convex function on a disk. *Appl. Math. Comput.* **2014**, *235*, 118–123. [CrossRef]
32. Sarikaya, M.Z.; Saglam, A.; Yildirim, H. On some Hadamard-type inequalities for h-convex functions. *J. Math. Inequal.* **2008**, *2*, 335–341. [CrossRef]
33. Sarikaya, M.Z.; Set, E.; Özdemir, M.E. On some new inequalities of Hadamard-type involving h-convex functions. *Acta Math. Univ. Comenian LXXIX* **2010**, *2*, 265–272.
34. Awan, M.U.; Noor, M.A.; Noor, K.I.; Khan, A.G. Some new classes of convex functions and inequalities. *Miskolc Math. Notes* **2018**, *19*, 77–94. [CrossRef]
35. Noor, M.A.; Noor, K.I.; Awan, M.U. A new Hermite-Hadamard type inequality for h-convex functions. *Creat. Math. Inf.* **2015**, *24*, 191–197.

36. Breckner, W.W. Stetigkeitsaussagen für eine Klasse verallgemeinerter konvexer funktionen in topologischen linearen Räumen. *Pupl. Inst. Math.* **1978**, *23*, 13–20.
37. Osuna-Gómez, R.; Jimenez-Gaméro, M.D.; Chalco-Cano, Y.; Rojas-Medar, M.A. Hadamard and Jensen inequalities for s-convex fuzzy processes. In *Soft Methodology and Random Information Systems*; Springer: Berlin/Heidelberg, Germany, 2004; pp. 645–652.

mathematics

MDPI

Article

New Integral Inequalities via the Katugampola Fractional Integrals for Functions Whose Second Derivatives Are Strongly η-Convex

Seth Kermausuor [1], Eze R. Nwaeze [2,*] and Ana M. Tameru [2]

1 Department of Mathematics and Computer Science, Alabama State University, Montgomery, AL 36101, USA; skermausuor@alasu.edu

2 Department of Mathematics, Tuskegee University, Tuskegee, AL 36088, USA; atameru@tuskegee.edu

* Correspondence: enwaeze@tuskegee.edu

Received: 28 December 2018; Accepted: 12 February 2019; Published: 15 February 2019

Abstract: In this paper, we introduced some new integral inequalities of the Hermite–Hadamard type for functions whose second derivatives in absolute values at certain powers are strongly η-convex functions via the Katugampola fractional integrals.

Keywords: Hermite–Hadamard type inequality; strongly η-convex functions; Hölder's inequality; Power mean inequality; Katugampola fractional integrals; Riemann–Liouville fractional integrals; Hadamard fractional integrals

2010 MSC: 26A33; 26A51; 26D10; 26D15

---

1. Introduction

Let I be an interval in $\mathbb{R}$. A function $f : I \to \mathbb{R}$ is said to be convex on I if

$$f(tx + (1-t)y) \leq tf(x) + (1-t)f(y)$$

for all $x, y \in I$ and $t \in [0,1]$. The following inequalities which hold for convex functions is known in the literature as the Hermite–Hadamard type inequality.

Theorem 1 ([1]). *If $f : [a,b] \to \mathbb{R}$ is convex on $[a,b]$ with $a < b$, then*

$$f\Big(\frac{a+b}{2}\Big) \leq \frac{1}{b-a}\int_a^b f(x)dx \leq \frac{f(a)+f(b)}{2}.$$

Many authors have studied and generalized the Hermite–Hadamard inequality in several ways via different classes of convex functions. For some recent results related to the Hermite–Hadamard inequality, we refer the interested reader to the papers [2–11].

In 2016, Gordji et al. [12] introduced the concept of η-convexity as follows:

Definition 1 ([12]). *A function $f : I \to \mathbb{R}$ is said to be η-convex with respect to the bifunction $\eta : \mathbb{R} \times \mathbb{R} \to \mathbb{R}$ if*

$$f(tx + (1-t)y) \leq f(y) + t\eta(f(x), f(y))$$

for all $x, y \in I$ and $t \in [0,1]$.

Remark 1. *If we take $\eta(x,y) = x - y$ in Definition 1, then we recover the classical definition of convex functions.*

In 2017, Awan et al. [13] extended the class of η-convex functions to the class of strongly η-convex functions as follows:

Definition 2 ([13])**.** *A function $f : I \to \mathbb{R}$ is said to be strongly η-convex with respect to the bifunction $\eta : \mathbb{R} \times \mathbb{R} \to \mathbb{R}$ with modulus $\mu \geq 0$ if*

$$f(tx + (1-t)y) \leq f(y) + t\eta(f(x), f(y)) - \mu t(1-t)(x-y)^2$$

for all $x, y \in I$ and $t \in [0,1]$.

Remark 2. *If $\eta(x,y) = x - y$ in Definition 2, then we have the class of strongly convex functions.*

For some recent results related to the class of η-convex functions, we refer the interested reader to the papers [8,12–16].

Definition 3 ([17])**.** *The left- and right-sided Riemann–Liouville fractional integrals of order $\alpha > 0$ of f are defined by*

$$J_{a+}^{\alpha} f(x) := \frac{1}{\Gamma(\alpha)} \int_a^x (x-t)^{\alpha-1} f(t) dt$$

and

$$J_{b-}^{\alpha} f(x) := \frac{1}{\Gamma(\alpha)} \int_x^b (t-x)^{\alpha-1} f(t) dt$$

with $a < x < b$ and $\Gamma(\cdot)$ is the gamma function given by

$$\Gamma(x) := \int_0^{\infty} t^{x-1} e^{-t} dt, \quad Re(x) > 0$$

with the property that $\Gamma(x+1) = x\Gamma(x)$.

Definition 4 ([18])**.** *The left- and right-sided Hadamard fractional integrals of order $\alpha > 0$ of f are defined by*

$$H_{a+}^{\alpha} f(x) := \frac{1}{\Gamma(\alpha)} \int_a^x \left(\ln \frac{x}{t}\right)^{\alpha-1} \frac{f(t)}{t} dt$$

and

$$H_{b-}^{\alpha} f(x) := \frac{1}{\Gamma(\alpha)} \int_x^b \left(\ln \frac{t}{x}\right)^{\alpha-1} \frac{f(t)}{t} dt.$$

Definition 5. *$X_c^p(a,b)$ $(c \in \mathbb{R}, 1 \leq p \leq \infty)$ denotes the space of all complex-valued Lebesgue measurable functions f for which $\|f\|_{X_c^p} < \infty$, where the norm $\|\cdot\|_{X_c^p}$ is defined by*

$$\|f\|_{X_c^p} = \left(\int_a^b |t^c f(t)|^p \frac{dt}{t}\right)^{1/p} \quad (1 \leq p < \infty)$$

and for $p = \infty$

$$\|f\|_{X_c^\infty} = ess \sup_{a \leq t \leq b} |t^c f(t)|.$$

In 2011, Katugampola [19] introduced a new fractional integral operator which generalizes the Riemann–Liouville and Hadamard fractional integrals as follows:

Definition 6. *Let* $[a, b] \subset \mathbb{R}$ *be a finite interval. Then, the left- and right-sided Katugampola fractional integrals of order* $\alpha > 0$ *of* $f \in X_c^p(a, b)$ *are defined by*

$$^{\rho}I_{a+}^{\alpha} f(x) := \frac{\rho^{1-\alpha}}{\Gamma(\alpha)} \int_a^x \frac{t^{\rho-1}}{(x^\rho - t^\rho)^{1-\alpha}} f(t) dt$$

and

$$^{\rho}I_{b-}^{\alpha} f(x) := \frac{\rho^{1-\alpha}}{\Gamma(\alpha)} \int_x^b \frac{t^{\rho-1}}{(t^\rho - x^\rho)^{1-\alpha}} f(t) dt$$

with $a < x < b$ *and* $\rho > 0$*, if the integrals exist.*

Remark 3. *It is shown in [19] that the Katugampola fractional integral operators are well-defined on* $X_c^p(a, b)$.

Theorem 2 ([19])**.** *Let* $\alpha > 0$ *and* $\rho > 0$*. Then for* $x > a$

1. $\lim_{\rho \to 1} {}^{\rho}I_{a+}^{\alpha} f(x) = J_{a+}^{\alpha} f(x)$,
2. $\lim_{\rho \to 0^+} {}^{\rho}I_{a+}^{\alpha} f(x) = H_{a+}^{\alpha} f(x)$.

Similar results also hold for right-sided operators.

For more information about the Katugampola fractional integrals and related results, we refer the interested reader to the papers [19–21]. Recently, Chen and Katugampola [20] introduced several integral inequalities of Hermite–Hadamard type for functions whose first derivatives in absolute value are convex functions via the Katugampola fractional integrals. We present two of their results here for the purpose of our discussion. The first result of importance to us employs the following lemma.

Lemma 1 ([20])**.** *Let* $\alpha > 0$*,* $\rho > 0$ *and* $f : [a^\rho, b^\rho] \to \mathbb{R}$ *be a differentiable mapping on* (a^ρ, b^ρ) *with* $0 \leq a < b$*. Then the following equality holds if the fractional integrals exist:*

$$\begin{aligned} &\frac{f(a^\rho) + f(b^\rho)}{\alpha\rho} - \frac{\rho^{\alpha-1}\Gamma(\alpha)}{(b^\rho - a^\rho)^\alpha} \left[{}^{\rho}I_{a+}^{\alpha} f(b^\rho) + {}^{\rho}I_{b-}^{\alpha} f(a^\rho) \right] \\ &= \frac{b^\rho - a^\rho}{\alpha} \int_0^1 t^{\rho(\alpha+1)-1} \left[f'((1-t^\rho)a^\rho + t^\rho b^\rho) - f'(t^\rho a^\rho + (1-t^\rho)b^\rho) \right] dt. \end{aligned} \tag{1}$$

By using Lemma 1, the authors proved the following result.

Theorem 3 ([20])**.** *Let* $f : [a^\rho, b^\rho] \to \mathbb{R}$ *be a differentiable mapping on* (a^ρ, b^ρ) *with* $0 \leq a < b$*. If* $|f'|$ *is convex on* $[a^\rho, b^\rho]$*, then the following inequality holds:*

$$\left| \frac{f(a^\rho) + f(b^\rho)}{2} - \frac{\rho^\alpha \Gamma(\alpha+1)}{2(b^\rho - a^\rho)^\alpha} \left[{}^{\rho}I_{a+}^{\alpha} f(b^\rho) + {}^{\rho}I_{b-}^{\alpha} f(a^\rho) \right] \right| \leq \frac{b^\rho - a^\rho}{2(\alpha+1)} \left[|f'(a^\rho)| + |f'(b^\rho)| \right].$$

The second result of importance to us also uses the following lemma.

Lemma 2 ([20]). *Let $\alpha > 0$, $\rho > 0$ and $f : [a^\rho, b^\rho] \to \mathbb{R}$ be a differentiable mapping on (a^ρ, b^ρ) with $0 \le a < b$. Then the following equality holds if the fractional integrals exist:*

$$\begin{aligned}&\frac{f(a^\rho)+f(b^\rho)}{2} - \frac{\rho^\alpha \Gamma(\alpha+1)}{2(b^\rho - a^\rho)^\alpha}\left[{}^\rho I^\alpha_{a+} f(b^\rho) + {}^\rho I^\alpha_{b-} f(a^\rho)\right] \\ &= \frac{b^\rho - a^\rho}{2}\int_0^1 [(1-t^\rho)^\alpha - t^{\rho\alpha}] t^{\rho-1} f'(t^\rho a^\rho + (1-t^\rho) b^\rho) dt. \end{aligned} \tag{2}$$

By using Lemma 2, the authors proved the following result.

Theorem 4 ([20]). *Let $f : [a^\rho, b^\rho] \to \mathbb{R}$ be a differentiable mapping on (a^ρ, b^ρ) with $0 \le a < b$. If $|f'|$ is convex on $[a^\rho, b^\rho]$, then the following inequality holds:*

$$\begin{aligned}&\left|\frac{f(a^\rho)+f(b^\rho)}{2} - \frac{\rho^\alpha \Gamma(\alpha+1)}{2(b^\rho - a^\rho)^\alpha}\left[{}^\rho I^\alpha_{a+} f(b^\rho) + {}^\rho I^\alpha_{b-} f(a^\rho)\right]\right| \\ &\qquad \le \frac{b^\rho - a^\rho}{2\rho(\alpha+1)}\left(1 - \frac{1}{2^\alpha}\right)\left[|f'(a^\rho)| + |f'(b^\rho)|\right]. \end{aligned}$$

Remark 4. *It is important to note that Lemmas 1 and 2 are corrected versions of [20] (Lemma 2.4 and Equation (14)).*

Our purpose in this paper is to provide some new estimates for the right hand side of the inequalities in Theorems 3 and 4 for functions whose second derivatives in absolute value at some powers are strongly η-convex.

2. Main Results

To prove the main results of this paper, we need the following lemmas which are extensions of Lemmas 1 and 2 for the second derivative case of the function f.

Lemma 3. *Let $\alpha > 0$, $\rho > 0$ and $f : [a^\rho, b^\rho] \to \mathbb{R}$ be a twice differentiable mapping on (a^ρ, b^ρ) with $0 \le a < b$. Then the following equality holds if the fractional integrals exist:*

$$\begin{aligned}&\frac{f(a^\rho)+f(b^\rho)}{\alpha\rho} - \frac{\rho^{\alpha-1} \Gamma(\alpha)}{(b^\rho - a^\rho)^\alpha}\left[{}^\rho I^\alpha_{a+} f(b^\rho) + {}^\rho I^\alpha_{b-} f(a^\rho)\right] \\ &= \frac{(b^\rho - a^\rho)^2}{\alpha(\alpha+1)}\left[\int_0^1 \left[1 - t^{\rho(\alpha+1)}\right] t^{\rho-1} f''((1-t^\rho)a^\rho + t^\rho b^\rho) dt \right. \\ &\qquad \left. - \int_0^1 t^{\rho(\alpha+2)-1} f''(t^\rho a^\rho + (1-t^\rho) b^\rho) dt\right]. \end{aligned} \tag{3}$$

Proof. Let

$$I_1 = \int_0^1 \left[1 - t^{\rho(\alpha+1)}\right] t^{\rho-1} f''((1-t^\rho)a^\rho + t^\rho b^\rho) dt$$

and

$$I_2 = \int_0^1 t^{\rho(\alpha+2)-1} f''(t^\rho a^\rho + (1-t^\rho) b^\rho) dt.$$

By using integration by parts we have that

$$
\begin{aligned}
I_1 &= \int_0^1 \left[1 - t^{\rho(\alpha+1)}\right] t^{\rho-1} f''((1-t^\rho)a^\rho + t^\rho b^\rho)dt \\
&= \frac{1}{\rho(b^\rho - a^\rho)}[1 - t^{\rho(\alpha+1)}]f'((1-t^\rho)a^\rho + t^\rho b^\rho)\Big|_0^1 \\
&+ \frac{\rho(\alpha+1)}{\rho(b^\rho - a^\rho)} \int_0^1 t^{\rho(\alpha+1)-1} f'((1-t^\rho)a^\rho + t^\rho b^\rho)dt \\
&= -\frac{1}{\rho(b^\rho - a^\rho)} f'(a^\rho) + \frac{(\alpha+1)}{(b^\rho - a^\rho)} \int_0^1 t^{\rho(\alpha+1)-1} f'((1-t^\rho)a^\rho + t^\rho b^\rho)dt. \qquad (4)
\end{aligned}
$$

By a similar argument, one gets:

$$
I_2 = -\frac{1}{\rho(b^\rho - a^\rho)} f'(a^\rho) + \frac{(\alpha+1)}{(b^\rho - a^\rho)} \int_0^1 t^{\rho(\alpha+1)-1} f'(t^\rho a^\rho + (1-t^\rho)b^\rho)dt. \qquad (5)
$$

Using (4) and (5), we have

$$
I_1 - I_2 = \frac{(\alpha+1)}{(b^\rho - a^\rho)} \int_0^1 t^{\rho(\alpha+1)-1} \left[f'((1-t^\rho)a^\rho + t^\rho b^\rho) - f'(t^\rho a^\rho + (1-t^\rho)b^\rho)\right] dt. \qquad (6)
$$

The desired identity in (3) follows from (6) by using (1) and rearranging the terms. □

Lemma 4. *Let $\alpha > 0, \rho > 0$ and $f : [a^\rho, b^\rho] \to \mathbb{R}$ be a twice differentiable mapping on (a^ρ, b^ρ) with $0 \le a < b$. Then the following equality holds if the fractional integrals exist:*

$$
\begin{aligned}
&\frac{f(a^\rho) + f(b^\rho)}{2} - \frac{\rho^\alpha \Gamma(\alpha+1)}{2(b^\rho - a^\rho)^\alpha}\left[{}^\rho I^\alpha_{a+} f(b^\rho) + {}^\rho I^\alpha_{b-} f(a^\rho)\right] \\
&= \frac{(b^\rho - a^\rho)^2}{2(\alpha+1)} \int_0^1 [1 - (1-t^\rho)^{\alpha+1} - t^{\rho(\alpha+1)}] t^{\rho-1} f''(t^\rho a^\rho + (1-t^\rho)b^\rho)dt. \qquad (7)
\end{aligned}
$$

Proof. We start by considering the following computation which is a direct application of integration by parts.

$$
\begin{aligned}
&\int_0^1 [1 - (1-t^\rho)^{\alpha+1} - t^{\rho(\alpha+1)}] t^{\rho-1} f''(t^\rho a^\rho + (1-t^\rho)b^\rho)dt \\
&= \frac{1}{\rho(a^\rho - b^\rho)}[1 - (1-t^\rho)^{\alpha+1} - t^{\rho(\alpha+1)}] f'(t^\rho a^\rho + (1-t^\rho)b^\rho)\Big|_0^1 \\
&- \frac{\rho(\alpha+1)}{\rho(a^\rho - b^\rho)} \int_0^1 [(1-t^\rho)^\alpha - t^{\rho\alpha}] t^{\rho-1} f'(t^\rho a^\rho + (1-t^\rho)b^\rho)dt \\
&= \frac{(\alpha+1)}{(b^\rho - a^\rho)} \int_0^1 [(1-t^\rho)^\alpha - t^{\rho\alpha}] t^{\rho-1} f'(t^\rho a^\rho + (1-t^\rho)b^\rho)dt. \qquad (8)
\end{aligned}
$$

The intended identity in (7) follows from (8) by using (2) and rearranging the terms. □

We are now in a position to prove our main results.

Theorem 5. *Let $\alpha > 0$, $\rho > 0$ and $f : [a^\rho, b^\rho] \to \mathbb{R}$ be a twice differentiable mapping on (a^ρ, b^ρ) with $0 \le a < b$. If $|f''|^q$ is strongly η-convex with modulus $\mu \ge 0$ for $q \ge 1$, then the following inequality holds:*

$$\begin{aligned}
&\left|\frac{f(a^\rho)+f(b^\rho)}{2} - \frac{\rho^\alpha \Gamma(\alpha+1)}{2(b^\rho-a^\rho)^\alpha}\left[{}^\rho I^\alpha_{a+} f(b^\rho) + {}^\rho I^\alpha_{b-} f(a^\rho)\right]\right| \\
&\le \frac{(b^\rho-a^\rho)^2}{2(\alpha+1)}\left[\left(\frac{\alpha+1}{\alpha+2}\right)^{1-\frac{1}{q}}\left(\frac{\alpha+1}{\alpha+2}|f''(a^\rho)|^q\right.\right. \\
&\left.+\frac{\alpha+1}{2(\alpha+3)}\eta\left(|f''(b^\rho)|^q, |f''(a^\rho)|^q\right) - \frac{\mu(b^\rho-a^\rho)^2\left[(\alpha+1)^2+5(\alpha+1)\right]}{6(\alpha+4)(\alpha+3)}\right)^{\frac{1}{q}} \\
&+\left(\frac{1}{\alpha+2}\right)^{1-\frac{1}{q}}\left(\frac{1}{\alpha+2}|f''(b^\rho)|^q + \frac{1}{\alpha+3}\eta\left(|f''(a^\rho)|^q, |f''(b^\rho)|^q\right)\right. \\
&\left.\left.-\frac{\mu(b^\rho-a^\rho)^2}{(\alpha+3)(\alpha+4)}\right)^{\frac{1}{q}}\right].
\end{aligned}$$

Proof. Using Lemma 3, the well-known power mean inequality and the strong η-convexity of $|f''|^q$, we obtain

$$\begin{aligned}
&\left|\frac{f(a^\rho)+f(b^\rho)}{\alpha\rho} - \frac{\rho^{\alpha-1}\Gamma(\alpha)}{(b^\rho-a^\rho)^\alpha}\left[{}^\rho I^\alpha_{a+} f(b^\rho) + {}^\rho I^\alpha_{b-} f(a^\rho)\right]\right| \\
&\le \frac{(b^\rho-a^\rho)^2}{\alpha(\alpha+1)}\left[\int_0^1 \left[1-t^{\rho(\alpha+1)}\right] t^{\rho-1}\left|f''((1-t^\rho)a^\rho + t^\rho b^\rho)\right| dt\right. \\
&\qquad \left.+\int_0^1 t^{\rho(\alpha+2)-1}\left|f''(t^\rho a^\rho + (1-t^\rho)b^\rho)\right| dt\right] \\
&\le \frac{(b^\rho-a^\rho)^2}{\alpha(\alpha+1)}\left[\left(\int_0^1 \left[1-t^{\rho(\alpha+1)}\right] t^{\rho-1} dt\right)^{1-\frac{1}{q}}\right. \\
&\qquad \times\left(\int_0^1 \left[1-t^{\rho(\alpha+1)}\right] t^{\rho-1}\left|f''((1-t^\rho)a^\rho + t^\rho b^\rho)\right|^q dt\right)^{\frac{1}{q}} \\
&\qquad \left.+\left(\int_0^1 t^{\rho(\alpha+2)-1} dt\right)^{1-\frac{1}{q}}\left(\int_0^1 t^{\rho(\alpha+2)-1}\left|f''(t^\rho a^\rho + (1-t^\rho)b^\rho)\right| dt\right)^{\frac{1}{q}}\right] \\
&\le \frac{(b^\rho-a^\rho)^2}{\alpha(\alpha+1)}\left[\left(\int_0^1 \left[1-t^{\rho(\alpha+1)}\right] t^{\rho-1} dt\right)^{1-\frac{1}{q}}\right. \\
&\qquad \times\left(\int_0^1 \left[1-t^{\rho(\alpha+1)}\right] t^{\rho-1}\left(|f''(a^\rho)|^q + t^\rho \eta\left(|f''(b^\rho)|^q, |f''(a^\rho)|^q\right)\right.\right. \\
&\qquad\qquad \left.\left.- \mu t^\rho (1-t^\rho)(b^\rho-a^\rho)^2\right) dt\right)^{\frac{1}{q}} \\
&\qquad +\left(\int_0^1 t^{\rho(\alpha+2)-1} dt\right)^{1-\frac{1}{q}}\left(\int_0^1 t^{\rho(\alpha+2)-1}\left(|f''(b^\rho)|^q\right.\right. \\
&\qquad \left.\left.\left.+ t^\rho \eta\left(|f''(a^\rho)|^q, |f''(b^\rho)|^q\right) - \mu t^\rho(1-t^\rho)(b^\rho-a^\rho)^2\right) dt\right)^{\frac{1}{q}}\right] \\
&= \frac{(b^\rho-a^\rho)^2}{\alpha(\alpha+1)}\left[\left(\int_0^1 \left[1-t^{\rho(\alpha+1)}\right] t^{\rho-1} dt\right)^{1-\frac{1}{q}}\left(|f''(a^\rho)|^q \int_0^1 \left[1-t^{\rho(\alpha+1)}\right] t^{\rho-1} dt\right.\right. \\
&\qquad +\eta\left(|f''(b^\rho)|^q, |f''(a^\rho)|^q\right)\int_0^1 \left[1-t^{\rho(\alpha+1)}\right] t^{2\rho-1} dt \\
&\qquad \left.-\mu(b^\rho-a^\rho)^2 \int_0^1 \left[1-t^{\rho(\alpha+1)}\right] t^{2\rho-1}(1-t^\rho) dt\right)^{\frac{1}{q}}
\end{aligned}$$

$$+\left(\int_0^1 t^{\rho(\alpha+2)-1}dt\right)^{1-\frac{1}{q}}\left(|f''(b^\rho)|^q\int_0^1 t^{\rho(\alpha+2)-1}dt\right.$$
$$+\eta\left(|f''(a^\rho)|^q,|f''(b^\rho)|^q\right)\int_0^1 t^{\rho(\alpha+3)-1}dt$$
$$\left.\left.-\mu(b^\rho-a^\rho)^2\int_0^1 t^{\rho(\alpha+3)-1}(1-t^\rho)dt\right)^{\frac{1}{q}}\right].$$

The desired inequality follows from the above estimation and observing that:

$$\int_0^1\left[1-t^{\rho(\alpha+1)}\right]t^{\rho-1}dt=\frac{\alpha+1}{\rho(\alpha+2)},\ \int_0^1\left[1-t^{\rho(\alpha+1)}\right]t^{2\rho-1}dt=\frac{\alpha+1}{2\rho(\alpha+3)},$$

$$\int_0^1\left[1-t^{\rho(\alpha+1)}\right]t^{2\rho-1}(1-t^\rho)dt=\frac{(\alpha+1)^2+5(\alpha+1)}{6\rho(\alpha+4)(\alpha+3)},\ \int_0^1 t^{\rho(\alpha+2)-1}dt=\frac{1}{\rho(\alpha+2)}$$

$$\int_0^1 t^{\rho(\alpha+3)-1}dt=\frac{1}{\rho(\alpha+3)}\ \text{and}\ \int_0^1 t^{\rho(\alpha+3)-1}(1-t^\rho)dt=\frac{1}{\rho(\alpha+3)(\alpha+4)}.$$

This completes the proof of Theorem 5. □

Corollary 1. *Let $\alpha>0$, $\rho>0$ and $f:[a^\rho,b^\rho]\to\mathbb{R}$ be a twice differentiable mapping on (a^ρ,b^ρ) with $0\le a<b$. If $|f''|^q$ is convex for $q\ge1$, then the following inequality holds:*

$$\left|\frac{f(a^\rho)+f(b^\rho)}{2}-\frac{\rho^\alpha\Gamma(\alpha+1)}{2(b^\rho-a^\rho)^\alpha}\left[{}^\rho I^\alpha_{a+}f(b^\rho)+{}^\rho I^\alpha_{b-}f(a^\rho)\right]\right|$$
$$\le\frac{(b^\rho-a^\rho)^2}{2(\alpha+1)}\left[\left(\frac{\alpha+1}{\alpha+2}\right)^{1-\frac{1}{q}}\left(\frac{(\alpha+1)(\alpha+4)}{2(\alpha+2)(\alpha+3)}|f''(a^\rho)|^q+\frac{\alpha+1}{2(\alpha+3)}|f''(b^\rho)|^q\right.\right.$$
$$\left.\left.+\left(\frac{1}{\alpha+2}\right)^{1-\frac{1}{q}}\left(\frac{1}{(\alpha+2)(\alpha+3)}|f''(b^\rho)|^q+\frac{1}{\alpha+3}|f''(a^\rho)|^q\right)^{\frac{1}{q}}\right]\right..$$

Proof. The result follows directly from Theorem 5 if we take $\eta(x,y)=x-y$ and $\mu=0$. □

Theorem 6. *Let $\alpha>0$, $\rho>0$ and $f:[a^\rho,b^\rho]\to\mathbb{R}$ be a twice differentiable mapping on (a^ρ,b^ρ) with $0\le a<b$. If $|f''|^q$ is strongly η-convex with modulus $\mu\ge0$ for $q>1$, then the following inequalities hold:*

$$\left|\frac{f(a^\rho)+f(b^\rho)}{2}-\frac{\rho^\alpha\Gamma(\alpha+1)}{2(b^\rho-a^\rho)^\alpha}\left[{}^\rho I^\alpha_{a+}f(b^\rho)+{}^\rho I^\alpha_{b-}f(a^\rho)\right]\right|$$
$$\le\frac{\rho(b^\rho-a^\rho)^2}{2(\alpha+1)}\left[\left(\frac{1}{\rho}\int_0^1\left[1-u^{\alpha+1}\right]^s du\right)^{\frac{1}{s}}\left(\frac{1}{\rho}|f''(a^\rho)|^q+\frac{1}{2\rho}\eta\left(|f''(b^\rho)|^q,|f''(a^\rho)|^q\right)\right.\right.$$
$$\left.-\frac{\mu(b^\rho-a^\rho)^2}{6\rho}\right)^{\frac{1}{q}}+\left(\frac{1}{s\rho(\alpha+2)-s+1}\right)^{\frac{1}{s}}\left(|f''(b^\rho)|^q+\frac{1}{\rho+1}\eta\left(|f''(a^\rho)|^q,|f''(b^\rho)|^q\right)\right.$$
$$\left.\left.-\frac{\mu\rho(b^\rho-a^\rho)^2}{(\rho+1)(2\rho+1)}\right)^{\frac{1}{q}}\right]$$

$$\leq \frac{\rho(b^\rho - a^\rho)^2}{2(\alpha+1)}\Bigg[\left(\frac{s(\alpha+1)}{\rho(s(\alpha+1)+1)}\right)^{\frac{1}{s}}\left(\frac{1}{\rho}|f''(a^\rho)|^q + \frac{1}{2\rho}\eta\left(|f''(b^\rho)|^q, |f''(a^\rho)|^q\right)\right.$$
$$\left. - \frac{\mu(b^\rho - a^\rho)^2}{6\rho}\right)^{\frac{1}{q}} + \left(\frac{1}{s\rho(\alpha+2)-s+1}\right)^{\frac{1}{s}}\left(|f''(b^\rho)|^q + \frac{1}{\rho+1}\eta\left(|f''(a^\rho)|^q, |f''(b^\rho)|^q\right)\right.$$
$$\left. - \frac{\mu\rho(b^\rho - a^\rho)^2}{(\rho+1)(2\rho+1)}\right)^{\frac{1}{q}}\Bigg],$$

where $\frac{1}{s} + \frac{1}{q} = 1$.

Proof. Using Lemma 3, the Hölder's inequality and the strong η-convexity of $|f''|^q$, we obtain

$$\left|\frac{f(a^\rho)+f(b^\rho)}{\alpha\rho} - \frac{\rho^{\alpha-1}\Gamma(\alpha)}{(b^\rho-a^\rho)^\alpha}\left[{}^\rho I^\alpha_{a+} f(b^\rho) + {}^\rho I^\alpha_{b-} f(a^\rho)\right]\right|$$
$$\leq \frac{(b^\rho-a^\rho)^2}{\alpha(\alpha+1)}\Bigg[\left(\int_0^1 \left|1 - t^{\rho(\alpha+1)}\right|^s t^{\rho-1} dt\right)^{\frac{1}{s}}\left(\int_0^1 t^{\rho-1}\left|f''((1-t^\rho)a^\rho + t^\rho b^\rho)\right|^q dt\right)^{\frac{1}{q}}$$
$$+ \left(\int_0^1 t^{s\rho(\alpha+2)-s} dt\right)^{\frac{1}{s}}\left(\int_0^1 \left|f''(t^\rho a^\rho + (1-t^\rho)b^\rho)\right|^q dt\right)^{\frac{1}{q}}\Bigg]$$
$$\leq \frac{(b^\rho-a^\rho)^2}{\alpha(\alpha+1)}\Bigg[\left(\int_0^1 \left|1 - t^{\rho(\alpha+1)}\right|^s t^{\rho-1} dt\right)^{\frac{1}{s}}\left(\int_0^1 t^{\rho-1}\Big(|f''(a^\rho)|^q\right.$$
$$\left. + t^\rho\eta\left(|f''(b^\rho)|^q, |f''(a^\rho)|^q\right) - \mu t^\rho(1-t^\rho)(b^\rho-a^\rho)^2\Big)dt\right)^{\frac{1}{q}}$$
$$+ \left(\int_0^1 t^{s\rho(\alpha+2)-s} dt\right)^{\frac{1}{s}}\left(\int_0^1 \Big(|f''(b^\rho)|^q\right.$$
$$\left. + t^\rho\eta\left(|f''(a^\rho)|^q, |f''(b^\rho)|^q\right) - \mu t^\rho(1-t^\rho)(b^\rho-a^\rho)^2\Big)dt\right)^{\frac{1}{q}}\Bigg]$$
$$= \frac{(b^\rho-a^\rho)^2}{\alpha(\alpha+1)}\Bigg[\left(\int_0^1 \left|1 - t^{\rho(\alpha+1)}\right|^s t^{\rho-1} dt\right)^{\frac{1}{s}}\left(|f''(a^\rho)|^q\int_0^1 t^{\rho-1} dt\right.$$
$$\left. + \eta\left(|f''(b^\rho)|^q, |f''(a^\rho)|^q\right)\int_0^1 t^{2\rho-1} dt - \mu(b^\rho-a^\rho)^2\int_0^1 t^{2\rho-1}(1-t^\rho) dt\right)^{\frac{1}{q}}$$
$$+ \left(\int_0^1 t^{s\rho(\alpha+2)-s} dt\right)^{\frac{1}{s}}\left(|f''(b^\rho)|^q\int_0^1 1\,dt + \eta\left(|f''(a^\rho)|^q, |f''(b^\rho)|^q\right)\int_0^1 t^\rho dt\right.$$
$$\left. - \mu(b^\rho-a^\rho)^2\int_0^1 t^\rho(1-t^\rho) dt\right)^{\frac{1}{q}}\Bigg]$$
$$= \frac{(b^\rho-a^\rho)^2}{\alpha(\alpha+1)}\Bigg[\left(\frac{1}{\rho}\int_0^1 \left[1-u^{\alpha+1}\right]^s du\right)^{\frac{1}{s}}\left(\frac{1}{\rho}|f''(a^\rho)|^q + \frac{1}{2\rho}\eta\left(|f''(b^\rho)|^q, |f''(a^\rho)|^q\right)\right.$$
$$\left. - \frac{\mu(b^\rho - a^\rho)^2}{6\rho}\right)^{\frac{1}{q}} + \left(\frac{1}{s\rho(\alpha+2)-s+1}\right)^{\frac{1}{s}}\left(|f''(b^\rho)|^q + \frac{1}{\rho+1}\eta\left(|f''(a^\rho)|^q, |f''(b^\rho)|^q\right)\right.$$
$$\left. - \frac{\mu\rho(b^\rho - a^\rho)^2}{(\rho+1)(2\rho+1)}\right)^{\frac{1}{q}}\Bigg].$$

This proves the first inequality. To prove the second inequality, we observe that for any $A > B \geq 0$ and $s \geq 1$, we have $(A-B)^s \leq A^s - B^s$. Thus, it follows that $\left[1-u^{\alpha+1}\right]^s \leq 1-u^{s(\alpha+1)}$ for all $u \in [0,1]$. Hence, we have that

$$\int_0^1 \left[1-u^{\alpha+1}\right]^s du \leq \int_0^1 1-u^{s(\alpha+1)} du = \frac{s(\alpha+1)}{s(\alpha+1)+1}.$$

This completes the proof. □

Corollary 2. *Let $\alpha > 0$, $\rho > 0$ and $f : [a^\rho, b^\rho] \to \mathbb{R}$ be a twice differentiable mapping on (a^ρ, b^ρ) with $0 < a < b$. If $|f''|^q$ is convex for $q > 1$, then the following inequalities hold:*

$$\begin{aligned}
&\left|\frac{f(a^\rho)+f(b^\rho)}{2} - \frac{\rho^\alpha\Gamma(\alpha+1)}{2(b^\rho-a^\rho)^\alpha}\left[{}^\rho I^\alpha_{a+} f(b^\rho) + {}^\rho I^\alpha_{b-} f(a^\rho)\right]\right| \\
&\leq \frac{\rho(b^\rho-a^\rho)^2}{2(\alpha+1)}\Bigg[\left(\frac{1}{\rho}\int_0^1 \left|1-u^{\alpha+1}\right|^s du\right)^{\frac{1}{s}}\left(\frac{1}{2\rho}|f''(a^\rho)|^q + \frac{1}{2\rho}|f''(b^\rho)|^q\right)^{\frac{1}{q}} \\
&\quad + \left(\frac{1}{s\rho(\alpha+2)-s+1}\right)^{\frac{1}{s}}\left(\frac{\rho}{\rho+1}|f''(b^\rho)|^q + \frac{1}{\rho+1}|f''(a^\rho)|^q\right)^{\frac{1}{q}}\Bigg] \\
&\leq \frac{\rho(b^\rho-a^\rho)^2}{2(\alpha+1)}\Bigg[\left(\frac{s(\alpha+1)}{\rho(s(\alpha+1)+1)}\right)^{\frac{1}{s}}\left(\frac{1}{2\rho}|f''(a^\rho)|^q + \frac{1}{2\rho}|f''(b^\rho)|^q\right)^{\frac{1}{q}} \\
&\quad + \left(\frac{1}{s\rho(\alpha+2)-s+1}\right)^{\frac{1}{s}}\left(\frac{\rho}{\rho+1}|f''(b^\rho)|^q + \frac{1}{\rho+1}|f''(a^\rho)|^q\right)^{\frac{1}{q}}\Bigg],
\end{aligned}$$

where $\frac{1}{s}+\frac{1}{q}=1$.

Proof. The result follows directly from Theorem 6 if we take $\eta(x,y) = x-y$ and $\mu = 0$. □

Theorem 7. *Let $\alpha > 0$, $\rho > 0$ and $f : [a^\rho, b^\rho] \to \mathbb{R}$ be a twice differentiable mapping on (a^ρ, b^ρ) with $0 \leq a < b$. If $|f''|^q$ is a strongly η-convex function on $[a^\rho, b^\rho]$ with modulus $\mu \geq 0$ for $q \geq 1$, then the following inequality holds:*

$$\begin{aligned}
&\left|\frac{f(a^\rho)+f(b^\rho)}{2} - \frac{\rho^\alpha\Gamma(\alpha+1)}{2(b^\rho-a^\rho)^\alpha}\left[{}^\rho I^\alpha_{a+} f(b^\rho) + {}^\rho I^\alpha_{b-} f(a^\rho)\right]\right| \\
&\leq \frac{(b^\rho-a^\rho)^2}{2\rho(\alpha+1)}\left(\frac{\alpha}{\alpha+2}\right)^{1-\frac{1}{q}}\Bigg[\frac{\alpha}{\alpha+2}|f''(b^\rho)|^q \\
&\quad + \left(\frac{\alpha+1}{2(\alpha+3)} - \mathrm{B}(2,\alpha+2)\right)\eta\left(|f''(a^\rho)|^q, |f''(b^\rho)|^q\right) \\
&\quad - \mu(b^\rho-a^\rho)^2\left(\frac{1}{6} - 2\mathrm{B}(2,\alpha+3)\right)\Bigg]^{\frac{1}{q}},
\end{aligned}$$

where $\mathrm{B}(\cdot,\cdot)$ denotes the beta function defined by $\mathrm{B}(x,y) = \int_0^1 t^{x-1}(1-t)^{y-1}dt$.

Proof. Using Lemma 4, the power mean inequality and the strong η-convexity of $|f''|^q$, we obtain

$$
\begin{aligned}
&\left|\frac{f(a^\rho)+f(b^\rho)}{2}-\frac{\rho^\alpha\Gamma(\alpha+1)}{2(b^\rho-a^\rho)^\alpha}\left[{}^\rho I^\alpha_{a+}f(b^\rho)+{}^\rho I^\alpha_{b-}f(a^\rho)\right]\right|\\
&\le \frac{(b^\rho-a^\rho)^2}{2(\alpha+1)}\int_0^1\left|1-(1-t^\rho)^{\alpha+1}-t^{\rho(\alpha+1)}\right|t^{\rho-1}\left|f''(t^\rho a^\rho+(1-t^\rho)b^\rho)\right|dt\\
&\le \frac{(b^\rho-a^\rho)^2}{2(\alpha+1)}\left(\int_0^1\left[1-(1-t^\rho)^{\alpha+1}-t^{\rho(\alpha+1)}\right]t^{\rho-1}dt\right)^{1-\frac{1}{q}}\\
&\qquad\times\left(\int_0^1\left[1-(1-t^\rho)^{\alpha+1}-t^{\rho(\alpha+1)}\right]t^{\rho-1}\left|f''(t^\rho a^\rho+(1-t^\rho)b^\rho)\right|^q dt\right)^{\frac{1}{q}}\\
&\le \frac{(b^\rho-a^\rho)^2}{2(\alpha+1)}\left(\int_0^1\left[1-(1-t^\rho)^{\alpha+1}-t^{\rho(\alpha+1)}\right]t^{\rho-1}dt\right)^{1-\frac{1}{q}}\\
&\qquad\times\left(\int_0^1\left[1-(1-t^\rho)^{\alpha+1}-t^{\rho(\alpha+1)}\right]t^{\rho-1}\left(|f''(b^\rho)|^q+t^\rho\eta\left(|f''(a^\rho)|^q,|f''(b^\rho)|^q\right)\right.\right.\\
&\qquad\qquad\left.\left.-\mu t^\rho(1-t^\rho)(b^\rho-a^\rho)^2\right)dt\right)^{\frac{1}{q}}\\
&\le \frac{(b^\rho-a^\rho)^2}{2(\alpha+1)}\left(\int_0^1\left[1-(1-t^\rho)^{\alpha+1}-t^{\rho(\alpha+1)}\right]t^{\rho-1}dt\right)^{1-\frac{1}{q}}\\
&\qquad\times\left(|f''(b^\rho)|^q\int_0^1\left[1-(1-t^\rho)^{\alpha+1}-t^{\rho(\alpha+1)}\right]t^{\rho-1}dt\right.\\
&\qquad\qquad+\eta\left(|f''(a^\rho)|^q,|f''(b^\rho)|^q\right)\int_0^1\left[1-(1-t^\rho)^{\alpha+1}-t^{\rho(\alpha+1)}\right]t^{2\rho-1}dt\\
&\qquad\qquad\left.-\mu(b^\rho-a^\rho)^2\int_0^1\left[1-(1-t^\rho)^{\alpha+1}-t^{\rho(\alpha+1)}\right]t^{2\rho-1}(1-t^\rho)dt\right)^{\frac{1}{q}}.
\end{aligned}
$$

The desired result follows from the above inequality and using the following computations:

$$
\begin{aligned}
\int_0^1\left[1-(1-t^\rho)^{\alpha+1}-t^{\rho(\alpha+1)}\right]t^{\rho-1}dt&=\frac{1}{\rho}\int_0^1\left[1-(1-u)^{\alpha+1}-u^{\alpha+1}\right]du\\
&=\frac{\alpha}{\rho(\alpha+2)},
\end{aligned}
$$

$$
\begin{aligned}
\int_0^1\left[1-(1-t^\rho)^{\alpha+1}-t^{\rho(\alpha+1)}\right]t^{2\rho-1}dt&=\frac{1}{\rho}\int_0^1\left[1-(1-u)^{\alpha+1}-u^{\alpha+1}\right]u\,du\\
&=\frac{1}{\rho}\left(\frac{1}{2}-\mathrm{B}(2,\alpha+2)-\frac{1}{\alpha+3}\right)\\
&=\frac{1}{\rho}\left(\frac{\alpha+1}{2(\alpha+3)}-\mathrm{B}(2,\alpha+2)\right)
\end{aligned}
$$

and

$$
\begin{aligned}
\int_0^1\left[1-(1-t^\rho)^{\alpha+1}-t^{\rho(\alpha+1)}\right]t^{2\rho-1}(1-t^\rho)dt&=\frac{1}{\rho}\int_0^1\left[1-(1-u)^{\alpha+1}-u^{\alpha+1}\right]u(1-u)\,du\\
&=\frac{1}{\rho}\left(\frac{1}{6}-2\mathrm{B}(2,\alpha+3)\right).
\end{aligned}
$$

This completes the proof of the theorem. □

Corollary 3. *Let $\alpha > 0$, $\rho > 0$ and $f : [a^\rho, b^\rho] \to \mathbb{R}$ be a twice differentiable mapping on (a^ρ, b^ρ) with $0 \le a < b$. If $|f''|^q$ is convex for $q \ge 1$, then the following inequality holds:*

$$\left|\frac{f(a^\rho)+f(b^\rho)}{2} - \frac{\rho^\alpha \Gamma(\alpha+1)}{2(b^\rho - a^\rho)^\alpha}\left[{}^\rho I^\alpha_{a+} f(b^\rho) + {}^\rho I^\alpha_{b-} f(a^\rho)\right]\right|$$
$$\le \frac{(b^\rho - a^\rho)^2}{2\rho(\alpha+1)}\left(\frac{\alpha}{\alpha+2}\right)^{1-\frac{1}{q}}\left[\left(\mathrm{B}(2,\alpha+2) - \frac{\alpha^2+3\alpha-2}{2(\alpha+2)(\alpha+3)}\right)|f''(b^\rho)|^q\right.$$
$$\left. + \left(\frac{\alpha+1}{2(\alpha+3)} - \mathrm{B}(2,\alpha+2)\right)|f''(a^\rho)|^q\right]^{\frac{1}{q}}.$$

Proof. The result follows directly from Theorem 7 if we take $\eta(x,y) = x - y$ and $\mu = 0$. □

Theorem 8. *Let $\alpha > 0$, $\rho > 0$ and $f : [a^\rho, b^\rho] \to \mathbb{R}$ be a twice differentiable mapping on (a^ρ, b^ρ) with $0 \le a < b$. If $|f''|^q$ is a strongly η-convex function on $[a^\rho, b^\rho]$ with modulus $\mu \ge 0$ for $q > 1$, then the following inequalities hold:*

$$\left|\frac{f(a^\rho)+f(b^\rho)}{2} - \frac{\rho^\alpha \Gamma(\alpha+1)}{2(b^\rho - a^\rho)^\alpha}\left[{}^\rho I^\alpha_{a+} f(b^\rho) + {}^\rho I^\alpha_{b-} f(a^\rho)\right]\right|$$
$$\le \frac{(b^\rho - a^\rho)^2}{2\rho(\alpha+1)}\left(\int_0^1 \left[1-(1-u)^{\alpha+1} - u^{\alpha+1}\right]^s du\right)^{\frac{1}{s}}$$
$$\times\left(|f''(b^\rho)|^q + \frac{1}{2}\eta\left(|f''(a^\rho)|^q, |f''(b^\rho)|^q\right) - \frac{\mu\rho(b^\rho - a^\rho)^2}{2(\rho+1)(2\rho+1)}\right)^{\frac{1}{q}}$$
$$\le \frac{(b^\rho - a^\rho)^2}{2\rho(\alpha+1)}\left(\frac{s(\alpha+1)-1}{s(\alpha+1)+1}\right)^{\frac{1}{s}}\left(|f''(b^\rho)|^q + \frac{1}{2}\eta\left(|f''(a^\rho)|^q, |f''(b^\rho)|^q\right)\right.$$
$$\left. - \frac{\mu\rho(b^\rho - a^\rho)^2}{2(\rho+1)(2\rho+1)}\right)^{\frac{1}{q}},$$

where $\frac{1}{s} + \frac{1}{q} = 1$.

Proof. Using Lemma 4, the Hölder's inequality and the strong η-convexity of $|f''|^q$, we obtain

$$\left|\frac{f(a^\rho)+f(b^\rho)}{2} - \frac{\rho^\alpha \Gamma(\alpha+1)}{2(b^\rho - a^\rho)^\alpha}\left[{}^\rho I^\alpha_{a+} f(b^\rho) + {}^\rho I^\alpha_{b-} f(a^\rho)\right]\right|$$
$$\le \frac{(b^\rho - a^\rho)^2}{2(\alpha+1)}\int_0^1 \left|1-(1-t^\rho)^{\alpha+1} - t^{\rho(\alpha+1)}\right| t^{\rho-1}\left|f''(t^\rho a^\rho + (1-t^\rho)b^\rho)\right| dt$$
$$\le \frac{(b^\rho - a^\rho)^2}{2(\alpha+1)}\left(\int_0^1 \left[1-(1-t^\rho)^{\alpha+1} - t^{\rho(\alpha+1)}\right]^s t^{\rho-1} dt\right)^{\frac{1}{s}}$$
$$\times\left(\int_0^1 t^{\rho-1}\left|f''(t^\rho a^\rho + (1-t^\rho)b^\rho)\right|^q dt\right)^{\frac{1}{q}}$$
$$\le \frac{(b^\rho - a^\rho)^2}{2(\alpha+1)}\left(\int_0^1 \left[1-(1-t^\rho)^{\alpha+1} - t^{\rho(\alpha+1)}\right]^s t^{\rho-1} dt\right)^{\frac{1}{s}}$$
$$\times\left(\int_0^1 t^{\rho-1}\left(|f''(b^\rho)|^q + t^\rho \eta\left(|f''(a^\rho)|^q, |f''(b^\rho)|^q\right)\right.\right.$$
$$\left.\left. - \mu t^\rho(1-t^\rho)(b^\rho - a^\rho)^2\right) dt\right)^{\frac{1}{q}}$$

$$
\begin{aligned}
&\leq \frac{(b^\rho - a^\rho)^2}{2(\alpha+1)} \left(\int_0^1 \left[1 - (1-t^\rho)^{\alpha+1} - t^{\rho(\alpha+1)}\right]^s t^{\rho-1} dt \right)^{\frac{1}{s}} \\
&\quad \times \left(|f''(b^\rho)|^q \int_0^1 t^{\rho-1} dt + \eta\left(|f''(a^\rho)|^q, |f''(b^\rho)|^q\right) \int_0^1 t^{2\rho-1} dt \right. \\
&\quad \left. - \mu(b^\rho - a^\rho)^2 \int_0^1 t^{2\rho-1}(1-t^\rho) dt \right)^{\frac{1}{q}},
\end{aligned}
$$

where

$$
\int_0^1 \left[1 - (1-t^\rho)^{\alpha+1} - t^{\rho(\alpha+1)}\right]^s t^{\rho-1} dt = \frac{1}{\rho} \int_0^1 \left[1 - (1-u)^{\alpha+1} - u^{\alpha+1}\right]^s du,
$$

$$
\int_0^1 t^{\rho-1}\, dt = \frac{1}{\rho}, \quad \int_0^1 t^{2\rho-1} dt = \frac{1}{2\rho} \text{ and } \int_0^1 t^{2\rho-1}(1-t^\rho) dt = \frac{1}{2(\rho+1)(2\rho+1)}.
$$

This proves the first inequality. Using a similar argument as in the proof of Theorem 6, we obtain

$$
\begin{aligned}
\int_0^1 \left[1 - (1-u)^{\alpha+1} - u^{\alpha+1}\right]^s du &\leq \int_0^1 1 - (1-u)^{s(\alpha+1)} - u^{s(\alpha+1)} du \\
&= 1 - \frac{2}{s(\alpha+1)+1} \\
&= \frac{s(\alpha+1)-1}{s(\alpha+1)+1}.
\end{aligned}
$$

This completes the proof of the theorem. □

Corollary 4. *Let $\alpha > 0$, $\rho > 0$ and $f : [a^\rho, b^\rho] \to \mathbb{R}$ be a twice differentiable mapping on (a^ρ, b^ρ) with $0 \leq a < b$. If $|f''|^q$ is convex for $q > 1$, then the following inequalities hold:*

$$
\begin{aligned}
&\left| \frac{f(a^\rho) + f(b^\rho)}{2} - \frac{\rho^\alpha \Gamma(\alpha+1)}{2(b^\rho - a^\rho)^\alpha} \left[{}^\rho I^\alpha_{a+} f(b^\rho) + {}^\rho I^\alpha_{b-} f(a^\rho) \right] \right| \\
&\leq \frac{(b^\rho - a^\rho)^2}{2\rho(\alpha+1)} \left(\int_0^1 \left[1 - (1-u)^{\alpha+1} - u^{\alpha+1}\right]^s du \right)^{\frac{1}{s}} \\
&\qquad \times \left(\frac{1}{2}|f''(b^\rho)|^q + \frac{1}{2}|f''(a^\rho)|^q \right)^{\frac{1}{q}} \\
&\leq \frac{(b^\rho - a^\rho)^2}{2\rho(\alpha+1)} \left(\frac{s(\alpha+1)-1}{s(\alpha+1)+1} \right)^{\frac{1}{s}} \\
&\qquad \times \left(\frac{1}{2}|f''(b^\rho)|^q + \frac{1}{2}|f''(a^\rho)|^q \right)^{\frac{1}{q}},
\end{aligned}
$$

where $\frac{1}{s} + \frac{1}{q} = 1$.

Proof. The result follows directly from Theorem 8 if we take $\eta(x, y) = x - y$ and $\mu = 0$. □

3. Conclusions

Four main results related to the Hermite–Hadamard inequality via the Katugampola fractional integrals involving strongly η-convex functions have been introduced. Similar results via the Riemann–Liouville and Hadamard fractional integrals could be derived as particular cases by taking $\rho \to 1$ and $\rho \to 0^+$, respectively. Several other interesting results can be obtained by considering different bifunctions η and/or the modulus μ as well as different values for the parameters α and ρ.

Author Contributions: S.K., E.R.N and A.M.T contributed equally to this work.

Funding: This research received no external funding.

Acknowledgments: We kindly appreciate the efforts of the anonymous referees for their valuable comments and suggestions.

Conflicts of Interest: The authors declare no conflict of interest.

References

1. Hadamard, J. Etude sur les properties des fonctions entries et an particular d'une fonction considree par, Riemann. *J. Math. Pures. Appl.* **1893**, *58*, 171–215.
2. Alomari, M.; Darus, M.; Dragomir, S.S. New inequalities of Hermite–Hadamard's type for functions whose second derivatives absolute values are quasiconvex. *Tamkang J. Math.* **2010**, *41*, 353–359.
3. Chun, L.; Qi, F. Integral inequalities of Hermite–Hadamard type for functions whose third derivatives are convex. *J. Inequal. Appl.* **2013**, *2013*, 451. [CrossRef]
4. Chun, L.; Qi, F. Integral inequalities of Hermite–Hadamard type for functions whose 3rd derivatives are *s*-convex. *Appl. Math.* **2012**, *3*, 1680–1685. [CrossRef]
5. Dragomir, S.S. Two mappings in connection to Hadamard's inequalities. *J. Math. Anal. Appl.* **1992**, *167*, 49–56. [CrossRef]
6. Dragomir, S.S.; Agarwal, R.P. Two inequalities for differentiable mappings and their applications to special means for real numbers and to trapezoidal formula. *Appl. Math. Lett.* **1998**, *11*, 91–95. [CrossRef]
7. Farid, G.; Rehman, A.U.; Zahra, M. On Hadamard-type inequalities for *k*-fractional integrals. *Konulrap J. Math.* **2016**, *4*, 79–86.
8. Khan, M.A.; Khurshid, Y.; Ali, T. Hermite–Hadamard inequality for fractional integrals via η-convex functions. *Acta Math. Univ. Comen.* **2017**, *LXXXVI* (1), 153–164.
9. Nwaeze, E.R. Inequalities of the Hermite–Hadamard type for Quasi-convex functions via the (k,s)-Riemann–Liouville fractional integrals. *Fract. Differ. Calc.* **2018**, *8*, 327–336. [CrossRef]
10. Nwaeze, E.R.; Kermausuor, S.; Tameru, A.M. Some new *k*-Riemann–Liouville Fractional integral inequalities associated with the strongly η-quasiconvex functions with modulus $\mu \geq 0$. *J. Inequal. Appl.* **2018**, *2018*, 139. [CrossRef] [PubMed]
11. Sarikaya, M.Z.; Set, E.; Yaldiz, H.; Basak, N. Hermite–Hadamard's inequalities for fractional integrals and related fractional inequalities. *Math. Comput. Model.* **2013**, *57*, 2403–2407. [CrossRef]
12. Gordji, M.E.; Delavar, M.R.; de la Sen, M. On φ-convex functions. *J. Math. Inequal.* **2016**, *10*, 173–183. [CrossRef]
13. Awan, M.U.; Noor, M.A.; Noor, K.I.; Safdar, F. On strongly generalized convex functions. *Filomat* **2017**, *31*, 5783–5790. [CrossRef]
14. Gordji, M.E.; Delavar, M.R.; Dragomir, S.S. Some inequalities related to η-convex functions. *Preprint RGMIA Res. Rep. Coll.* **2015**, *18*, 1–14
15. Gordji, M.E.; Dragomir, S.S.; Delavar, M.R. An inequality related to η-convex functions (II). *Int. J. Nonlinear Anal. Appl.* **2015**, *6*, 27–33.
16. Nwaeze, E.R.; Torres, D.F.M. Novel results on the Hermite–Hadamard kind inequality for η-convex funxtions by means of the (k,r)-fractional integral operators. In *Advances in Mathematical Inequalities and Applications (AMIA)*; Trends in Mathematics; Dragomir, S.S., Agarwal, P., Jleli, M., Samet, B., Eds.; Birkhäuser: Singapore, 2018; pp. 311–321.
17. Podlubny, I. *Fractional Differential Equations: Mathematics in Science and Engineering;* Academic Press: San Diego, CA, USA, 1999.

18. Samko, S.G.; Kilbas, A.A.; Marichev, O.I. *Fractional Integrals and Derivatives: Theory and Applications*; Gordon and Breach: Amsterdam, The Netherlands, 1993.
19. Katugampola, U.N. New approach to a generalized fractional integral. *Appl. Math. Comput.* **2011**, *218*, 860–865. [CrossRef]
20. Chen, H.; Katugampola, U.N. Hermite–Hadamard and Hermite–Hadamard–Fejér type inequalities for generalized fractional integrals. *J. Math. Anal. Appl.* **2017**, *446*, 1274–1291. [CrossRef]
21. Katugampola, U.N. A new approach to generalized fractional derivatives. *Bull. Math. Anal. Appl.* **2014**, *6*, 1–15.

mathematics

Article

Some Quantum Estimates of Hermite-Hadamard Inequalities for Quasi-Convex Functions

Hefeng Zhuang [1,†], Wenjun Liu [1,*,†] and Jaekeun Park [2]

1 College of Mathematics and Statistics, Nanjing University of Information Science and Technology, Nanjing 210044, China; hfzhuang11@163.com
2 Department of Mathematics, Hanseo University, Chungnam-do, Seosan-si 356-706, Korea; jkpark@hanseo.ac.kr
* Correspondence: wjliu@nuist.edu.cn; Tel.: +86-25-5873-1160
† These authors contributed equally to this work.

Received: 18 December 2018; Accepted: 2 February 2019; Published: 5 February 2019

Abstract: In this paper, we develop some quantum estimates of Hermite-Hadamard type inequalities for quasi-convex functions. In some special cases, these quantum estimates reduce to the known results.

Keywords: quantum estimates; Hermite-Hadamard type inequalities; quasi-convex

---

1. Introduction

1.1. Current State of Hermite-Hadamard Inequalities

Many important inequalities are established for the class of convex functions [1], but one of the most famous is the so-called Hermite-Hadamard inequality, which was first discovered by Hermite in 1881, and is stated as follows: Let $f: I \subseteq \mathbb{R} \to \mathbb{R}$ be a convex function, where $a, b \in I$ with $a < b$. Then

$$f\left(\frac{a+b}{2}\right) \leq \frac{1}{b-a}\int_a^b f(x)dx \leq \frac{f(a)+f(b)}{2}.$$

This famous result can be considered as a necessary and sufficient condition for a function to be convex. Hermite-Hadamard's inequality has raised many scholars' attention, and a variety of refinements and generalizations have been found (see [1–20]).

In [16], Özdemir used the following lemma and established some estimates on it via quasi-convex functions.

Lemma 1. *([16], Lemma 1) Let $f: I \subset \mathbb{R} \to \mathbb{R}$ be a twice differentiable mapping on I^0, $a, b \in I$ with $a < b$ and f'' be integrable on $[a, b]$. Then the following equality holds:*

$$\frac{f(a)+f(b)}{2} - \frac{1}{b-a}\int_a^b f(x)dx = \frac{(b-a)^2}{2}\int_0^1 s(1-s)f''(sa+(1-s)b)ds. \tag{1}$$

Theorem 1. *([16], Theorem 2) Let $f: I^0 \subset [0,\infty) \to \mathbb{R}$ be a twice differentiable mapping on I^0, such that $f'' \in L[a,b]$, $a, b \in I$ with $a < b$. If $|f''|^r$ is quasi-convex on $[a,b]$ for $r \geq 1$, then the following inequality holds:*

$$\left|\frac{f(a)+f(b)}{2} - \frac{1}{b-a}\int_a^b f(x)dx\right| \leq \frac{(b-a)^2}{4}\left(\frac{2}{(r+1)(r+2)}\right)^{\frac{r-1}{r}}\left(\sup\{|f''(a)|^r, |f''(b)|^r\}\right)^{\frac{1}{r}}. \tag{2}$$

Theorem 2. *([16], Theorem 3) Let $f : I^0 \subset [0,\infty) \to \mathbb{R}$ be a twice differentiable mapping on I^0, such that $f'' \in L[a,b], a,b \in I$ with $a < b$. If $|f''|^r$ is quasi-convex on $[a,b]$ for $r > 1$, then the following inequality holds:*

$$\left|\frac{f(a)+f(b)}{2} - \frac{1}{b-a}\int_a^b f(x)dx\right| \leq \frac{(b-a)^2}{2^{1+\frac{1}{r}}}\left(\beta(2,p+1)\right)^{\frac{1}{p}}\left(\sup\{|f''(a)|^r, |f''(b)|^r\}\right)^{\frac{1}{r}}, \quad (3)$$

where $\frac{1}{p}+\frac{1}{r}=1$ and $\beta(,)$ is Euler Beta Function:

$$\beta(x,y) = \int_0^1 t^{x-1}(1-t)^{y-1}dt, \quad x,y>0.$$

In [2], Alomari et al. established the following inequalities through Lemma 1.

Theorem 3. *([2], Theorem 3) Let $f : I \subset \mathbb{R} \to \mathbb{R}$ be a twice differentiable mapping on I^0, $a,b \in I$ with $a < b$ and f'' be integrable on $[a,b]$. If $|f''|$ is quasi-convex on $[a,b]$, then the following inequality holds:*

$$\left|\frac{f(a)+f(b)}{2} - \frac{1}{b-a}\int_a^b f(x)dx\right| \leq \frac{(b-a)^2}{12}\sup\{|f''(a)|, |f''(b)|\}. \quad (4)$$

Theorem 4. *([2], Theorem 4) Let $f : I \subset \mathbb{R} \to \mathbb{R}$ be a twice differentiable mapping on I^0, $a,b \in I$ with $a < b$ and f'' be integrable on $[a,b]$. If $|f''|^{p/(p-1)}$ is quasi-convex on $[a,b]$ for $p > 1$, then the following inequality holds:*

$$\left|\frac{f(a)+f(b)}{2} - \frac{1}{b-a}\int_a^b f(x)dx\right| \leq \frac{(b-a)^2}{8}\left(\frac{\sqrt{\pi}}{2}\right)^{\frac{1}{p}}\left(\frac{\Gamma(1+p)}{\Gamma(\frac{3}{2}+p)}\right)^{\frac{1}{p}}\left(\sup\{|f''(a)|^r, |f''(b)|^r\}\right)^{\frac{1}{r}}, \quad (5)$$

where $r = p/(p-1)$.

Theorem 5. *([2], Theorem 5) Let $f : I \subset \mathbb{R} \to \mathbb{R}$ be a twice differentiable mapping on I^0, $a,b \in I$ with $a < b$ and f'' be integrable on $[a,b]$. If $|f''|^r$ is quasi-convex on $[a,b]$ for $q \geq 1$, then the following inequality holds:*

$$\left|\frac{f(a)+f(b)}{2} - \frac{1}{b-a}\int_a^b f(x)dx\right| \leq \frac{(b-a)^2}{12}\left(\sup\{|f''(a)|^r, |f''(b)|^r\}\right)^{\frac{1}{r}}. \quad (6)$$

1.2. Motivation of Quantum Estimates

In recent years, many researchers have shown their interest in studying and investigating quantum calculus. Quantum analysis has large applications in many mathematical areas such as number theory ([21]), special functions ([22]), quantum mechanics ([23]) and mathematical inequalities. At present, q-analogues of many identities and inequalities have been established ([13–15,19,20,24]).

The Hermite-Hadamard inequality has been extended by considering its quantum estimates. For example, in [13], Noor et al. established the following lemma and developed some quantum estimates for it.

Lemma 2. *([13], Lemma 3.1) Let $f : I = [a,b] \subset \mathbb{R} \to \mathbb{R}$ be a q-differentiable function on I^0 (the interior of I) with ${}_aD_q$ be continuous and integrable on I where $0 < q < 1$, then*

$$\frac{1}{b-a}\int_a^b f(x)_a d_q x - \frac{qf(a)+f(b)}{1+q} = \frac{q(b-a)}{1+q}\int_0^1 (1-(1+q)t)_a D_q f((1-t)a+tb)_0 d_q t.$$

Theorem 6. *([13], Theorem 3.2) Let $f : I = [a, b] \subset \mathbb{R} \to \mathbb{R}$ be a q-differentiable function on I^0 (the interior of I) with ${}_aD_q$ be continuous and integrable on I where $0 < q < 1$. If $|{}_aD_qf|^r, r \geq 1$ is a convex function, then*

$$\left|\frac{1}{b-a}\int_a^b f(x)_a d_q x - \frac{qf(a)+f(b)}{1+q}\right|$$
$$\leq \frac{q(b-a)}{1+q}\left(\frac{2q}{(1+q)^2}\right)^{1-\frac{1}{r}}\left(\frac{q(1+3q^2+2q^3)}{(1+q+q^2)(1+q)^3}|{}_aD_qf(a)|^r + \frac{q(1+4q+q^2)}{(1+q+q^2)(1+q)^3}|{}_aD_qf(b)|^r\right)^{\frac{1}{r}}.$$

Theorem 7. *([13], Theorem 3.3) Let $f : I = [a, b] \subset \mathbb{R} \to \mathbb{R}$ be a q-differentiable function on I^0 (the interior of I) with ${}_aD_q$ be continuous and integrable on I where $0 < q < 1$. If $|{}_aD_qf|^r$ is a convex function where $p, r > 1, \frac{1}{p} + \frac{1}{r} = 1$, then*

$$\left|\frac{1}{b-a}\int_a^b f(x)_a d_q x - \frac{qf(a)+f(b)}{1+q}\right|$$
$$\leq \frac{q(b-a)}{1+q}\left(\frac{2q}{(1+q)^2}\right)^{\frac{1}{p}}\left(\frac{q(1+3q^2+2q^3)}{(1+q+q^2)(1+q)^3}|{}_aD_qf(a)|^r + \frac{q(1+4q+q^2)}{(1+q+q^2)(1+q)^3}|{}_aD_qf(b)|^r\right)^{\frac{1}{r}}.$$

The main purpose of this paper is to use a new quantum integral identity established in [11] to develop some quantum estimates of Hermite-Hadamard type inequalities for quasi-convex functions (Section 3). These quantum estimates of Hermite-Hadamard type inequalities reduces to Theorems 1–5 as $q \to 1$.

1.3. Possible Applications of the Estimates

Quantum calculus has large applications in many mathematical areas. We expect these new quantum estimates for Hermite-Hadamard type inequalities to have potential applications in the fields of integral inequalities, approximation theory, special means theory, optimization theory, information theory and numerical analysis.

2. Preliminaries

In this section, we first recall some previously known concepts on q-calculus which will be used in this paper.

Let $J = [a, b] \subseteq \mathbb{R}$ be an interval and $0 < q < 1$ be a constant.

Definition 1. *[19] Assume $f : J \to \mathbb{R}$ is a continuous function and let $x \in J$. Then q-derivative on J of function f at x is defined as*

$$ {}_aD_qf(x) = \frac{f(x) - f(qx + (1-q)a)}{(1-q)(x-a)}, x \neq a, \quad {}_aD_qf(a) = \lim_{x\to a} {}_aD_qf(x). \tag{7}$$

We say that f is q-differentiable on J provided ${}_aD_qf(x)$ exists for all $x \in J$. Note that if $a = 0$ in (2.1), then ${}_0D_qf = D_qf$, where D_q is the well-known q-derivative of the function $f(x)$ defined by

$$D_qf(x) = \frac{f(x) - f(qx)}{(1-q)x}. \tag{8}$$

Definition 2. *[19] Let $f : J \to \mathbb{R}$ be a continuous function. We define the second-order q-derivative on interval J, which denoted as ${}_aD_q^2f$, provided ${}_aD_qf$ is q-differentiable on J with ${}_aD_q^2f = {}_aD_q({}_aD_qf) : J \to \mathbb{R}$. Similarly, we define higher order q-derivative on J, ${}_aD_q^n : J_k \to \mathbb{R}$.*

Definition 3. [19] *Let $f : J \subset \mathbb{R} \to \mathbb{R}$ be a continuous function. Then q-integral on J is defined by*

$$\int_a^x f(t)\, {}_a d_q t = (1-q)(x-a) \sum_{n=0}^{\infty} q^n f(q^n x + (1-q^n) a) \tag{9}$$

for $x \in J$. Moreover, if $c \in (a, x)$ then the definite q-integral on J is defined by

$$\begin{aligned}\int_c^x f(t)_a d_q t &= \int_a^x f(t)_a d_q t - \int_a^c f(t)_a d_q t \\ &= (1-q)(x-a) \sum_{n=0}^{\infty} q^n f(q^n x + (1-q^n)a) - (1-q)(c-a) \sum_{n=0}^{\infty} q^n f(q^n c + (1-q^n)a).\end{aligned}$$

Note that if $a = 0$, then we have the classical q-integral, which is defined by

$$\int_0^x f(t)\, {}_0 d_q t = (1-q)\, x \sum_{n=0}^{\infty} q^n f(q^n x) \tag{10}$$

for $x \in [0, +\infty)$.

Theorem 8. [19] *Assume that $f, g : J \to \mathbb{R}$ are continuous functions, $\alpha \in \mathbb{R}$. Then, for $x \in J$,*

$$\begin{aligned}&\int_a^x [f(t) + g(t)]\, {}_a d_q t = \int_a^x f(t)_a d_q t + \int_a^x g(t)_a d_q t; \\ &\int_a^x (\alpha f)(t)_a d_q t = \alpha \int_a^x f(t)_a d_q t.\end{aligned}$$

In addition, we introduce the q-analogues of a and $(x-a)^n$ and the definition of q-Beta function.

Definition 4. [22] *For any real number a,*

$$[a]_q = \frac{q^a - 1}{q - 1} \tag{11}$$

is called the q-analogue of a. In particular, if $n \in \mathbb{Z}^+$, we denote

$$[n] = \frac{q^n - 1}{q - 1} = q^{n-1} + \cdots + q + 1.$$

Definition 5. [22] *If n is an integer, the q-analogue of $(x-a)^n$ is the polynomial*

$$(x-a)_q^n = \begin{cases} 1, & \text{if } \; n = 0, \\ (x-a)(x-qa)\cdots(x-q^{n-1}a), & \text{if } \; n \geq 1. \end{cases} \tag{12}$$

Definition 6. [22] *For any $t, s > 0$,*

$$\beta_q(t,s) = \int_0^1 x^{t-1}(1-qx)_q^{s-1}\, {}_0 d_q x \tag{13}$$

is called the q-Beta function. Note that

$$\beta_q(t,1) = \int_0^1 x^{t-1}\, {}_0 d_q x = \frac{1}{[t]}, \tag{14}$$

where $[t]$ is the q-analogue of t.

At last, we present four simple calculations that will be used in this paper.

Lemma 3. *Let $f(x) = 1$, then we have*

$$\int_0^1 {}_0d_qx = (1-q)\sum_{n=0}^{\infty} q^n = 1.$$

Lemma 4. *Let $f(x) = x$ for $x \in [a,b]$, then we have*

$$\int_0^1 x_0d_qx = (1-q)\sum_{n=0}^{\infty} q^{2n} = \frac{1}{1+q}.$$

Lemma 5. *Let $f(x) = 1 - qx$ for $x \in [0,1]$ where $0 < q < 1$ be a constant , then we have*

$$\int_0^1 (1-qx)_0d_qx = \int_0^1 {}_0d_qx - q\int_0^1 x_0d_qx = \frac{1}{1+q}.$$

Lemma 6. *Let $f(x) = x(1-qx)$ for $x \in [0,1]$ where $0 < q < 1$ be a constant , then we have*

$$\begin{aligned}\int_0^1 x(1-qx)_0d_qx &= \int_0^1 (x-qx^2)_0d_qx = \int_0^1 x_0d_qx - q\int_0^1 x^2{}_0d_qx \\ &= \frac{1}{1+q} - q(1-q)\sum_{n=0}^{\infty} q^{3n} = \frac{1}{1+q} - q\frac{1}{1+q+q^2} \\ &= \frac{1}{(1+q)(1+q+q^2)}.\end{aligned}$$

In [6], we can find the notion of quasi-convex functions generalizes the notion of convex functions. More exactly, a function $f : [a,b] \to \mathbb{R}$ is said to be quasi-convex on $[a,b]$ if

$$f((1-\lambda)x + \lambda y) \le \sup\{f(x), f(y)\} \tag{15}$$

holds for any $x, y \in [a,b]$ and $\lambda \in [0,1]$. It's obviously that any convex function is a quasi-convex function. Furthermore, there exist quasi-convex functions which are not convex.

In [11], we have established the following q-integral identity and used it to prove some quantum estimates of Hermite-Hadamard type inequalities for convex functions.

Lemma 7. *([11], Lemma 4.1) Let $f : I = [a,b] \subset \mathbb{R} \to \mathbb{R}$ be a twice q-differentiable function on I^0 with ${}_aD_q^2f$ be continuous and integrable on I where $0 < q < 1$. Then the following identity holds:*

$$\frac{qf(a)+f(b)}{1+q} - \frac{1}{b-a}\int_a^b f(x)_ad_qx = \frac{q^2(b-a)^2}{1+q}\int_0^1 t(1-qt)_aD_q^2f((1-t)a+tb)_0d_qt. \tag{16}$$

Remark 1. *If $q \to 1$ and substitute $(1-t)a + tb$ for $sa + (1-s)b$, then (16) reduces to identity (1) in Lemma 1.*

3. Hermite-Hadamard Inequalities for Quasi-Convex Functions

In this section, we will give some estimates for the left-hand side of the result of (16) through quasi-convex functions.

Theorem 9. *Let $f : I = [a,b] \subset \mathbb{R} \to \mathbb{R}$ be a twice q-differentiable function on I^0 with ${}_aD_q^2 f$ be continuous and integrable on I where $0 < q < 1$. If $\left|{}_aD_q^2 f\right|^r$ is quasi-convex on $[a,b]$ for $r \geq 1$, then the following inequality holds:*

$$\left|\frac{qf(a)+f(b)}{1+q} - \frac{1}{b-a}\int_a^b f(x)_a d_q x\right| \leq \frac{q^2(b-a)^2}{1+q}\left(\frac{1}{1+q}\right)^{1-\frac{1}{r}}\left(h_1 \sup\{\left|{}_aD_q^2 f(a)\right|^r, \left|{}_aD_q^2 f(b)\right|^r\}\right)^{\frac{1}{r}}, \tag{17}$$

where

$$h_1 = (1-q)\sum_{n=0}^{\infty} q^{2n}(1-q^{n+1})^r.$$

Proof. Using Lemma 7, Hölder's inequality and the fact that $\left|{}_aD_q^2 f\right|^r$ is a quasi-convex function, we have

$$\begin{aligned}
&\left|\frac{qf(a)+f(b)}{1+q} - \frac{1}{b-a}\int_a^b f(x)_a d_q x\right| \\
=&\left|\frac{q^2(b-a)^2}{1+q}\int_0^1 t(1-qt)_a D_q^2 f((1-t)a+tb)_0 d_q t\right| \\
\leq&\frac{q^2(b-a)^2}{1+q}\int_0^1 t(1-qt)\left|{}_aD_q^2 f((1-t)a+tb)\right|_0 d_q t \\
\leq&\frac{q^2(b-a)^2}{1+q}\left(\int_0^1 t_0 d_q t\right)^{1-\frac{1}{r}}\left(\int_0^1 t(1-qt)^r\left|{}_aD_q^2 f((1-t)a+tb)\right|^r {}_0d_q t\right)^{\frac{1}{r}} \\
\leq&\frac{q^2(b-a)^2}{1+q}\left(\int_0^1 t_0 d_q t\right)^{1-\frac{1}{r}}\left(\sup\{\left|{}_aD_q^2 f(a)\right|^r, \left|{}_aD_q^2 f(b)\right|^r\}\int_0^1 t(1-qt)^r {}_0d_q t\right)^{\frac{1}{r}}
\end{aligned}$$

Applying Lemma 4, we have

$$\begin{aligned}
&\left|\frac{qf(a)+f(b)}{1+q} - \frac{1}{b-a}\int_a^b f(x)_a d_q x\right| \\
\leq&\frac{q^2(b-a)^2}{1+q}\left(\frac{1}{1+q}\right)^{1-\frac{1}{r}}\left(\sup\{\left|{}_aD_q^2 f(a)\right|^r, \left|{}_aD_q^2 f(b)\right|^r\}\int_0^1 t(1-qt)^r {}_0d_q t\right)^{\frac{1}{r}} \\
=&\frac{q^2(b-a)^2}{1+q}\left(\frac{1}{1+q}\right)^{1-\frac{1}{r}}\left(h_1 \sup\{\left|{}_aD_q^2 f(a)\right|^r, \left|{}_aD_q^2 f(b)\right|^r\}\right)^{\frac{1}{r}}.
\end{aligned}$$

It is easy to check that

$$h_1 = \int_0^1 t(1-qt)^r {}_0d_q t = (1-q)\sum_{n=0}^{\infty} q^{2n}(1-q^{n+1})^r,$$

thus, we get (17). □

Remark 2. *If $q \to 1$, then*

$$h_1 = \int_0^1 t(1-t)^r dt = \frac{1}{(r+1)(r+2)}.$$

Inequality (17) reduces to inequality (2) in Theorem 1 due to the fact that

$$\begin{aligned}&\frac{(b-a)^2}{2}\left(\frac{1}{2}\right)^{1-\frac{1}{r}}\left(\frac{1}{(r+1)(r+2)}\right)^{\frac{1}{r}}\left(\sup\{|f''(a)|^r,|f''(b)|^r\}\right)^{\frac{1}{r}}\\=&\frac{(b-a)^2}{4}\left(\frac{2}{(r+1)(r+2)}\right)^{\frac{1}{r}}\left(\sup\{|f''(a)|^r,|f''(b)|^r\}\right)^{\frac{1}{r}}.\end{aligned}$$

Corollary 1. *In Theorem 9, if r is a positive integer , then*

$$(1-qt)^r \le (1-qt)_q^r,$$

and (17) reduces to

$$\begin{aligned}&\left|\frac{qf(a)+f(b)}{1+q}-\frac{1}{b-a}\int_a^b f(x)_a d_q x\right|\\\le&\frac{q^2(b-a)^2}{1+q}\left(\frac{1}{1+q}\right)^{1-\frac{1}{r}}\left(\beta_q(2,r+1)\sup\{\left|{}_aD_q^2f(a)\right|^r,\left|{}_aD_q^2f(b)\right|^r\}\right)^{\frac{1}{r}}.\end{aligned}$$

Theorem 10. *Let $f: I=[a,b]\subset\mathbb{R}\to\mathbb{R}$ be a twice q-differentiable function on I^0 with ${}_aD_q^2 f$ be continuous and integrable on I where $0<q<1$. If $\left|{}_aD_q^2 f\right|^r$ is quasi-convex on $[a,b]$ where $p,r>1$, $\frac{1}{p}+\frac{1}{r}=1$, then*

$$\left|\frac{qf(a)+f(b)}{1+q}-\frac{1}{b-a}\int_a^b f(x)_a d_q x\right|\le\frac{q^2(b-a)^2}{1+q}(l_1)^{\frac{1}{p}}\left(\frac{\sup\{\left|{}_aD_q^2f(a)\right|^r,\left|{}_aD_q^2f(b)\right|^r\}}{1+q}\right)^{\frac{1}{r}},\tag{18}$$

where

$$l_1=(1-q)\sum_{n=0}^{\infty}q^{2n}(1-q^{n+1})^p.$$

Proof. Using Lemma 7, Hölder's inequality and the fact that $\left|{}_aD_q^2 f\right|^r$ is a quasi-convex function, we have

$$\begin{aligned}&\left|\frac{qf(a)+f(b)}{1+q}-\frac{1}{b-a}\int_a^b f(x)_a d_q x\right|\\=&\left|\frac{q^2(b-a)^2}{1+q}\int_0^1 t(1-qt)_aD_q^2f((1-t)a+tb)_0d_qt\right|\\\le&\frac{q^2(b-a)^2}{1+q}\int_0^1 t(1-qt)\left|{}_aD_q^2f((1-t)a+tb)\right|{}_0d_qt\\\le&\frac{q^2(b-a)^2}{1+q}\left(\int_0^1 t(1-qt)^p{}_0d_qt\right)^{\frac{1}{p}}\left(\int_0^1 t\left|{}_aD_q^2f((1-t)a+tb)\right|^r{}_0d_qt\right)^{\frac{1}{r}}\\\le&\frac{q^2(b-a)^2}{1+q}\left(\int_0^1 t(1-qt)^p{}_0d_qt\right)^{\frac{1}{p}}\left(\sup\{\left|{}_aD_q^2f(a)\right|^r,\left|{}_aD_q^2f(b)\right|^r\}\int_0^1 t_0d_qt\right)^{\frac{1}{r}}\end{aligned}$$

Applying Lemma 4, we have

$$\begin{aligned}&\left|\frac{qf(a)+f(b)}{1+q}-\frac{1}{b-a}\int_a^b f(x)_a d_q x\right|\\ \leq&\frac{q^2(b-a)^2}{1+q}\left(\int_0^1 t(1-qt)^p{}_0d_qt\right)^{\frac{1}{p}}\left(\frac{\sup\{\left|{}_aD_q^2f(a)\right|^r,\left|{}_aD_q^2f(b)\right|^r\}}{1+q}\right)^{\frac{1}{r}}\\ =&\frac{q^2(b-a)^2}{1+q}(l_1)^{\frac{1}{p}}\left(\frac{\sup\{\left|{}_aD_q^2f(a)\right|^r,\left|{}_aD_q^2f(b)\right|^r\}}{1+q}\right)^{\frac{1}{r}}.\end{aligned}$$

It is easy to check that

$$l_1=\int_0^1 t(1-qt)^p{}_0d_qt=(1-q)\sum_{n=0}^{\infty}q^{2n}(1-q^{n+1})^p,$$

thus, we get (18). □

Remark 3. *If $q\to 1$, then*

$$l_1=\int_0^1 t(1-t)^p dt=\beta(2,p+1).$$

Inequality (18) reduces to inequality (3) in Theorem 2.

Corollary 2. *In Theorem 10, if p is a positive integer and $p>1$, then*

$$(1-qt)^p\leq(1-qt)_q^p,$$

and (18) reduces to

$$\left|\frac{qf(a)+f(b)}{1+q}-\frac{1}{b-a}\int_a^b f(x)_a d_q x\right|\leq\frac{q^2(b-a)^2}{1+q}\left(\beta_q(2,p+1)\right)^{\frac{1}{p}}\left(\frac{\sup\{\left|{}_aD_q^2f(a)\right|^r,\left|{}_aD_q^2f(b)\right|^r\}}{1+q}\right)^{\frac{1}{r}}.$$

Theorem 11. *Let $f:I=[a,b]\subset\mathbb{R}\to\mathbb{R}$ be a twice q-differentiable function on I^0 with ${}_aD_q^2f$ be continuous and integrable on I where $0<q<1$. If $\left|{}_aD_q^2f\right|^r$ is quasi-convex on $[a,b]$ where $p,r>1$, $\frac{1}{p}+\frac{1}{r}=1$, then the following inequality holds:*

$$\left|\frac{qf(a)+f(b)}{1+q}-\frac{1}{b-a}\int_a^b f(x)_a d_q x\right|\leq\frac{q^2(b-a)^2}{1+q}(s_1)^{\frac{1}{p}}\left(\sup\{\left|{}_aD_q^2f(a)\right|^r,\left|{}_aD_q^2f(b)\right|^r\}\right)^{\frac{1}{r}},\tag{19}$$

where

$$s_1=(1-q)\sum_{n=0}^{\infty}(q^n)^{p+1}(1-q^{n+1})^p.$$

Proof. Using Lemma 7, Hölder's inequality and the fact that $\left|{}_aD_q^2 f\right|^r$ is a quasi-convex function, we have

$$\begin{aligned}
&\left|\frac{qf(a)+f(b)}{1+q}-\frac{1}{b-a}\int_a^b f(x)_a d_q x\right| \\
=&\left|\frac{q^2(b-a)^2}{1+q}\int_0^1 t(1-qt)_a D_q^2 f((1-t)a+tb)_0 d_q t\right| \\
\leq&\frac{q^2(b-a)^2}{1+q}\int_0^1 t(1-qt)\left|{}_aD_q^2 f((1-t)a+tb)\right|{}_0 d_q t \\
\leq&\frac{q^2(b-a)^2}{1+q}\left(\int_0^1 t^p(1-qt)^p{}_0 d_q t\right)^{\frac{1}{p}}\left(\int_0^1 \left|{}_aD_q^2 f((1-t)a+tb)\right|^r{}_0 d_q t\right)^{\frac{1}{r}} \\
\leq&\frac{q^2(b-a)^2}{1+q}\left(\int_0^1 t^p(1-qt)^p{}_0 d_q t\right)^{\frac{1}{p}}\left(\sup\{\left|{}_aD_q^2 f(a)\right|^r,\left|{}_aD_q^2 f(b)\right|^r\}\int_0^1 {}_0 d_q t\right)^{\frac{1}{r}}
\end{aligned}$$

Applying Lemma 3, we have

$$\begin{aligned}
&\left|\frac{qf(a)+f(b)}{1+q}-\frac{1}{b-a}\int_a^b f(x)_a d_q x\right| \\
\leq&\frac{q^2(b-a)^2}{1+q}(s_1)^{\frac{1}{p}}\left(\sup\{\left|{}_aD_q^2 f(a)\right|^r,\left|{}_aD_q^2 f(b)\right|^r\}\right)^{\frac{1}{r}}.
\end{aligned}$$

It is easy to check that

$$s_1=\int_0^1 t^p(1-qt)^p{}_0 d_q t=(1-q)\sum_{n=0}^{\infty}(q^n)^{p+1}(1-q^{n+1})^p,$$

thus, we get (19). □

Remark 4. *If $q\to 1$, then*

$$s_1=\int_0^1 t^p(1-t)^p dt=\beta(p+1,p+1).$$

Using the properties of Beta function, that is, $\beta(x,x)=2^{1-2x}\beta\left(\frac{1}{2},x\right)$ and $\beta(x,y)=\frac{\Gamma(x)\Gamma(y)}{\Gamma(xy)}$, we can obtain that

$$\beta(p+1,p+1)=2^{1-2(p+1)}\beta\left(\frac{1}{2},p+1\right)=2^{-2p-1}\frac{\Gamma\left(\frac{1}{2}\right)\Gamma(p+1)}{\Gamma(\frac{3}{2}+p)},$$

where $\Gamma(\frac{1}{2})=\sqrt{\pi}$ and $\Gamma(t)$ is Gamma function:

$$\Gamma(t)=\int_0^\infty x^{t-1}e^{-x}dx,\quad t>0.$$

Inequality (19) reduces to inequality (5) in Theorem 4 due to the fact that

$$\begin{aligned}
&\frac{(b-a)^2}{2}\left(2^{-2p-1}\frac{\Gamma(\frac{1}{2})\Gamma(p+1)}{\Gamma(\frac{3}{2}+p)}\right)^{\frac{1}{p}}\left(\sup\{|f''(a)|^r,|f''(b)|^r\}\right)^{\frac{1}{r}} \\
=&\frac{(b-a)^2}{8}\left(\frac{\sqrt{\pi}}{2}\right)^{\frac{1}{p}}\left(\frac{\Gamma(1+p)}{\Gamma(\frac{3}{2}+p)}\right)^{\frac{1}{p}}\left(\sup\{|f''(a)|^r,|f''(b)|^r\}\right)^{\frac{1}{r}}.
\end{aligned}$$

Corollary 3. *In Theorem 11, if p is a positive integer, $p > 1$, then*

$$(1-qt)^p \leq (1-qt)_q^p,$$

and (19) reduces to

$$\left|\frac{qf(a)+f(b)}{1+q} - \frac{1}{b-a}\int_a^b f(x)_a d_q x\right| \leq \frac{q^2(b-a)^2}{1+q}\left(\beta_q(p+1,p+1)\right)^{\frac{1}{p}}\left(\sup\{\left|{}_aD_q^2 f(a)\right|^r, \left|{}_aD_q^2 f(b)\right|^r\}\right)^{\frac{1}{r}}.$$

Theorem 12. *Let $f : I = [a,b] \subset \mathbb{R} \to \mathbb{R}$ be a twice q-differentiable function on I^0 with ${}_aD_q^2 f$ be continuous and integrable on I where $0 < q < 1$. If $\left|{}_aD_q^2 f\right|^r$ is quasi-convex on $[a,b]$ where $p, r > 1$, $\frac{1}{p} + \frac{1}{r} = 1$, then the following inequality holds:*

$$\begin{aligned} &\left|\frac{qf(a)+f(b)}{1+q} - \frac{1}{b-a}\int_a^b f(x)_a d_q x\right| \\ \leq &\frac{q^2(b-a)^2}{1+q}\left(\frac{1}{[p+1]}\right)^{\frac{1}{p}}\left(m_1 \sup\{\left|{}_aD_q^2 f(a)\right|^r, \left|{}_aD_q^2 f(b)\right|^r\}\right)^{\frac{1}{r}}, \end{aligned} \tag{20}$$

where

$$m_1 = (1-q)\sum_{n=0}^{\infty} q^n (1-q^{n+1})^r$$

and $[p+1]$ is the q-analogue of $p+1$.

Proof. Using Lemma 7, Hölder's inequality and the fact that $\left|{}_aD_q^2 f\right|^r$ is a quasi-convex function, we have

$$\begin{aligned} &\left|\frac{qf(a)+f(b)}{1+q} - \frac{1}{b-a}\int_a^b f(x)_a d_q x\right| \\ =&\left|\frac{q^2(b-a)^2}{1+q}\int_0^1 t(1-qt)_a D_q^2 f((1-t)a+tb)_0 d_q t\right| \\ \leq&\frac{q^2(b-a)^2}{1+q}\int_0^1 t(1-qt)\left|{}_aD_q^2 f((1-t)a+tb)\right|_0 d_q t \\ \leq&\frac{q^2(b-a)^2}{1+q}\left(\int_0^1 t^p{}_0d_q t\right)^{\frac{1}{p}}\left(\int_0^1 (1-qt)^r\left|{}_aD_q^2 f((1-t)a+tb)\right|^r{}_0d_q t\right)^{\frac{1}{r}} \\ \leq&\frac{q^2(b-a)^2}{1+q}\left(\int_0^1 t^p{}_0d_q t\right)^{\frac{1}{p}}\left(\sup\{\left|{}_aD_q^2 f(a)\right|^r, \left|{}_aD_q^2 f(b)\right|^r\}\int_0^1 (1-qt)^r{}_0d_q t\right)^{\frac{1}{r}} \end{aligned}$$

Applying (14) in Definition 6, we have

$$\begin{aligned} &\left|\frac{qf(a)+f(b)}{1+q} - \frac{1}{b-a}\int_a^b f(x)_a d_q x\right| \\ \leq&\frac{q^2(b-a)^2}{1+q}\left(\frac{1}{[p+1]}\right)^{\frac{1}{p}}\left(\sup\{\left|{}_aD_q^2 f(a)\right|^r, \left|{}_aD_q^2 f(b)\right|^r\}\int_0^1 (1-qt)^r{}_0d_q t\right)^{\frac{1}{r}} \\ =&\frac{q^2(b-a)^2}{1+q}\left(\frac{1}{[p+1]}\right)^{\frac{1}{p}}\left(m_1 \sup\{\left|{}_aD_q^2 f(a)\right|^r, \left|{}_aD_q^2 f(b)\right|^r\}\right)^{\frac{1}{r}}. \end{aligned}$$

It is easy to check that

$$m_1 = \int_0^1 (1-qt)^r{}_0d_q t = (1-q)\sum_{n=0}^{\infty} q^n (1-q^{n+1})^r,$$

thus, we get (20). □

Remark 5. *If $q \to 1$, then*

$$m_1 = \int_0^1 (1-t)^r dt = \frac{1}{r+1},$$

and (20) reduces to

$$\left|\frac{f(a)+f(b)}{2} - \frac{1}{b-a}\int_a^b f(x)dx\right| \le \frac{(b-a)^2}{2}\left(\frac{1}{p+1}\right)^{\frac{1}{p}}\left(\frac{\sup\{|f''(a)|^r, |f''(b)|^r\}}{r+1}\right)^{\frac{1}{r}}. \tag{21}$$

Corollary 4. *In Theorem 12, if r is a positive integer, $r > 1$, then*

$$(1-qt)^r \le (1-qt)_q^r,$$

and (20) reduces to

$$\begin{aligned}&\left|\frac{qf(a)+f(b)}{1+q} - \frac{1}{b-a}\int_a^b f(x)_a d_q x\right| \\ \le &\frac{q^2(b-a)^2}{1+q}\left(\frac{1}{[p+1]}\right)^{\frac{1}{p}}\left(\beta_q(1,r+1)\sup\{\left|{}_aD_q^2 f(a)\right|^r, \left|{}_aD_q^2 f(b)\right|^r\}\right)^{\frac{1}{r}}.\end{aligned}$$

Theorem 13. *Let $f : I = [a,b] \subset \mathbb{R} \to \mathbb{R}$ be a twice q-differentiable function on I^0 with ${}_aD_q^2 f$ be continuous and integrable on I where $0 < q < 1$. If $\left|{}_aD_q^2 f\right|^r$ is quasi-convex on $[a,b]$ where $p, r > 1$, $\frac{1}{p} + \frac{1}{r} = 1$, then the following inequality holds:*

$$\begin{aligned}&\left|\frac{qf(a)+f(b)}{1+q} - \frac{1}{b-a}\int_a^b f(x)_a d_q x\right| \\ \le &\frac{q^2(b-a)^2}{1+q}(n_1)^{\frac{1}{p}}\left(\frac{\sup\{\left|{}_aD_q^2 f(a)\right|^r, \left|{}_aD_q^2 f(b)\right|^r\}}{[r+1]}\right)^{\frac{1}{r}},\end{aligned} \tag{22}$$

where

$$n_1 = (1-q)\sum_{n=0}^{\infty} q^n (1-q^{n+1})^p$$

and $[r+1]$ is the q-analogue of $r+1$.

Proof. Using Lemma 7, Hölder's inequality and the fact that $\left|{}_aD_q^2 f\right|^r$ is a quasi-convex function, we have

$$\begin{aligned}&\left|\frac{qf(a)+f(b)}{1+q} - \frac{1}{b-a}\int_a^b f(x)_a d_q x\right| \\ =&\left|\frac{q^2(b-a)^2}{1+q}\int_0^1 t(1-qt)_a D_q^2 f((1-t)a+tb)_0 d_q t\right| \\ \le&\frac{q^2(b-a)^2}{1+q}\int_0^1 t(1-qt)\left|{}_aD_q^2 f((1-t)a+tb)\right|{}_0 d_q t \\ \le&\frac{q^2(b-a)^2}{1+q}\left(\int_0^1 (1-qt)^p{}_0 d_q t\right)^{\frac{1}{p}}\left(\int_0^1 t^r\left|{}_aD_q^2 f((1-t)a+tb)\right|^r{}_0 d_q t\right)^{\frac{1}{r}} \\ \le&\frac{q^2(b-a)^2}{1+q}\left(\int_0^1 (1-qt)^p{}_0 d_q t\right)^{\frac{1}{p}}\left(\sup\{\left|{}_aD_q^2 f(a)\right|^r, \left|{}_aD_q^2 f(b)\right|^r\}\int_0^1 t^r{}_0 d_q t\right)^{\frac{1}{r}}\end{aligned}$$

Applying (14) in Definition 6, we have

$$\left|\frac{qf(a)+f(b)}{1+q}-\frac{1}{b-a}\int_a^b f(x)_a d_q x\right|$$
$$\leq \frac{q^2(b-a)^2}{1+q}\left(\int_0^1 (1-qt)^p{}_0 d_q t\right)^{\frac{1}{p}}\left(\frac{\sup\{\left|{}_aD_q^2 f(a)\right|^r, \left|{}_aD_q^2 f(b)\right|^r\}}{[r+1]}\right)^{\frac{1}{r}}$$
$$= \frac{q^2(b-a)^2}{1+q}(n_1)^{\frac{1}{p}}\left(\frac{\sup\{\left|{}_aD_q^2 f(a)\right|^r, \left|{}_aD_q^2 f(b)\right|^r\}}{[r+1]}\right)^{\frac{1}{r}}.$$

It is easy to check that

$$n_1 = \int_0^1 (1-qt)^p{}_0 d_q t = (1-q)\sum_{n=0}^{\infty} q^n(1-q^{n+1})^p,$$

thus, we get (22). □

Remark 6. *If $q \to 1$, then*

$$n_1 = \int_0^1 (1-t)^p dt = \frac{1}{p+1},$$

and (22) reduces to (21) in Remark 5.

Corollary 5. *In Theorem 13, if p is a positive integer, $p > 1$, then*

$$(1-qt)^p \leq (1-qt)_q^p,$$

and (22) reduces to

$$\left|\frac{qf(a)+f(b)}{1+q}-\frac{1}{b-a}\int_a^b f(x)_a d_q x\right|$$
$$\leq \frac{q^2(b-a)^2}{1+q}\left(\beta_q(1,p+1)\right)^{\frac{1}{p}}\left(\frac{\sup\{\left|{}_aD_q^2 f(a)\right|^r, \left|{}_aD_q^2 f(b)\right|^r\}}{[r+1]}\right)^{\frac{1}{r}}.$$

Theorem 14. *Let $f : I = [a,b] \subset \mathbb{R} \to \mathbb{R}$ be a twice q-differentiable function on I^0 with ${}_aD_q^2 f$ be continuous and integrable on I where $0 < q < 1$. If $\left|{}_aD_q^2 f\right|^r$ is quasi-convex on $[a,b]$ for $r \geq 1$, then*

$$\left|\frac{qf(a)+f(b)}{1+q}-\frac{1}{b-a}\int_a^b f(x)_a d_q x\right| \leq \frac{q^2(b-a)^2}{1+q}(\mu_1)^{\frac{1}{r}}\left(\sup\{\left|{}_aD_q^2 f(a)\right|^r, \left|{}_aD_q^2 f(b)\right|^r\}\right)^{\frac{1}{r}}, \tag{23}$$

where

$$\mu_1 = (1-q)\sum_{n=0}^{\infty}(q^n)^{r+1}(1-q^{n+1})^r.$$

Proof. Using Lemma 7, Hölder's inequality and the fact that $\left|{}_aD_q^2 f\right|^r$ is a quasi-convex function, we have

$$\begin{aligned}&\left|\frac{qf(a)+f(b)}{1+q}-\frac{1}{b-a}\int_a^b f(x){}_ad_qx\right|\\ =&\left|\frac{q^2(b-a)^2}{1+q}\int_0^1 t(1-qt){}_aD_q^2f((1-t)a+tb){}_0d_qt\right|\\ \leq&\frac{q^2(b-a)^2}{1+q}\left(\int_0^1 {}_0d_qt\right)^{1-\frac{1}{r}}\left(\int_0^1 |t(1-qt)|^r\left|{}_aD_q^2f((1-t)a+tb)\right|^r {}_0d_qt\right)^{\frac{1}{r}}\\ \leq&\frac{q^2(b-a)^2}{1+q}\left(\int_0^1 {}_0d_qt\right)^{1-\frac{1}{r}}\left(\sup\{\left|{}_aD_q^2f(a)\right|^r,\left|{}_aD_q^2f(b)\right|^r\}\right)^{\frac{1}{r}}\left(\int_0^1 |t(1-qt)|^r {}_0d_qt\right)^{\frac{1}{r}}\end{aligned}$$

Applying Lemma 3, we have

$$\begin{aligned}&\left|\frac{qf(a)+f(b)}{1+q}-\frac{1}{b-a}\int_a^b f(x){}_ad_qx\right|\\ \leq&\frac{q^2(b-a)^2}{1+q}\left(\sup\{\left|{}_aD_q^2f(a)\right|^r,\left|{}_aD_q^2f(b)\right|^r\}\right)^{\frac{1}{r}}\left(\int_0^1 |t(1-qt)|^r {}_0d_qt\right)^{\frac{1}{r}}\\ =&\frac{q^2(b-a)^2}{1+q}\left(\sup\{\left|{}_aD_q^2f(a)\right|^r,\left|{}_aD_q^2f(b)\right|^r\}\right)^{\frac{1}{r}}(\mu_1)^{\frac{1}{r}}.\end{aligned}$$

It is easy to check that

$$\mu_1=\int_0^1 t^r(1-qt)^r {}_0d_qt=(1-q)\sum_{n=0}^{\infty}(q^n)^{r+1}(1-q^{n+1})^r,$$

thus, we get (23). □

Remark 7. *If $q\to 1$, then*

$$\mu_1=\int_0^1 t^r(1-t)^r {}_0d_qt=\beta(r+1,r+1),$$

and (23) *reduces to*

$$\left|\frac{f(a)+f(b)}{2}-\frac{1}{b-a}\int_a^b f(x)dx\right|\leq\frac{(b-a)^2}{2}(\beta(r+1,r+1))^{\frac{1}{r}}\left(\sup\{|f''(a)|^r,|f''(b)|^r\}\right)^{\frac{1}{r}}.$$

Corollary 6. *In Theorem 14, if r is a positive integer, then*

$$(1-qt)^r\leq(1-qt)_q^r,$$

and (23) *reduces to*

$$\left|\frac{qf(a)+f(b)}{1+q}-\frac{1}{b-a}\int_a^b f(x){}_ad_qx\right|\leq\frac{q^2(b-a)^2}{1+q}(\beta_q(r+1,r+1))^{\frac{1}{r}}\left(\sup\{\left|{}_aD_q^2f(a)\right|^r,\left|{}_aD_q^2f(b)\right|^r\}\right)^{\frac{1}{r}}.$$

Theorem 15. *Let $f : I = [a,b] \subset \mathbb{R} \to \mathbb{R}$ be a twice q-differentiable function on I^0 with ${}_aD_q^2 f$ be continuous and integrable on I where $0 < q < 1$. If $\left|{}_aD_q^2 f\right|^r$ is quasi-convex on $[a,b]$ for $r \geq 1$, then*

$$\left|\frac{qf(a)+f(b)}{1+q} - \frac{1}{b-a}\int_a^b f(x)_a d_q x\right| \leq \frac{q^2(b-a)^2}{1+q}\left(\frac{1}{1+q}\right)^{1-\frac{1}{r}}\left(\beta_q(r+1,2)\sup\{\left|{}_aD_q^2 f(a)\right|^r, \left|{}_aD_q^2 f(b)\right|^r\}\right)^{\frac{1}{r}}. \tag{24}$$

Proof. Using Lemma 7, Hölder's inequality and the fact that $\left|{}_aD_q^2 f\right|^r$ is a quasi-convex function, we have

$$\begin{aligned}
&\left|\frac{qf(a)+f(b)}{1+q} - \frac{1}{b-a}\int_a^b f(x)_a d_q x\right| \\
=&\left|\frac{q^2(b-a)^2}{1+q}\int_0^1 t(1-qt)_a D_q^2 f((1-t)a+tb)_0 d_q t\right| \\
\leq&\frac{q^2(b-a)^2}{1+q}\int_0^1 t(1-qt)\left|{}_aD_q^2 f((1-t)a+tb)\right|{}_0 d_q t \\
\leq&\frac{q^2(b-a)^2}{1+q}\left(\int_0^1 (1-qt)_0 d_q t\right)^{1-\frac{1}{r}}\left(\int_0^1 (1-qt)t^r\left|{}_aD_q^2 f((1-t)a+tb)\right|^r\right)^{\frac{1}{r}} \\
\leq&\frac{q^2(b-a)^2}{1+q}\left(\int_0^1 (1-qt)_0 d_q t\right)^{1-\frac{1}{r}}\left(\sup\{\left|{}_aD_q^2 f(a)\right|^r, \left|{}_aD_q^2 f(b)\right|^r\}\int_0^1 (1-qt)t^r{}_0 d_q t\right)^{\frac{1}{r}}
\end{aligned}$$

Applying Lemma 5 and the fact that $(1-qt) = (1-qt)_q^1$, we have

$$\begin{aligned}
&\left|\frac{qf(a)+f(b)}{1+q} - \frac{1}{b-a}\int_a^b f(x)_a d_q x\right| \\
\leq&\frac{q^2(b-a)^2}{1+q}\left(\frac{1}{1+q}\right)^{1-\frac{1}{r}}\left(\sup\{\left|{}_aD_q^2 f(a)\right|^r, \left|{}_aD_q^2 f(b)\right|^r\}\int_0^1 t^r(1-qt)_q^1{}_0 d_q t\right)^{\frac{1}{r}} \\
=&\frac{q^2(b-a)^2}{1+q}\left(\frac{1}{1+q}\right)^{1-\frac{1}{r}}\left(\beta_q(r+1,2)\sup\{\left|{}_aD_q^2 f(a)\right|^r, \left|{}_aD_q^2 f(b)\right|^r\}\right)^{\frac{1}{r}},
\end{aligned}$$

thus, we gett (24). □

Remark 8. *If $q \to 1$, then*

$$\beta(r+1,2) = \int_0^1 t^r(1-t)_0 d_q t = \frac{1}{(r+1)(r+2)},$$

and (24) reduces to inequality (2) in Theorem 1.

Theorem 16. *Let $f : I = [a,b] \subset \mathbb{R} \to \mathbb{R}$ be a twice q-differentiable function on I^0 with ${}_aD_q^2 f$ be continuous and integrable on I where $0 < q < 1$. If $\left|{}_aD_q^2 f\right|^r$ is quasi-convex on $[a,b]$ where $p, r > 1$, $\frac{1}{p} + \frac{1}{r} = 1$, then*

$$\left|\frac{qf(a)+f(b)}{1+q} - \frac{1}{b-a}\int_a^b f(x)_a d_q x\right| \leq \frac{q^2(b-a)^2}{1+q}\left(\beta_q(p+1,2)\right)^{\frac{1}{p}}\left(\frac{\sup\{\left|{}_aD_q^2 f(a)\right|^r, \left|{}_aD_q^2 f(b)\right|^r\}}{1+q}\right)^{\frac{1}{r}}. \tag{25}$$

Proof. Using Lemma 7, Hölder's inequality and the fact that $\left|{}_aD_q^2 f\right|^r$ is a quasi-convex function, we have

$$\begin{aligned}
&\left|\frac{qf(a)+f(b)}{1+q}-\frac{1}{b-a}\int_a^b f(x)_a d_q x\right| \\
=&\left|\frac{q^2(b-a)^2}{1+q}\int_0^1 t(1-qt)_a D_q^2 f((1-t)a+tb)_0 d_q t\right| \\
\leq&\frac{q^2(b-a)^2}{1+q}\int_0^1 t(1-qt)\left|{}_aD_q^2 f((1-t)a+tb)\right|{}_0 d_q t \\
\leq&\frac{q^2(b-a)^2}{1+q}\left(\int_0^1 t^p(1-qt)_0 d_q t\right)^{\frac{1}{p}}\left(\int_0^1 (1-qt)\left|{}_aD_q^2 f((1-t)a+tb)\right|^r {}_0 d_q t\right)^{\frac{1}{r}} \\
\leq&\frac{q^2(b-a)^2}{1+q}\left(\int_0^1 t^p(1-qt)_0 d_q t\right)^{\frac{1}{p}}\left(\sup\left\{\left|{}_aD_q^2 f(a)\right|^r,\left|{}_aD_q^2 f(b)\right|^r\right\}\int_0^1 (1-qt)_0 d_q t\right)^{\frac{1}{r}}
\end{aligned}$$

Applying Lemma 5 and the fact that $(1-qt) = (1-qt)_q^1$, we have

$$\begin{aligned}
&\left|\frac{qf(a)+f(b)}{1+q}-\frac{1}{b-a}\int_a^b f(x)_a d_q x\right| \\
\leq&\frac{q^2(b-a)^2}{1+q}\left(\int_0^1 t^p(1-qt)_q^1{}_0 d_q t\right)^{\frac{1}{p}}\left(\frac{\sup\left\{\left|{}_aD_q^2 f(a)\right|^r,\left|{}_aD_q^2 f(b)\right|^r\right\}}{1+q}\right)^{\frac{1}{r}} \\
=&\frac{q^2(b-a)^2}{1+q}\left(\beta_q(p+1,2)\right)^{\frac{1}{p}}\left(\frac{\sup\left\{\left|{}_aD_q^2 f(a)\right|^r,\left|{}_aD_q^2 f(b)\right|^r\right\}}{1+q}\right)^{\frac{1}{r}},
\end{aligned}$$

thus, we get (25). □

Remark 9. *If $q \to 1$, then*

$$\beta(p+1,2) = \int_0^1 t^p(1-t)dt = \int_0^1 s(1-s)^p ds = \beta(2,p+1).$$

Inequality (25) reduces to inequality (3) in Theorem 2.

Theorem 17. *Let $f : I = [a,b] \subset \mathbb{R} \to \mathbb{R}$ be a twice q-differentiable function on I^0 with ${}_aD_q^2 f$ be continuous and integrable on I where $0 < q < 1$. If $\left|{}_aD_q^2 f\right|$ is quasi-convex on $[a,b]$, then*

$$\left|\frac{qf(a)+f(b)}{1+q}-\frac{1}{b-a}\int_a^b f(x)_a d_q x\right| \leq \frac{q^2(b-a)^2\sup\left\{\left|{}_aD_q^2 f(a)\right|,\left|{}_aD_q^2 f(b)\right|\right\}}{(1+q)^2(1+q+q^2)}. \tag{26}$$

Proof. Using Lemma 7, Hölder's inequality and the fact that $\left|{}_aD_q^2 f\right|$ is a quasi-convex function, we have

$$\begin{aligned}
&\left|\frac{qf(a)+f(b)}{1+q} - \frac{1}{b-a}\int_a^b f(x)_a d_q x\right| \\
=&\left|\frac{q^2(b-a)^2}{1+q}\int_0^1 t(1-qt)_a D_q^2 f((1-t)a+tb)_0 d_q t\right| \\
\leq&\frac{q^2(b-a)^2}{1+q}\int_0^1 t(1-qt)\left|{}_aD_q^2 f((1-t)a+tb)\right|{}_0 d_q t \\
\leq&\frac{q^2(b-a)^2}{1+q}\sup\{\left|{}_aD_q^2 f(a)\right|,\left|{}_aD_q^2 f(b)\right|\}\int_0^1 t(1-qt)_0 d_q t
\end{aligned}$$

Applying Lemma 6, we have

$$\begin{aligned}
&\left|\frac{qf(a)+f(b)}{1+q} - \frac{1}{b-a}\int_a^b f(x)_a d_q x\right| \\
\leq&\frac{q^2(b-a)^2\sup\{\left|{}_aD_q^2 f(a)\right|,\left|{}_aD_q^2 f(b)\right|\}}{(1+q)^2(1+q+q^2)},
\end{aligned}$$

thus, we get (26). □

Remark 10. *If $q \to 1$, then inequality (26) reduces to inequality (4) in Theorem 3.*

Theorem 18. *Let $f : I = [a,b] \subset \mathbb{R} \to \mathbb{R}$ be a twice q-differentiable function on I^0 with ${}_aD_q^2 f$ be continuous and integrable on I where $0 < q < 1$. If $\left|{}_aD_q^2 f\right|^r$ is quasi-convex on $[a,b]$ for $r \geq 1$, then the following inequality holds:*

$$\left|\frac{qf(a)+f(b)}{1+q} - \frac{1}{b-a}\int_a^b f(x)_a d_q x\right| \leq \frac{q^2(b-a)^2}{(1+q)^2(1+q+q^2)}\left(\sup\{\left|{}_aD_q^2 f(a)\right|^r,\left|{}_aD_q^2 f(b)\right|^r\}\right)^{\frac{1}{r}}. \tag{27}$$

Proof. Using Lemma 7, Hölder's inequality and the fact that $\left|{}_aD_q^2 f\right|^r$ is a quasi-convex function, we have

$$\begin{aligned}
&\left|\frac{qf(a)+f(b)}{1+q} - \frac{1}{b-a}\int_a^b f(x)_a d_q x\right| \\
=&\left|\frac{q^2(b-a)^2}{1+q}\int_0^1 t(1-qt)_a D_q^2 f((1-t)a+tb)_0 d_q t\right| \\
\leq&\frac{q^2(b-a)^2}{1+q}\int_0^1 t(1-qt)\left|{}_aD_q^2 f((1-t)a+tb)\right|{}_0 d_q t \\
\leq&\frac{q^2(b-a)^2}{1+q}\left(\int_0^1 t(1-qt)_0 d_q t\right)^{1-\frac{1}{r}}\left(\int_0^1 t(1-qt)\left|{}_aD_q^2 f((1-t)a+tb)\right|^r {}_0 d_q t\right)^{\frac{1}{r}} \\
\leq&\frac{q^2(b-a)^2}{1+q}\left(\int_0^1 t(1-qt)_0 d_q t\right)^{1-\frac{1}{r}}\left(\sup\{\left|{}_aD_q^2 f(a)\right|^r,\left|{}_aD_q^2 f(b)\right|^r\}\int_0^1 t(1-qt)_0 d_q t\right)^{\frac{1}{r}}
\end{aligned}$$

Applying Lemma 6, we have

$$\begin{aligned}
&\left|\frac{qf(a)+f(b)}{1+q}-\frac{1}{b-a}\int_a^b f(x)_a d_q x\right| \\
\leq&\frac{q^2(b-a)^2}{1+q}\left(\frac{1}{(1+q)(1+q+q^2)}\right)^{1-\frac{1}{r}}\left(\frac{\sup\{\left|{}_aD_q^2 f(a)\right|^r,\left|{}_aD_q^2 f(b)\right|^r\}}{(1+q)(1+q+q^2)}\right)^{\frac{1}{r}} \\
=&\frac{q^2(b-a)^2}{(1+q)^2(1+q+q^2)}\left(\sup\{\left|{}_aD_q^2 f(a)\right|^r,\left|{}_aD_q^2 f(b)\right|^r\}\right)^{\frac{1}{r}},
\end{aligned}$$

thus, we get (27). □

Remark 11. *If $q \to 1$, then inequality (27) reduces to inequality (6) in Theorem 5.*

4. Discussion of New Perspectives

Currently, the Hermite-Hadamard inequality plays a significant role in the development of all fields of Mathematics. It has sgnificant applications in a variety of applied Mathematics, such as integral inequalities, approximation theory, special means theory, optimization theory, information theory and numerical analysis. In recent years, a number of authors have discovered new Hermite-Hadamard-type inequalities for convex, s-convex functions, logarithmic convex functions, h-convex functions, quasi-convex functions, m-convex functions, (K, m)-convex functions, co-ordinated convex functions, and the Godunova-Levin function, P-function, and so on. In this paper, we use a new quantum integral identity established in [11] (Lemma 4.1) to develop some quantum estimates for Hermite-Hadamard type inequalities in which some quasi-convex functions are involved.

Since quantum calculus has large applications in many mathematical areas such as number theory, special functions, quantum mechanics and mathematical inequalities, we hope interested readers will continue to explore more quantum estimates of Hermite-Hadamard type inequalities for other kinds of convex functions, and, furthermore, to find applications in the above-mentioned mathematical areas.

Author Contributions: The work presented here was carried out in collaboration between all authors. All authors contributed equally and significantly in writing this article. All authors have read and approved the final manuscript.

Funding: This research was funded by the National Natural Science Foundation of China (11771216), the Natural Science Foundation of Jiangsu Province (BK20151523), the Six Talent Peaks Project in Jiangsu Province (2015-XCL-020), and the Qing Lan Project of Jiangsu Province.

Acknowledgments: The authors thank the referees for their valuable suggestions and remarks.

Conflicts of Interest: The authors declare no conflict of interest.

References

1. Pečarić, J.E.; Proschan, F.; Tong, Y.L. *Convex Functions, Partial Orderings, and Statistical Applications*; Mathematics in Science and Engineering, 187; Academic Press, Inc.: Boston, MA, USA, 1992.
2. Alomari, M.; Darus, M.; Dragomir, S.S. New inequalities of Hermite-Hadamard type for functions whose second derivatives absolute values are quasi-convex. *Tamkang J. Math.* **2010**, *41*, 353–359.
3. Dragomir, S.S.; Agarwal, R.P. Two inequalities for differentiable mappings and applications to special means of real numbers and to trapezoidal formula. *Appl. Math. Lett.* **1998**, *11*, 91–95. [CrossRef]
4. Dragomir, S.S. On some new inequalities of Hermite-Hadamard type for m-convex functions. *Tamkang J. Math.* **2002**, *33*, 55–65.
5. Dragomir, S.S.; Fitzpatrick, S. The Hadamard inequalities for s-convex functions in the second sense. *Demonstratio Math.* **1999**, *32*, 687–696. [CrossRef]
6. Ion, D.A. Some estimates on the Hermite-Hadamard inequality through quasi-convex functions. *Ann. Univ. Craiova Ser. Mat. Inform.* **2007**, *34*, 83–88.

7. Liu, W.J. New integral inequalities via (α, m)-convexity and quasi-convexity. *Hacet. J. Math. Stat.* **2013**, *42*, 289–297.
8. Liu, W.J. Some Simpson type inequalities for h-convex and (α, m)-convex functions. *J. Comput. Anal. Appl.* **2014**, *16*, 1005–1012.
9. Liu, W.J. Ostrowski type fractional integral inequalities for MT-convex functions. *Miskolc Math. Notes* **2015**, *16*, 249–256. [CrossRef]
10. Liu, W.J.; Wen, W.S.; Park, J.K. Hermite-Hadamard type inequalities for MT-convex functions via classical integrals or fractional integrals. *J. Nonlinear Sci. Appl.* **2016**, *9*, 766–777. [CrossRef]
11. Liu, W.J.; Zhuang, H.F. Some quantum estimates of Hermite-Hadamard inequalities for convex functions. *J. Appl. Anal. Comput.* **2017**, *7*, 501–522.
12. Liu, Z. Generalization and improvement of some Hadamard type inequalities for Lipschitzian mappings. *J. Pure Appl. Math. Adv. Appl.* **2009**, *1*, 175–181.
13. Noor, M.A.; Noor, K.I.; Awan, M.U. Some quantum estimates for Hermite-Hadamard inequalities. *Appl. Math. Comput.* **2015**, *251*, 675–679. [CrossRef]
14. Noor, M.A.; Noor, K.I.; Awan, M.U. Some quantum integral inequalities via preinvex functions. *Appl. Math. Comput.* **2015**, *269*, 242–251. [CrossRef]
15. Noor, M.A.; Noor, K.I.; Awan, M.U. Quantum analogues of Hermite-Hadamard type inequalities for generalized convexity. In *Computation, Cryptography and Network Security*; Daras, N., Rassias, M.T., Eds.; Springer: Cham, Switzerland, 2015; pp. 413–439.
16. Özdemir, M.E. On Iyengar-type inequalities via quasi-convexity and quasi-concavity. *Miskolc Math. Notes* **2014**, *15*, 171–181. [CrossRef]
17. Sarikaya, M.Z.; Set, E.; Özdemir, M.E. On some new inequalities of Hadamard type involving h-convex functions. *Acta Math. Univ. Comen. (N.S.)* **2010**, *79*, 265–272.
18. Sudsutad, W.; Ntouyas, S.K.; Tariboon, J. Quantum integral inequalities for convex functions. *J. Math. Inequal.* **2015**, *9*, 781–793. [CrossRef]
19. Tariboon, J.; Ntouyas, S.K. Quantum integral inequalities on finite intervals. *J. Inequal. Appl.* **2014**, *2014*, 121. [CrossRef]
20. Tariboon, J.; Ntouyas, S.K. Quantum calculus on finite intervals and applications to impulsive difference equations. *Adv. Differ. Equ.* **2013**, *2013*, 282. [CrossRef]
21. Al-Salam, W.A. q-Bernoulli numbers and polynomials. *Math. Nachr.* **1959**, *17*, 239–260. [CrossRef]
22. Kac, V.; Cheung, P. *Quantum Calculus*; Universitext; Springer: New York, NY, USA, 2002.
23. Von Neumann, J. *Mathematical Foundations of Quantum Mechanics*, new ed.; translated from the German and with a preface by Robert T. Beyer; Princeton University Press: Princeton, NJ, USA, 2018.
24. Alp, N.; Sarikaya, M.Z.; Kunt, M.; Iscan, I. q-Hermite Hadamard inequalities and quantum estimates for midpoint type inequalities via convex and quasi-convex functions. *J. King Saud Univ. Sci.* **2018**, *30*, 193–203. [CrossRef]

mathematics

MDPI

Article

New Refinement of the Operator Kantorovich Inequality

Hamid Reza Moradi [1,*], Shigeru Furuichi [2] and Zahra Heydarbeygi [1]

1 Department of Mathematics, Payame Noor University (PNU), P.O. Box 19395-4697, Tehran, Iran; zheydarbeygi@yahoo.com
2 Department of Information Science, College of Humanities and Sciences, Nihon University, Tokyo 102-0074, Japan; furuichi@chs.nihon-u.ac.jp
* Correspondence: hrmoradi@mshdiau.ac.ir or hrmoradi.68@gmail.com; Tel.: +98-915-529-5869

Received: 7 December 2018; Accepted: 24 January 2019; Published: 1 February 2019

Abstract: We focus on the improvement of operator Kantorovich type inequalities. Among the consequences, we improve the main result of the paper [H.R. Moradi, I.H. Gümüş, Z. Heydarbeygi, A glimpse at the operator Kantorovich inequality, Linear Multilinear Algebra, doi:10.1080/03081087.2018.1441799].

Keywords: operator inequality; positive linear map; operator Kantorovich inequality; geometrically convex function

MSC: Primary 47A63; Secondary 46L05; 47A60

---

1. Notation and Preliminaries

At the beginning of this paper, we cite the following inequality which is called the operator Kantorovich inequality [1]:

$$\Phi\left(A^{-1}\right) \le \frac{(M+m)^2}{4Mm}\Phi(A)^{-1} \tag{1}$$

where Φ is a normalized positive linear map from $\mathcal{B}(\mathcal{H})$ to $\mathcal{B}(\mathcal{K})$, (we represent $\mathcal{H}$ and $\mathcal{K}$ as complex Hilbert spaces throughout the paper) and A is a positive operator with spectrum contained in $[m, M]$ with $0 < m < M$. This is a non-commutative analogue of the classical inequality [2],

$$\langle Ax, x\rangle \left\langle A^{-1}x, x\right\rangle \le \frac{(M+m)^2}{4Mm}$$

where $x \in \mathcal{H}$ is a unit vector.

In recent years, various attempts have been made by many authors to improve and generalize the operator Kantorovich inequality. One may see the basic references [3–5] and the excellent survey [6] on this topic. In [7], it was shown that

$$\Phi\left(A^{-1}\right) \le \Phi\left(m^{\frac{A-MI}{M-m}} M^{\frac{mI-A}{M-m}}\right) \le \frac{(M+m)^2}{4Mm}\Phi(A)^{-1}. \tag{2}$$

The main aim of the present short paper is to improve both inequalities in (2). Actually, we prove that

$$\begin{aligned}\Phi\left(A^{-1}\right) &\le \Phi\left(\left(A-\left(\sqrt{m}-\sqrt{M}\right)^{2}r(A)\right)^{-1}\right)\\ &\le \Phi\left(\left(m^{\frac{A-MI}{M-m}}M^{\frac{mI-A}{M-m}}\right)^{-1}\right)\\ &\le \frac{(M+m)^2}{4Mm}\Phi(A)^{-1}-\left(\frac{\left(\sqrt{M}-\sqrt{m}\right)^2}{Mm}\right)r(A)\end{aligned}$$

where $r(A)=\min\left\{\frac{MI-A}{M-m},\frac{A-mI}{M-m}\right\}=\frac{1}{2}I-\frac{1}{M-m}\left|A-\frac{M+m}{2}I\right|$.

In what follows, an operator means a bounded linear one acting on a complex Hilbert space $\mathcal{H}$. As customary, we reserve m, M for scalars and I for the identity operator. A self-adjoint operator A is said to be positive if $\langle Ax,x\rangle\ge 0$ holds for all $x\in\mathcal{H}$. A linear map Φ is positive if $\Phi(A)\ge 0$ whenever $A\ge 0$. It is said to be normalized if $\Phi(I)=I$. We denote by $\sigma(A)$ the spectrum of the operator A.

2. Main Results

Before we present the proof of our theorems, we begin with a general observation. We say that a non-negative function f on $[0,\infty)$ is geometrically convex [8] when

$$f\left(a^{1-v}b^{v}\right)\le f(a)^{1-v}f(b)^{v} \tag{3}$$

for all $a,b>0$ and $v\in[0,1]$. Equivalently, a function f is geometrically convex if and only if the associated function $F(y)=\log(f(e^y))$ is convex.

Example 1 ([9] Example 2.12). *Given real numbers $c_i\ge 0$ and $p_i\in(-\infty,0]\cup[1,\infty)$ for $i=1,\cdots,n$, the function $f(t)=\sum_{i=1}^{n}c_it^{p_i}$ is geometrically convex on $(0,\infty)$.*

Kittaneh and Manasrah [10] Theorem 2.1 obtained a refinement of the weighted arithmetic-geometric mean inequality as follows:

$$a^{1-v}b^{v}\le(1-v)a+vb-r\left(\sqrt{a}-\sqrt{b}\right)^2 \tag{4}$$

where $r=\min\{v,1-v\}$.

Now, if f is a decreasing geometrically convex function, then

$$\begin{aligned}f((1-v)a+vb) &\le f\left(((1-v)a+vb)-r\left(\sqrt{a}-\sqrt{b}\right)^2\right)\\ &\le f\left(a^{1-v}b^{v}\right)\\ &\le f(a)^{1-v}f(b)^{v}\\ &\le (1-v)f(a)+vf(b)-r\left(\sqrt{f(a)}-\sqrt{f(b)}\right)^2\\ &\le (1-v)f(a)+vf(b)\end{aligned} \tag{5}$$

where the first inequality follows from the inequality $(1-v)\,a+vb-r\left(\sqrt{a}-\sqrt{b}\right)^2 \le (1-v)\,a+vb$ and the fact that f is decreasing function, in the second inequality we used (4), the third inequality is obvious by (3), and the fourth inequality again follows from (4) by interchanging a by $f(a)$ and b by $f(b)$.

Of course, each decreasing geometrically convex function is also convex. However, the converse does not hold in general.

The inequality (5) applied to $a=m, b=M, 1-v=\frac{M-t}{M-m}$, and $v=\frac{t-m}{M-m}$ gives

$$\begin{aligned} f(t) &\le f\left(t-\left(\sqrt{m}-\sqrt{M}\right)^2 r(t)\right) \\ &\le f\left(m^{\frac{M-t}{M-m}} M^{\frac{t-m}{M-m}}\right) \\ &\le f(m)^{\frac{M-t}{M-m}} f(M)^{\frac{t-m}{M-m}} \\ &\le \frac{M-t}{M-m} f(m) + \frac{t-m}{M-m} f(M) - \left(\sqrt{f(m)}-\sqrt{f(M)}\right)^2 r(t) \\ &\le \frac{M-t}{M-m} f(m) + \frac{t-m}{M-m} f(M) \end{aligned} \tag{6}$$

with $r(t)=\min\left\{\frac{t-m}{M-m}, \frac{M-t}{M-m}\right\}=\frac{1}{2}-\frac{1}{M-m}\left|t-\frac{M+m}{2}\right|$ whenever $t\in[m,M]$.

In order to establish our promised refinement of the operator Kantorovich inequality, we also use the well-known monotonicity principle for bounded self-adjoint operators on Hilbert space (see, e.g., [6] (p. 3)): If $A\in\mathcal{B}(\mathcal{H})$ is a self-adjoint operator, then

$$f(t)\le g(t),\ t\in\sigma(A)\ \Rightarrow\ f(A)\le g(A) \tag{7}$$

provided that f and g are real-valued continuous functions. Under the same assumptions, $h(t)=|t|$ implies $h(A)=|A|$.

Now, we are in a position to state and prove our main results. We remark that the following theorem can be regarded as an extension of [5] Remark 4.14 to the context of geometrical convex functions.

Theorem 1. *Let $A\in\mathcal{B}(\mathcal{H})$ be a self-adjoint operator with $\sigma(A)\subseteq[m,M]$ for some scalars m, M with $0<m<M$ and Φ be a normalized positive linear map from $\mathcal{B}(\mathcal{H})$ to $\mathcal{B}(\mathcal{K})$. If f is strictly positive decreasing geometrically convex function, then*

$$\begin{aligned} \Phi\left(f\left(A-\left(\sqrt{m}-\sqrt{M}\right)^2 r(A)\right)\right) &\le \Phi\left(f\left(m^{\frac{MI-A}{M-m}} M^{\frac{A-mI}{M-m}}\right)\right) \\ &\le \mu(m,M,f)\, f(\Phi(A)) - \left(\sqrt{f(m)}-\sqrt{f(M)}\right)^2 \Phi(r(A)) \end{aligned}$$

where $r(A)=\min\left\{\frac{A-mI}{M-m}, \frac{MI-A}{M-m}\right\}=\frac{1}{2}I-\frac{1}{M-m}\left|A-\frac{M+m}{2}I\right|$ and

$$\mu(m,M,f)=\max\left\{\frac{1}{f(t)}\left(\frac{M-t}{M-m}f(m)+\frac{t-m}{M-m}f(M)\right):\ t\in[m,M]\right\}.$$

Proof. On account of the assumptions, from parts of (6), we have

$$f\left(t-\left(\sqrt{m}-\sqrt{M}\right)^{2}r(t)\right) \le f\left(m^{\frac{M-t}{M-m}}M^{\frac{t-m}{M-m}}\right) \le L(t)-\left(\sqrt{f(m)}-\sqrt{f(M)}\right)^{2}r(t) \tag{8}$$

where

$$L(t)=\frac{M-t}{M-m}f(m)+\frac{t-m}{M-m}f(M).$$

Note that inequality (8) holds for all $t \in [m, M]$. On the other hand, $\sigma(A) \subseteq [m, M]$, which, by virtue of monotonicity principle (7) for operator functions, yields the series of inequalities

$$f\left(A-\left(\sqrt{m}-\sqrt{M}\right)^{2}r(A)\right) \le f\left(m^{\frac{MI-A}{M-m}}M^{\frac{A-mI}{M-m}}\right) \le L(A)-\left(\sqrt{f(m)}-\sqrt{f(M)}\right)^{2}r(A).$$

It follows from the linearity and the positivity of the map Φ that

$$\Phi\left(f\left(A-\left(\sqrt{m}-\sqrt{M}\right)^{2}r(A)\right)\right) \le \Phi\left(f\left(m^{\frac{MI-A}{M-m}}M^{\frac{A-mI}{M-m}}\right)\right) \le \Phi(L(A))-\left(\sqrt{f(m)}-\sqrt{f(M)}\right)^{2}\Phi(r(A)).$$

Now, by using [5] Corollary 4.12 we get

$$\begin{aligned}\Phi\left(f\left(A-\left(\sqrt{m}-\sqrt{M}\right)^{2}r(A)\right)\right) &\le \Phi\left(f\left(m^{\frac{MI-A}{M-m}}M^{\frac{A-mI}{M-m}}\right)\right)\\ &\le \Phi(L(A))-\left(\sqrt{f(m)}-\sqrt{f(M)}\right)^{2}\Phi(r(A))\\ &\le \mu(m,M,f)\,f(\Phi(A))-\left(\sqrt{f(m)}-\sqrt{f(M)}\right)^{2}\Phi(r(A)).\end{aligned}$$

This completes the proof. □

As discussed extensively in [6] Cahpter 2, for $f(t)=t^p$, we have

$$\begin{aligned}\mu(m,M,t^p) &= \max\left\{\frac{1}{t^p}\left(\frac{M-t}{M-m}m^p+\frac{t\quad m}{M-m}M^p\right) : t\in[m,M]\right\}\\ &= \frac{(mM^p-Mm^p)}{(p-1)(M-m)}\left(\frac{p-1}{p}\frac{M^p-m^p}{mM^p-Mm^p}\right)^p.\end{aligned}$$

Now, the following fact can be easily deduced from Theorem 1 and Example 1.

Corollary 1. *Let $A \in \mathcal{B}(\mathcal{H})$ be a positive operator with $\sigma(A) \subseteq [m, M]$ for some scalars m, M with $0 < m < M$ and Φ be a normalized positive linear map from $\mathcal{B}(\mathcal{H})$ to $\mathcal{B}(\mathcal{K})$. Then for any $p < 0$,*

$$\begin{aligned}\Phi(A^p) &\leq \Phi\left(\left(A - \left(\sqrt{m} - \sqrt{M}\right)^2 r(A)\right)^p\right) \\ &\leq \Phi\left(\left(m^{\frac{A-MI}{M-m}} M^{\frac{mI-A}{M-m}}\right)^p\right) \\ &< K(m, M, p)\, \Phi(A)^p - \left(m^{p/2} - M^{p/2}\right)^2 \Phi(r(A))\end{aligned}$$

where

$$K(m, M, p) = \frac{(mM^p - Mm^p)}{(p-1)(M-m)} \left(\frac{p-1}{p} \frac{M^p - m^p}{mM^p - Mm^p}\right)^p.$$

In particular,

$$\begin{aligned}\Phi\left(A^{-1}\right) &\leq \Phi\left(\left(A - \left(\sqrt{m} - \sqrt{M}\right)^2 r(A)\right)^{-1}\right) \\ &\leq \Phi\left(\left(m^{\frac{A-MI}{M-m}} M^{\frac{mI-A}{M-m}}\right)^{-1}\right) \\ &\leq \frac{(M+m)^2}{4Mm} \Phi(A)^{-1} - \left(\frac{\left(\sqrt{M} - \sqrt{m}\right)^2}{Mm}\right) \Phi(r(A)).\end{aligned}$$

We note that $K(m, M, -1) = \frac{(M+m)^2}{4Mm}$ is the original Kantorovich constant.

Theorem 2. *Let all the assumptions of Theorem 1 hold. Then*

$$\begin{aligned}f\left(\Phi(A) - \left(\sqrt{m} - \sqrt{M}\right)^2 r(\Phi(A))\right) &\leq f\left(m^{\frac{MI-\Phi(A)}{M-m}} M^{\frac{\Phi(A)-mI}{M-m}}\right) \\ &\leq \mu(m, M, f)\, \Phi(f(A)) - \left(\sqrt{f(m)} - \sqrt{f(M)}\right)^2 r(\Phi(A)).\end{aligned}$$

Proof. By applying a standard functional calculus for the operator $\Phi(A)$ such that $mI \leq \Phi(A) \leq MI$, we get from (8)

$$\begin{aligned}f\left(\Phi(A) - \left(\sqrt{m} - \sqrt{M}\right)^2 r(\Phi(A))\right) &\leq f\left(m^{\frac{MI-\Phi(A)}{M-m}} M^{\frac{\Phi(A)-mI}{M-m}}\right) \\ &\leq \Phi(L(A)) - \left(\sqrt{f(m)} - \sqrt{f(M)}\right)^2 r(\Phi(A)).\end{aligned}$$

We thus have

$$\begin{aligned}
f\left(\Phi(A) - \left(\sqrt{m}-\sqrt{M}\right)^2 r(\Phi(A))\right) &\le f\left(m^{\frac{MI-\Phi(A)}{M-m}} M^{\frac{\Phi(A)-mI}{M-m}}\right) \\
&\le L(\Phi(A)) - \left(\sqrt{f(m)} - \sqrt{f(M)}\right)^2 r(\Phi(A)) \\
&= \Phi(L(A)) - \left(\sqrt{f(m)} - \sqrt{f(M)}\right)^2 r(\Phi(A)) \\
&\le \mu(m, M, f)\, \Phi(f(A)) - \left(\sqrt{f(m)} - \sqrt{f(M)}\right)^2 r(\Phi(A))
\end{aligned}$$

where at the last step we used the basic inequality [5] Corollary 4.12.

Hence, the proof is complete. □

As a corollary of Theorem 2 we have:

Corollary 2. *Let all the assumptions of Corollary 1 hold. Then for any $p < 0$*

$$\begin{aligned}
\Phi(A)^p &\le \left(\Phi(A) - \left(\sqrt{m}-\sqrt{M}\right)^2 r(\Phi(A))\right)^p \\
&\le \left(m^{\frac{MI-\Phi(A)}{M-m}} M^{\frac{\Phi(A)-mI}{M-m}}\right)^p \\
&\le K(m, M, p)\, \Phi(A^p) - \left(\sqrt{m^p}-\sqrt{M^p}\right)^2 r(\Phi(A)).
\end{aligned}$$

Remark 1. *Notice that the inequalities in Corollary 2 are stronger than the inequalities obtained in [11] Corollary 2.1.*

Recall that if f is operator convex, the solidarities [12] or the perspective [13] of f is defined by

$$\mathcal{P}_f(A \mid B) = A^{\frac{1}{2}} f\left(A^{-\frac{1}{2}} B A^{-\frac{1}{2}}\right) A^{\frac{1}{2}}.$$

Using a series of inequalities (6) we have the upper bounds of the perspective for non-negative decreasing geometrically convex function (not necessary operator convex f). We use the same symbol $\mathcal{P}_f(A \mid B)$ for a simplicity.

Proposition 1. *Let $A, B > 0$ with $mA \le B \le MA$ for some scalars $0 < m < M$. For a non-negative decreasing geometrically convex function f, we have*

$$\begin{aligned}
\mathcal{P}_f(A \mid B) &\le A^{1/2} f\left(A^{-1/2} B A^{-1/2} - \left(\sqrt{m}-\sqrt{M}\right)^2 r(A, B)\right) A^{1/2} \\
&\le A^{1/2} f\left(m^{\frac{MI-A^{-1/2}BA^{-1/2}}{M-m}} M^{\frac{A^{-1/2}BA^{-1/2}-mI}{M-m}}\right) A^{1/2} \\
&\le A^{1/2} f(m)^{\frac{MI-A^{-1/2}BA^{-1/2}}{M-m}} f(M)^{\frac{A^{-1/2}BA^{-1/2}-mI}{M-m}} A^{1/2} \\
&\le \frac{Mf(m)-mf(M)}{M-m} A + \frac{f(M)-f(m)}{M-m} B - \left(\sqrt{f(m)}-\sqrt{f(M)}\right)^2 A^{1/2} r(A, B) A^{1/2} \\
&\le \frac{Mf(m)-mf(M)}{M-m} A + \frac{f(M)-f(m)}{M-m} B,
\end{aligned}$$

where

$$r(A,B) = \min\left\{ \frac{A^{-1/2}BA^{-1/2} - mI}{M-m}, \frac{MI - A^{-1/2}BA^{-1/2}}{M-m} \right\}$$
$$= \frac{1}{2}I - \frac{1}{M-m}\left| A^{-1/2}BA^{-1/2} - \frac{M+m}{2}I \right|.$$

Author Contributions: The work presented here was carried out in collaboration between all authors. The study was initiated by the first author. The first author played also the role of the corresponding author. All authors contributed equally and significantly in writing this article. All authors have read and approved the final manuscript.

Funding: This research was funded by JSPS KAKENHI Grant Number 16K05257.

Acknowledgments: The authors would like to express their hearty thanks to the referees for their valuable comments.

Conflicts of Interest: The authors declare no conflict of interest.

References

1. Marshall, A.W.; Olkin, I. Matrix versions of Cauchy and Kantorovich inequalities. *Aequ. Math.* **1990**, *40*, 89–93. [CrossRef]
2. Kantorovich, L.V. Functional analysis and applied mathematics. *Uspehi Matem. Nauk* **1948**, *3*, 89–185. (In Russian)
3. Bourin, J.C. Matrix versions of some classical inequalities. *Linear Algebra Appl.* **2006**, *416*, 890–907. [CrossRef]
4. Fujii, M.; Zuo, H.; Cheng, N. Generalization on Kantorovich inequality. *J. Math. Inequal.* **2013**, *7*, 517–522. [CrossRef]
5. Mićić, J.; Carić, J.P.; Seo, Y.; Tominaga, M. Inequalities for positive linear maps on Hermitian matrices. *J. Math. Inequal. Appl.* **2000**, *3*, 559–591.
6. Furuta, T.; Mićić, J.; Carić, J.P.; Seo, Y. *Mond–Pečarić Method in Operator Inequalities*; Element: Guernsey, France, 2005.
7. Moradi, H.R.; Gümüş, I.H.; Heydarbeygi, Z. A glimpse at the operator Kantorovich inequality. *Linear Multilinear Algebra* **2018**. [CrossRef]
8. Montel, P. Sur les functions convexes et les fonctions sousharmoniques. *J. Math.* **1928**, *9*, 29–60.
9. Bourin, J.C.; Hiai, F. Jensen and Minkowski inequalities for operator means and anti–norms. *Linear Algebra Appl.* **2014**, *456*, 22–53. [CrossRef]
10. Kittaneh, F.; Manasrah, Y. Improved Young and Heinz inequalities for matrices. *J. Math. Anal. Appl.* **2010**, *361*, 262–269. [CrossRef]
11. Sababheh, M.; Moradi, H.R.; Furuichi, S. Exponential inequalities for positive linear mappings. *J. Funct. Spaces* **2018**, *2018*, 5467413. [CrossRef]
12. Fujii, J.I.; Fujii, M.; Seo, Y. An extension of the Kubo–Ando theory: Soridarities. *Math. Japonica* **1990**, *35*, 387–396.
13. Ebadian, A.; Nikoufar, I.; Gordji, M.E. Perspectives of matrix convex functions. *Proc. Natl. Acad. Sci. USA* **2011**, *108*, 7313–7314. [CrossRef]

mathematics

MDPI

Article

Refinements of Majorization Inequality Involving Convex Functions via Taylor's Theorem with Mean Value form of the Remainder

Shanhe Wu [1,*], Muhammad Adil Khan [2] and Hidayat Ullah Haleemzai [2]

[1] Department of Mathematics, Longyan University, Longyan 364012, China
[2] Department of Mathematics, University of Peshawar, Peshawar 25000, Pakistan
* Correspondence: shanhewu@gmail.com

Received: 16 June 2019; Accepted: 22 July 2019; Published: 24 July 2019

Abstract: The aim of this paper is to establish some refined versions of majorization inequality involving twice differentiable convex functions by using Taylor theorem with mean-value form of the remainder. Our results improve several results obtained in earlier literatures. As an application, the result is used for deriving a new fractional inequality.

Keywords: majorization inequality; twice differentiable convex functions; refined inequality; Taylor theorem

MSC: 26A51; 26D15; 26D20

---

1. Introduction

The notion of majorization was introduced in the celebrated monograph [1] by Hardy, Littlewood and Pólya, which was used as a measure of the diversity of the components of an n-dimensional vector.

Let $\boldsymbol{\nu} = (\nu_1, \nu_2, \ldots, \nu_n)$ and $\boldsymbol{\vartheta} = (\vartheta_1, \vartheta_2, \ldots, \vartheta_n)$ be two n-tuples. The n-tuple $\boldsymbol{\nu}$ is said to be majorized by $\boldsymbol{\vartheta}$ (in symbols $\boldsymbol{\nu} \prec \boldsymbol{\vartheta}$) if $\sum_{i=1}^{k} \nu_{[i]} \leq \sum_{i=1}^{k} \vartheta_{[i]}$ for $k = 1, 2, \ldots, n-1$ and $\sum_{i=1}^{n} \nu_i = \sum_{i=1}^{n} \vartheta_i$, where $\nu_{[1]} \geq \nu_{[2]} \geq \cdots \geq \nu_{[n]}$ and $\vartheta_{[1]} \geq \vartheta_{[2]} \geq \cdots \geq \vartheta_{[n]}$ are rearrangements of $\boldsymbol{\nu}$ and $\boldsymbol{\vartheta}$ in a descending order.

The majorization has been found many applications in different fields of mathematics. A survey of the applications of majorization and relevant results can be found in the monograph of Marshall and Olkin [2]. Recently, the authors have given considerable attention to the generalizations and applications of the majorization and related inequalities, for details, we refer the reader to our papers [3–13].

In this paper we focus on a type of majorization inequality involving convex functions, which reveals the correlations among majorization, convex functions and inequalities. Now, let us recall briefly this type of majorization inequality.

The following classical majorization inequality can be found in the monographs of Marshall and Olkin [2] and Pečarić et al. [14].

Theorem 1. *Let $\boldsymbol{\nu} = (\nu_1, \nu_2, \ldots, \nu_n)$, $\boldsymbol{\vartheta} = (\vartheta_1, \vartheta_2, \ldots, \vartheta_n)$ be two n-tuples, $\nu_i, \vartheta_i \in I$ $(i = 1, 2, \ldots, n)$, I is an interval. Then*

$$\sum_{i=1}^{n} \Psi(\nu_i) \leq \sum_{i=1}^{n} \Psi(\vartheta_i) \tag{1}$$

holds for every continuous convex function $\Psi : I \to \mathbb{R}$ if and only if $\boldsymbol{\nu} \prec \boldsymbol{\vartheta}$ holds.

Fuchs [15] gave a weighted generalization of the majorization theorem, as follows:

Theorem 2. *Let $v = (v_1, v_2, \ldots, v_n)$, $\boldsymbol{\vartheta} = (\vartheta_1, \vartheta_2, \ldots, \vartheta_n)$ be two decreasing n-tuples, $v_i, \vartheta_i \in I$ $(i = 1, 2, \ldots, n)$, I is an interval. Suppose $\ell_1, \ell_2, \ldots, \ell_n$ are real numbers such that $\sum_{i=1}^{k} \ell_i v_i \leq \sum_{i=1}^{k} \ell_i \vartheta_i$ for $k = 1, 2, \ldots, n-1$ and $\sum_{i=1}^{n} \ell_i v_i = \sum_{i=1}^{n} \ell_i \vartheta_i$. Then*

$$\sum_{i=1}^{n} \ell_i \Psi(v_i) \leq \sum_{i=1}^{n} \ell_i \Psi(\vartheta_i) \tag{2}$$

holds for any continuous convex function $\Psi : I \to \mathbb{R}$.

Bullen, Vasić, and Stanković [16] presented a result similar to the above result, in which the condition of the tuples v, $\boldsymbol{\vartheta}$ is relaxed and the condition of the function Ψ is intensified.

Theorem 3. *Let $v = (v_1, v_2, \ldots, v_n)$, $\boldsymbol{\vartheta} = (\vartheta_1, \vartheta_2, \ldots, \vartheta_n)$ be two decreasing n-tuples, $v_i, \vartheta_i \in I$ $(i = 1, 2, \ldots, n)$, I is an interval. Suppose $\ell_1, \ell_2, \ldots, \ell_n$ are real numbers such that $\sum_{i=1}^{k} \ell_i v_i \leq \sum_{i=1}^{k} \ell_i \vartheta_i$ for $k = 1, 2, \ldots, n$. If $\Psi : I \to \mathbb{R}$ is a continuous increasing convex function, then*

$$\sum_{i=1}^{n} \ell_i \Psi(v_i) \leq \sum_{i=1}^{n} \ell_i \Psi(\vartheta_i). \tag{3}$$

The aim of this paper is to establish the refinements of majorization inequalities of Theorems 1–3. To achieve this, we will first establish an equality by using Taylor theorem with mean-value form of the remainder, which enables us to deduce the refined versions of majorization inequalities mentioned above.

2. Lemma

Lemma 1. *Let $v = (v_1, v_2, \ldots, v_n)$, $\boldsymbol{\vartheta} = (\vartheta_1, \vartheta_2, \ldots, \vartheta_n)$ be two n-tuples, $v_i, \vartheta_i \in (a, b)$ $(i = 1, 2, \ldots, n)$, and let $\ell_1, \ell_2, \ldots, \ell_n$ be real numbers. If $\Psi : [a, b] \to \mathbb{R}$ is a function such that $\Psi' \in C[a, b]$ and Ψ'' exists on (a, b), then there exists τ_i between v_i and ϑ_i satisfying*

$$\sum_{i=1}^{n} \ell_i \Psi(\vartheta_i) - \sum_{i=1}^{n} \ell_i \Psi(v_i) = \sum_{i=1}^{n} \Psi'(v_i) \ell_i (\vartheta_i - v_i) + \sum_{i=1}^{n} \frac{\Psi''(\tau_i)}{2} \ell_i (\vartheta_i - v_i)^2. \tag{4}$$

Proof. Using the Taylor's formula with the Lagrange remainder (mean-value form of the remainder) gives

$$\Psi(\vartheta_i) = \Psi(v_i) + \frac{\Psi'(v_i)}{1!}(\vartheta_i - v_i) + \frac{\Psi''(\tau_i)}{2!}(\vartheta_i - v_i)^2, \tag{5}$$

where $v_i, \vartheta_i \in (a, b)$, τ_i is a real number between v_i and ϑ_i $(i = 1, 2, \ldots, n)$.

Multiplying both sides of (5) by ℓ_i and taking summation over i $(i = 1, 2, \ldots, n)$, we get

$$\sum_{i=1}^{n} \ell_i \Psi(\vartheta_i) = \sum_{i=1}^{n} \ell_i \Psi(v_i) + \sum_{i=1}^{n} \Psi'(v_i) \ell_i (\vartheta_i - v_i) + \sum_{i=1}^{n} \frac{\Psi''(\tau_i)}{2} \ell_i (\vartheta_i - v_i)^2,$$

which is the desired equality (4). The proof of Lemma 1 is complete. □

3. Main Results

In this section, we establish some refinements of the majorization inequality.

Theorem 4. *Let $\nu = (\nu_1, \nu_2, \ldots, \nu_n)$, $\boldsymbol{\vartheta} = (\vartheta_1, \vartheta_2, \ldots, \vartheta_n)$ be two n-tuples, $\nu_i, \vartheta_i \in (a,b)$ $(i = 1,2,\ldots,n)$. If $\nu \prec \boldsymbol{\vartheta}$ and $\Psi : [a,b] \to \mathbb{R}$ is a twice differentiable convex function, then there exists a real number τ_i between $\nu_{[i]}$ and $\vartheta_{[i]}$ $(i = 1,2,\ldots,n)$ such that*

$$\sum_{i=1}^{n} \Psi(\vartheta_i) - \sum_{i=1}^{n} \Psi(\nu_i) \geq \sum_{i=1}^{n} \frac{\Psi''(\tau_i)}{2} (\vartheta_{[i]} - \nu_{[i]})^2. \tag{6}$$

where $\nu_{[1]} \geq \nu_{[2]} \geq \cdots \geq \nu_{[n]}$ and $\vartheta_{[1]} \geq \vartheta_{[2]} \geq \cdots \geq \vartheta_{[n]}$ are rearrangements of ν and $\boldsymbol{\vartheta}$ in a descending order.

Proof. Using Lemma 1 with $\ell_i = 1$, $\nu_i = \nu_{[i]}$, $\vartheta_i = \vartheta_{[i]}$ $(i = 1,2,\ldots,n)$, one has

$$\sum_{i=1}^{n} \Psi(\vartheta_{[i]}) - \sum_{i=1}^{n} \Psi(\nu_{[i]}) = \sum_{i=1}^{n} \Psi'(\nu_{[i]})(\vartheta_{[i]} - \nu_{[i]}) + \sum_{i=1}^{n} \frac{\Psi''(\tau_i)}{2} (\vartheta_{[i]} - \nu_{[i]})^2,$$

that is

$$\sum_{i=1}^{n} \Psi(\vartheta_i) - \sum_{i=1}^{n} \Psi(\nu_i) = \sum_{i=1}^{n} \Psi'(\nu_{[i]})(\vartheta_{[i]} - \nu_{[i]}) + \sum_{i=1}^{n} \frac{\Psi''(\tau_i)}{2} (\vartheta_{[i]} - \nu_{[i]})^2, \tag{7}$$

where $\nu_i, \vartheta_i \in (a,b)$, τ_i is a real number between $\nu_{[i]}$ and $\vartheta_{[i]}$ $(i = 1,2,\ldots,n)$.

Let

$$A_k = \sum_{i=1}^{k} \vartheta_{[i]}, \quad B_k = \sum_{i=1}^{k} \nu_{[i]} \quad (k = 1,2,\ldots,n), \quad A_0 = B_0 = 0.$$

Considering the first term in the right hand side of (7), we have

$$\begin{aligned}
\sum_{i=1}^{n} \Psi'(\nu_{[i]})(\vartheta_{[i]} - \nu_{[i]}) &= \sum_{i=1}^{n} \Psi'(\nu_{[i]})(A_i - A_{i-1} - B_i + B_{i-1}) \\
&= \sum_{i=1}^{n} \Psi'(\nu_{[i]})(A_i - B_i) - \sum_{i=1}^{n} \Psi'(\nu_{[i]})(A_{i-1} - B_{i-1}) \\
&= \Psi'(\nu_{[n]})(A_n - B_n) + \sum_{i=1}^{n-1} (\Psi'(\nu_{[i]}) - \Psi'(\nu_{[i+1]}))(A_i - B_i).
\end{aligned}$$

It follows from $\nu \prec \boldsymbol{\vartheta}$ that $A_n - B_n = 0$ and $A_i - B_i \geq 0$ for $i = 1,2,\ldots,n-1$.

Additionally, since Ψ is a continuous convex function on $[a,b]$, we deduce from $\nu_{[i]} \geq \nu_{[i+1]}$ $(i = 1,2,\ldots,n-1)$ that

$$\Psi'(\nu_{[i]}) - \Psi'(\nu_{[i+1]}) \geq 0 \quad \text{for} \quad i = 1,2,\ldots,n-1.$$

Hence

$$\sum_{i=1}^{n} \Psi'(\nu_{[i]})(\vartheta_{[i]} - \nu_{[i]}) \geq 0,$$

which, along with the equality (7), leads to the required inequality (6). This completes the proof of Theorem 4. □

Remark 1. *The inequality of Theorem 4 is a refinement of the inequality of Theorem 1, since the term $\sum_{i=1}^{n} \frac{\Psi''(\tau_i)}{2} (\vartheta_{[i]} - \nu_{[i]})^2$ in inequality (6) is nonnegative.*

In the following, we provide two refinements of majorization inequality by keeping one of the tuples decreasing (increasing).

Theorem 5. *Let $v = (v_1, v_2, \ldots, v_n)$, $\boldsymbol{\vartheta} = (\vartheta_1, \vartheta_2, \ldots, \vartheta_n)$ be two n-tuples, $v_i, \vartheta_i \in (a,b)$ $(i = 1,2,\ldots,n)$, let $\Psi : [a,b] \to \mathbb{R}$ be a twice differentiable convex function, and let $\ell_1, \ell_2, \ldots, \ell_n$ be real numbers such that $\sum_{i=1}^{k} \ell_i v_i \leq \sum_{i=1}^{k} \ell_i \vartheta_i$ for $k = 1,2,\ldots,n-1$ and $\sum_{i=1}^{n} \ell_i v_i = \sum_{i=1}^{n} \ell_i \vartheta_i$.*

(i) If v is a decreasing n-tuple, then there exists a real number τ_i between $v_{[i]}$ and $\vartheta_{[i]}$ $(i = 1,2,\ldots,n)$ such that

$$\sum_{i=1}^{n} \ell_i \Psi(\vartheta_i) - \sum_{i=1}^{n} \ell_i \Psi(v_i) \geq \sum_{i=1}^{n} \frac{\Psi''(\tau_i)}{2} \ell_i (\vartheta_i - v_i)^2. \tag{8}$$

(ii) If $\boldsymbol{\vartheta}$ is a increasing n-tuple, then there exists another real number σ_i between $v_{[i]}$ and $\vartheta_{[i]}$ $(i = 1,2,\ldots,n)$ such that

$$\sum_{i=1}^{n} \ell_i \Psi(v_i) - \sum_{i=1}^{n} \ell_i \Psi(\vartheta_i) \geq \sum_{i=1}^{n} \frac{\Psi''(\sigma_i)}{2} \ell_i (\vartheta_i - v_i)^2. \tag{9}$$

Proof. (i) It follows from Lemma 1 that

$$\sum_{i=1}^{n} \ell_i \Psi(\vartheta_i) - \sum_{i=1}^{n} \ell_i \Psi(v_i) = \sum_{i=1}^{n} \Psi'(v_i) \ell_i (\vartheta_i - v_i) + \sum_{i=1}^{n} \frac{\Psi''(\tau_i)}{2} \ell_i (\vartheta_i - v_i)^2, \tag{10}$$

where $v_i, \vartheta_i \in (a,b)$, τ_i is a real number between v_i and ϑ_i $(i = 1,2,\ldots,n)$. Let

$$A_k = \sum_{i=1}^{k} \ell_i \vartheta_i, \quad B_k = \sum_{i=1}^{k} \ell_i v_i \ \ (k = 1,2,\ldots,n), \quad A_0 = B_0 = 0.$$

Then, we have $A_i \geq B_i$ $(i = 1,2,\ldots,n-1)$, $A_n = B_n$, and

$$\sum_{i=1}^{n} \Psi'(v_i) \ell_i (\vartheta_i - v_i) = \sum_{i=1}^{n} \Psi'(v_i)(A_i - A_{i-1} - B_i + B_{i-1})$$

$$= \Psi'(v_n)(A_n - B_n) + \sum_{i=1}^{n-1} (\Psi'(v_i) - \Psi'(v_{i+1}))(A_i - B_i).$$

$$= \sum_{i=1}^{n-1} (\Psi'(v_i) - \Psi'(v_{i+1}))(A_i - B_i).$$

Noting that Ψ is a continuous convex function on $[a,b]$, and v is a decreasing n-tuple, we obtain $\Psi'(v_i) - \Psi'(v_{i+1}) \geq 0$ for $i = 1,2,\ldots,n-1$.

Hence

$$\sum_{i=1}^{n} \Psi'(v_i) \ell_i (\vartheta_i - v_i) \geq 0,$$

which, together with inequality (10), leads to the required inequality (8).

(ii) Similarly, we can prove the inequality (9) under the condition that $\boldsymbol{\vartheta}$ is an increasing n-tuple. The proof of Theorem 5 is complete. □

Remark 2. *The inequality (8) of Theorem 5 is a refinement of the inequality (2) of Theorem 2 in the case when $\ell_1, \ell_2, \ldots, \ell_n$ are positive numbers.*

Theorem 6. *Let $v = (v_1, v_2, \ldots, v_n)$, $\boldsymbol{\vartheta} = (\vartheta_1, \vartheta_2, \ldots, \vartheta_n)$ be two n-tuples, $v_i, \vartheta_i \in (a,b)$ $(i = 1,2,\ldots,n)$, let $\Psi : [a,b] \to \mathbb{R}$ be a twice differentiable and increasing convex function, and let $\ell_1, \ell_2, \ldots, \ell_n$ be real numbers such that $\sum_{i=1}^{k} \ell_i v_i \leq \sum_{i=1}^{k} \ell_i \vartheta_i$ for $k = 1,2,\ldots,n$. If v is a decreasing n-tuple, then there exists a real number τ_i between $v_{[i]}$ and $\vartheta_{[i]}$ $(i = 1,2,\ldots,n)$ such that*

$$\sum_{i=1}^{n} \ell_i \Psi(\vartheta_i) - \sum_{i=1}^{n} \ell_i \Psi(v_i) \geq \sum_{i=1}^{n} \frac{\Psi''(\tau_i)}{2} \ell_i (\vartheta_i - v_i)^2. \tag{11}$$

Proof. Let

$$A_k = \sum_{i=1}^{k} \ell_i \vartheta_i, \quad B_k = \sum_{i=1}^{k} \ell_i \nu_i \ (k = 1, 2, \ldots, n), \quad A_0 = B_0 = 0.$$

By Lemma 1, for any $\nu_i, \vartheta_i \in (a, b)$ $(i = 1, 2, \ldots, n)$, there exists a real number between ν_i and ϑ_i such that

$$\sum_{i=1}^{n} \ell_i \Psi(\vartheta_i) - \sum_{i=1}^{n} \ell_i \Psi(\nu_i) = \sum_{i=1}^{n} \Psi'(\nu_i) \ell_i (\vartheta_i - \nu_i) + \sum_{i=1}^{n} \frac{\Psi''(\tau_i)}{2} \ell_i (\vartheta_i - \nu_i)^2$$

$$= \sum_{i=1}^{n} \Psi'(\nu_i)(A_i - A_{i-1} - B_i + B_{i-1}) + \sum_{i=1}^{n} \frac{\Psi''(\tau_i)}{2} \ell_i (\vartheta_i - \nu_i)^2$$

$$= \Psi'(\nu_n)(A_n - B_n) + \sum_{i=1}^{n-1} (\Psi'(\nu_i) - \Psi'(\nu_{i+1}))(A_i - B_i) + \sum_{i=1}^{n} \frac{\Psi''(\tau_i)}{2} \ell_i (\vartheta_i - \nu_i)^2.$$

Since Ψ is a continuous convex function on $[a, b]$, and $\boldsymbol{\nu}$ is a decreasing n-tuple, we obtain $\Psi'(\nu_i) - \Psi'(\nu_{i+1}) \geq 0$ for $i = 1, 2, \ldots, n-1$. In addition, since Ψ is an increasing function on $[a, b]$, we get $\Psi'(\nu_n) \geq 0$. Now, by using the assumption conditions $A_i \geq B_i$ $(k = 1, 2, \ldots, n)$, we conclude that

$$\Psi'(\nu_n)(A_n - B_n) + \sum_{i=1}^{n-1} (\Psi'(\nu_i) - \Psi'(\nu_{i+1}))(A_i - B_i) \geq 0.$$

Therefore, we have

$$\sum_{i=1}^{n} \ell_i \Psi(\vartheta_i) - \sum_{i=1}^{n} \ell_i \Psi(\nu_i) \geq \sum_{i=1}^{n} \frac{\Psi''(\tau_i)}{2} \ell_i (\vartheta_i - \nu_i)^2.$$

The Theorem 6 is proved. □

Remark 3. *The inequality (11) of Theorem 6 is a refinement of the inequality (3) of Theorem 3 in the case when* $\ell_1, \ell_2, \ldots, \ell_n$ *are positive numbers.*

Theorem 7. *Let* $\boldsymbol{\nu} = (\nu_1, \nu_2, \ldots, \nu_n)$, $\boldsymbol{\vartheta} = (\vartheta_1, \vartheta_2, \ldots, \vartheta_n)$ *be two n-tuples,* $\nu_i, \vartheta_i \in (a, b)$ $(i = 1, 2, \ldots, n)$, *let* $\Psi : [a, b] \to \mathbb{R}$ *be a twice differentiable convex function, and let* $\ell_1, \ell_2, \ldots, \ell_n$ *be positive numbers. If* $\boldsymbol{\nu}$ *and* $\boldsymbol{\vartheta} - \boldsymbol{\nu}$ *are monotonic in the same sense, then there exists a real number* τ_i *between* $\nu_{[i]}$ *and* $\vartheta_{[i]}$ $(i = 1, 2, \ldots, n)$ *such that*

$$\sum_{i=1}^{n} \ell_i \Psi(\vartheta_i) - \sum_{i=1}^{n} \ell_i \Psi(\nu_i) \geq \frac{1}{\ell_1 + \ell_2 + \cdots + \ell_n} \sum_{i=1}^{n} \ell_i \Psi'(\nu_i) \sum_{i=1}^{n} \ell_i (\vartheta_i - \nu_i) + \sum_{i=1}^{n} \frac{\Psi''(\tau_i)}{2} \ell_i (\vartheta_i - \nu_i)^2. \tag{12}$$

Proof. Since Ψ is convex function, and tuple $\boldsymbol{\nu}$ and tuple $\boldsymbol{\vartheta} - \boldsymbol{\nu}$ are monotonic in the same sense, we conclude that $\Psi'(\boldsymbol{\nu})$ and $\boldsymbol{\vartheta} - \boldsymbol{\nu}$ are monotonic in the same sense.

Using the Chebyshev's inequality for weights $\ell_1, \ell_2, \ldots, \ell_n$, we obtain

$$(\sum_{i=1}^{n} \ell_i) \sum_{i=1}^{n} \ell_i \Psi'(\nu_i)(\vartheta_i - \nu_i) \geq \sum_{i=1}^{n} \ell_i \Psi'(\nu_i) \sum_{i=1}^{n} \ell_i (\vartheta_i - \nu_i).$$

On the other hand, by Lemma 1, for any $\nu_i, \vartheta_i \in (a, b)$ $(i = 1, 2, \ldots, n)$, there exists a real number τ_i between ν_i and ϑ_i such that

$$\sum_{i=1}^{n} \ell_i \Psi(\vartheta_i) - \sum_{i=1}^{n} \ell_i \Psi(\nu_i) = \sum_{i=1}^{n} \Psi'(\nu_i) \ell_i (\vartheta_i - \nu_i) + \sum_{i=1}^{n} \frac{\Psi''(\tau_i)}{2} \ell_i (\vartheta_i - \nu_i)^2.$$

Hence, we get

$$\sum_{i=1}^{n}\ell_i\Psi(\vartheta_i)-\sum_{i=1}^{n}\ell_i\Psi(\nu_i)\geq\frac{1}{\ell_1+\ell_2+\cdots+\ell_n}\sum_{i=1}^{n}\ell_i\Psi'(\nu_i)\sum_{i=1}^{n}\ell_i(\vartheta_i-\nu_i)$$
$$+\sum_{i=1}^{n}\frac{\Psi''(\tau_i)}{2}\ell_i(\vartheta_i-\nu_i)^2.$$

This proves the required inequality (12) in Theorem 7. □

Applying an additional condition $\sum_{i=1}^{n}\ell_i\nu_i \leq \sum_{i=1}^{n}\ell_i\vartheta_i$ to inequality (12), we obtain the following result.

Corollary 1. *Let $\boldsymbol{\nu}=(\nu_1,\nu_2,\ldots,\nu_n)$, $\boldsymbol{\vartheta}=(\vartheta_1,\vartheta_2,\ldots,\vartheta_n)$ be two n-tuples, $\nu_i,\vartheta_i\in(a,b)$ $(i=1,2,\ldots,n)$, let $\Psi:[a,b]\to\mathbb{R}$ be a twice differentiable and increasing convex function, and let $\ell_1,\ell_2,\ldots,\ell_n$ be positive numbers. If $\boldsymbol{\nu}$ and $\boldsymbol{\vartheta}-\boldsymbol{\nu}$ are monotonic in the same sense, and $\sum_{i=1}^{n}\ell_i\nu_i\leq\sum_{i=1}^{n}\ell_i\vartheta_i$, then there exists a real number τ_i between $\nu_{[i]}$ and $\vartheta_{[i]}$ $(i=1,2,\ldots,n)$ such that*

$$\sum_{i=1}^{n}\ell_i\Psi(\vartheta_i)-\sum_{i=1}^{n}\ell_i\Psi(\nu_i)\geq\sum_{i=1}^{n}\frac{\Psi''(\tau_i)}{2}\ell_i(\vartheta_i-\nu_i)^2. \tag{13}$$

4. An Application

In this section we establish a new fractional inequality to illustrate the application of our results.

Theorem 8. *Let ξ_1, ξ_2, ξ_3 be positive numbers and $\xi_1\geq\xi_2\geq\xi_3$. Then we have the inequality*

$$\frac{1}{2\xi_1}+\frac{1}{2\xi_2}+\frac{1}{2\xi_3}-\frac{1}{\xi_1+\xi_2}-\frac{1}{\xi_1+\xi_3}-\frac{1}{\xi_2+\xi_3}$$
$$\geq\frac{(\xi_1-\xi_2)^2}{2\xi_1(\xi_1+\xi_2)^2}+\frac{(2\xi_2-\xi_1-\xi_3)^2}{2\xi_2(\xi_1+\xi_3)^2}+\frac{(\xi_2-\xi_3)^2}{2\xi_3(\xi_2+\xi_3)^2}. \tag{14}$$

Proof. From the given condition $\xi_1\geq\xi_2\geq\xi_3$, it is easy to check that

$$\xi_1+\xi_2\geq\xi_1+\xi_3\geq\xi_2+\xi_3,\ \ 2\xi_1\geq2\xi_2\geq2\xi_3$$

and

$$(\xi_1+\xi_2,\xi_1+\xi_3,\xi_2+\xi_3)\prec(2\xi_1,2\xi_2,2\xi_3).$$

Using Theorem 4 and taking $\boldsymbol{\nu}=(\xi_1+\xi_2,\xi_1+\xi_3,\xi_2+\xi_3)$, $\boldsymbol{\vartheta}=(2\xi_1,2\xi_2,2\xi_3)$, $\Psi(x)=\frac{1}{x}$, $x\in(0,+\infty)$ in (6), we obtain that there exists a real number τ_i between $\nu_{[i]}$ and $\vartheta_{[i]}$ $(i=1,2,3)$ such that

$$\frac{1}{2\xi_1}+\frac{1}{2\xi_2}+\frac{1}{2\xi_3}-\frac{1}{\xi_1+\xi_2}-\frac{1}{\xi_1+\xi_3}-\frac{1}{\xi_2+\xi_3}$$
$$\geq\frac{1}{\tau_1^3}(\xi_1-\xi_2)^2+\frac{1}{\tau_2^3}(2\xi_2-\xi_1-\xi_3)^2+\frac{1}{\tau_3^3}(\xi_2-\xi_3)^2. \tag{15}$$

Further, by (5) we find that τ_1,τ_2,τ_3 satisfy

$$\frac{1}{2\xi_1}-\frac{1}{\xi_1+\xi_2}=-\frac{\xi_1-\xi_2}{(\xi_1+\xi_2)^2}+\frac{1}{\tau_1^3}(\xi_1-\xi_2)^2,$$

$$\frac{1}{2\xi_2}-\frac{1}{\xi_1+\xi_3}=-\frac{2\xi_2-\xi_1-\xi_3}{(\xi_1+\xi_3)^2}+\frac{1}{\tau_2^3}(2\xi_2-\xi_1-\xi_3)^2,$$

$$\frac{1}{2\xi_3} - \frac{1}{\xi_2+\xi_3} = -\frac{\xi_3-\xi_2}{(\xi_2+\xi_3)^2} + \frac{1}{\tau_3^3}(\xi_3-\xi_2)^2.$$

From the above equations, we have

$$\tau_1^3 = 2\xi_1(\xi_1+\xi_2)^2,\ \tau_2^3 = 2\xi_2(\xi_1+\xi_3)^2,\ \tau_3^3 = 2\xi_3(\xi_3+\xi_2)^2. \tag{16}$$

Combining (15) and (16) leads to the desired inequality (14). The proof of Theorem 8 is complete. □

Author Contributions: S.W. and M.A.K. finished the proofs of the main results and the writing work. H.U.H. gave lots of advice on the proofs of the main results and the writing work. All authors read and approved the final manuscript.

Funding: This work was supported by the Teaching Reform Project of Longyan University (Grant No. 2017JZ02) and the Teaching Reform Project of Fujian Provincial Education Department (Grant No. FBJG20180120).

Acknowledgments: The authors would like to express sincere appreciation to the anonymous reviewers for their helpful comments and suggestions.

Conflicts of Interest: The authors declare no conflict of interest.

References

1. Hardy, G.H.; Littlewood, J.E.; Pólya, G. *Inequalities*, 2nd ed.; Cambridge Univ. Press: Cambridge, UK, 1952.
2. Marshall, A.W.; Olkin, I. *Inequalities: The Theory of Majorization and Its Applicaitons*; Academic Press: New York, NY, USA, 1979.
3. Wu, S.H.; Adil Khan, M.; Basir, A.; Saadati, R. Some majorization integral inequalities for functions defined on rectangles. *J. Inequal. Appl.* **2018**, *2018*, 146. [CrossRef] [PubMed]
4. Adil Khan, M.; Wu, S.H.; Ullah, H.; Chu, Y.M. Discrete majorization type inequaliites for convex functions on rectangles. *J. Inequal. Appl.* **2019**, *2019*, 16. [CrossRef]
5. Wu, S.H.; Shi, H.N. A relation of weak majorization and its applications to certain inequalities for means. *Math. Slov.* **2009**, *61*, 561–570. [CrossRef]
6. Shi, H.N.; Wu, S.H. Majorized proof and refinement of the discrete Steffensen's inequality. *Taiwan. J. Math.* **2007**, *11*, 1203–1208. [CrossRef]
7. Zaheer Ullah, S.; Adil Khan, M.; Abdulhameed Khan, Z.; Chu, Y.M. Integral majorization type inequalities for the functions in the sense of strongly convexity. *J. Funct. Spaces* **2019**, *2019*, 9487823.
8. Adil Khan, M.; Khalid, S.; Pečarić, J.E. Refinements of some majorization type inequalities. *J. Math. Inequal.* **2013**, *7*, 73–92. [CrossRef]
9. Adil Khan, M.; Khan, J.; Pečarić, J.E. Generalization of Jensen's and Jensen-Steffensen's inequalities by generalized majorization theorem. *J. Math. Inequal.* **2017**, *11*, 1049–1074. [CrossRef]
10. Adil Khan, M.; Latif, N.; Pečarić, J.E. Generalization of majorization theorem. *J. Math. Inequal.* **2015**, *9*, 847–872. [CrossRef]
11. Adil Khan, M.; Latif, N.; Pečarić, J.E.; Perić, I. On Majorization for Matrices. *Math. Balk.* **2013**, *27*, 3–19.
12. Khan, J.; Adil Khan, M.; Pečarić, J. On Jensen's Type Inequalities via Generalized Majorization Inequalities. *Filomat* **2018**, *32*, 5719–5733. [CrossRef]
13. Zaheer Ullah, S.; Adil Khan, M.; Chu, Y.M. Majorization theorems for strongly convex functions. *J. Inequal. Appl.* **2019**, *2019*, 58. [CrossRef]
14. Pečarić, J.E.; Proschan, F.; Tong, Y.L. *Convex Functions, Partial Orderings and Statistical Applications*; Academic Press: New York, NY, USA, 1992.
15. Fuchs, L. A new proof of an inequality of Hardy-Littlewood Pólya. *Mat. Tidsskr.* B **1947**, 53–54; stable/24530530.
16. Bullen, P.S.; Vasić, P.M.; Stanković, L.J. *A Problem of A. Oppenheim*; University of Belgrade: Belgrade, Serbia, 1973.

mathematics MDPI

Article

Difference Mappings Associated with Nonsymmetric Monotone Types of Fejér's Inequality

Mohsen Rostamian Delavar [1,*] and Manuel De La Sen [2]

1 Department of Mathematics, Faculty of Basic Sciences, University of Bojnord, P.O. Box 1339, Bojnord 94531, Iran
2 Institute of Research and Development of Processes, University of Basque Country, Campus of Leioa (Bizkaia)—Aptdo. 644, 48080 Bilbao, Spain
* Correspondence: m.rostamian@ub.ac.ir

Received: 28 July 2019; Accepted: 26 August 2019; Published: 1 September 2019

Abstract: Two mappings L_w and P_w, in connection with Fejér's inequality, are considered for the convex and nonsymmetric monotone functions. Some basic properties and results along with some refinements for Fejér's inequality according to these new settings are obtained. As applications, some special means type inequalities are given.

Keywords: convex functions; Fejér's inequality; special means

MSC: 26D15, 26A51, 52A01

---

1. Introduction

In 1906, L. Fejér [1] proved the following integral inequalities known in the literature as Fejér's inequality:

$$f\Big(\frac{a+b}{2}\Big)\int_a^b g(x)dx \le \int_a^b f(x)g(x)dx \le \frac{f(a)+f(b)}{2}\int_a^b g(x)dx, \tag{1}$$

where $f : [a,b] \to \mathbb{R}$ is convex and $g : [a,b] \to \mathbb{R}^+ = [0,+\infty)$ is integrable and symmetric to $x = \frac{a+b}{2}\big(g(x) = g(a+b-x), \forall x \in [a,b]\big)$. If in (1) we consider $g \equiv 1$, we recapture the classic Hermite–Hadamard inequality [2,3]:

$$f\Big(\frac{a+b}{2}\Big) \le \frac{1}{b-a}\int_a^b f(x)dx \le \frac{f(a)+f(b)}{2}.$$

In [4], two difference mappings L and P associated with Hermite–Hadamard's inequality have been introduced as follows:

$$L : [a,b] \to \mathbb{R}, \qquad L(t) = \frac{f(a)+f(t)}{2}(t-a) - \int_a^t f(s)ds$$
$$P : [a,b] \to \mathbb{R}, \qquad P(t) = \int_a^t f(s)ds - (t-a)f\Big(\frac{a+t}{2}\Big).$$

Some properties for L and P, refinements for Hermite–Hadamard's inequality and some applications were raised in [4] as well:

Theorem 1 (Theorem 1 in [4]). *Let $f : I \subset \mathbb{R} \to \mathbb{R}$ be a convex mapping on the interval I and let $a < b$ be fixed in I°. Then, we have the following:*

(i) The mapping L is nonnegative, monotonically nondecreasing, and convex on $[a,b]$
(ii) The following refinement of Hadamard's inequality holds:

$$\frac{1}{b-a}\int_a^b f(s)ds \le \frac{1}{b-a}\int_y^b f(s)ds + \left(\frac{y-a}{b-a}\right)\frac{f(a)+f(y)}{2} \le \frac{f(a)+f(b)}{2},$$

for each $y \in [a,b]$.
(iii) The following inequality holds:

$$\begin{aligned}&\alpha\frac{f(t)+f(a)}{2}(t-a) + (1-\alpha)\frac{f(s)+f(a)}{2}(s-a) - \\ &\frac{f(\alpha t+(1-\alpha)s)+f(a)}{2}[\alpha t+(1-\alpha)s-\alpha] \ge \\ &\alpha\int_a^t f(u)du + (1-\alpha)\int_a^s f(u)du - \int_a^{\alpha t+(1-\alpha)s} f(u)du,\end{aligned}$$

for every $t, s \in [a,b]$ and each $\alpha \in [0,1]$.

Theorem 2 (Theorem 2 in [4]). *Let $f : I \subset \mathbb{R} \to \mathbb{R}$ be a convex mapping on the interval I and let $a < b$ be fixed in I°. Then, we have the following:*

(i) The mapping P is nonnegative and monotonically nondecreasing on $[a,b]$.
(ii) The following inequality holds:

$$0 \le P(t) \le L(t), \qquad \text{for all } t \in [a,b].$$

(iii) The following refinement of Hadamard's inequality holds:

$$\begin{aligned}&f\left(\frac{a+b}{2}\right) \le \left[(b-a)f\left(\frac{a+b}{2}\right) - (y-a)f\left(\frac{a+y}{2}\right)\right] + \\ &\frac{1}{b-a}\int_a^y f(s)ds \le \frac{1}{b-a}\int_a^b f(s)ds,\end{aligned}$$

for all $y \in [a,b]$.

The main results obtained in [4] (Theorems 1 and 2) are based on the facts that if $f : [a,b] \to \mathbb{R}$ is convex, then for all $x, y \in [a,b]$ with $x \neq y$ we have (see, [5,6]):

$$f\left(\frac{x+y}{2}\right) \le \frac{1}{y-x}\int_x^y f(s)ds < \frac{f(x)+f(y)}{2},$$

and

$$f(x) - f(y) \ge (x-y)f'_+(y),$$

where $f'_+(y)$ is the right-derivative of f at y.
Motivated by the above concepts, inequalities and results, we introduce two difference mappings, L_w and P_w, related to Fejér's inequality:

$$L_w : [a,b] \to \mathbb{R}, \qquad L_w(t) = \frac{f(a)+f(t)}{2}\int_a^t w(s)ds - \int_a^t f(x)w(x)dx,$$

$$P_w : [a,b] \to \mathbb{R}, \qquad P_w(t) = \int_a^t f(x)w(x)dx - f\left(\frac{a+t}{2}\right)\int_a^t w(x)dx.$$

In the case that $w \equiv 1$, the mappings L_w and P_w reduce to L and P, respectively.

In this paper we obtain some properties for L_w and P_w that imply some refinements for Fejér's inequality in the case that w is a nonsymmetric monotone function. Also, our results generalize Theorems 1 and 2 from Hermite–Hadamard's type to Fejér's type. Furthermore as applications, we find some numerical and special means type inequalities.

To obtain our respective results, we need the modified version of Theorem 5 in [7] which includes the left and right part of Fejér's inequality in the monotone nonsymmetric case.

Theorem 3. *Let $f : I \subset \mathbb{R} \to \mathbb{R}$ be a convex function on the interval I and differentiable on I°. Consider $a, b \in I^\circ$ with $a < b$ such that $w : [a,b] \to \mathbb{R}$ is a nonnegative, integrable and monotone function. Then*

(1) If $w'(x) \leq 0$ ($w'(x) \geq 0$), $a \leq x \leq b$ and $f(a) \leq f(b)$ ($f(a) \geq f(b)$), then

$$\int_a^b f(x)w(x)dx \leq \frac{f(a)+f(b)}{2}\int_a^b w(x)dx. \tag{2}$$

(2) If $w'(x) \geq 0$ ($w'(x) \leq 0$), $a \leq x \leq b$ and $f(a) \leq f(\frac{a+b}{2})$ ($f(a) \geq f(\frac{a+b}{2})$), then

$$f\Big(\frac{a+b}{2}\Big)\int_a^b f(x)w(x)dx \leq \int_a^b f(x)w(x)dx. \tag{3}$$

The main point in Theorem 3 (1) ($w'(x) \leq 0$), is that we have (2) for any $x, y \in [a,b]$ with $f(x) \leq f(y)$ without the need for w to be symmetric with respect to $\frac{x+y}{2}$. Also similar properties hold for other parts of the above theorem.

Example 1. *Consider $f(x) = \frac{1}{t}$ and $w(x) = \frac{1}{t^2}$ for $t > 0$. It is clear that f is convex and w is nonsymmetric and decreasing. If we consider $0 < x \leq y$, then from the fact that $(y-x)^2 \geq 0$ we obtain that*

$$\frac{2}{x+y} \leq \frac{x+y}{2xy}.$$

This inequality implies that

$$\frac{2}{x+y}\Big(\frac{y-x}{xy}\Big) \leq \frac{y^2-x^2}{2x^2y^2}.$$

It follows that

$$\frac{2}{x+y}\Big(\frac{1}{x}-\frac{1}{y}\Big) \leq \frac{1}{2x^2}-\frac{1}{2y^2}.$$

So

$$\Big(\frac{1}{\frac{x+y}{2}}\Big)\int_x^y \frac{1}{t^2}dt \leq \int_x^y \frac{1}{t^3}dt,$$

shows that f and w satisfy (3) on $[x,y]$, where w is not symmetric. Also, we can see that f and w satisfy (2).

2. Main Results

The first result of this section is about some properties of the mapping L_w where the function w is nonincreasing.

Theorem 4. *Let $f : I \subset \mathbb{R} \to \mathbb{R}$ be a convex function on the interval I and differentiable on I°. Consider $a, b \in I^\circ$ with $a < b$ such that $w : [a,b] \to \mathbb{R}$ is a nonnegative and differentiable function with $w'(x) \leq 0$ for all $a \leq x \leq b$. Then*

- *(i) The mapping L_w is nonnegative on $[a,b]$, if $f(a) \leq f(t)$ for all $t \in [a,b]$.*
- *(ii) The mapping L_w is convex on $[a,b]$, if f is nondecreasing. Also L_w is monotonically nondecreasing on $[a,b]$.*

(iii) *The following refinement of (2) holds:*

$$\begin{aligned}&\int_a^b f(x)w(x) \leq \\ &\int_y^b f(x)w(x)dx + \frac{f(a)+f(y)}{2}\int_a^y w(x)dx \leq \\ &\frac{f(a)+f(b)}{2}\int_a^b w(x)dx,\end{aligned} \tag{4}$$

for any $y \in [a,b]$ *with* $f(a) \leq f(y)$.

(iv) *If* f *is nondecreasing, then the following inequality holds:*

$$\begin{aligned}&t\int_a^u f(x)w(x)dx + (1-t)\int_a^v f(x)w(x)dx - \int_a^{tu+(1-t)v} f(x)w(x)dx \leq \\ &t\frac{f(u)+f(a)}{2}\int_a^u w(x)dx + (1-t)\frac{f(v)+f(u)}{2}\int_a^v w(x)dx \\ &-\frac{f(tu+(1-t)v)+f(a)}{2}\int_a^{tu+(1-t)v} w(x)dx,\end{aligned} \tag{5}$$

for any $u, v \in [a,b]$ *and each* $t \in [0,1]$.

(v) *If* $f' \in L([a,b])$, *then for each* $t \in [a,b]$ *we have*

$$|L_w(t)| \leq \frac{(t-a)^2}{2}\int_a^t w(x)|f'(x)|dx. \tag{6}$$

Furthermore when $|f'|$ *is convex on* $[a,b]$, *then:*

$$|L_w(t)| \leq \frac{t-a}{2}\left[|f'(a)|\int_a^t (t-x)w(x)dx + |f'(t)|\int_a^t (x-a)w(x)dx\right]. \tag{7}$$

Proof. (i) We need only the inequality

$$\int_a^t f(x)w(x)dx \leq \frac{f(a)+f(t)}{2}\int_a^t w(x)dx,$$

for all $t \in [a,b]$. This happens according to Theorem 3 (1).

(ii) Without loss of generality for $a \leq y < x < b$ consider the following identity:

$$\begin{aligned}&L_w(x) - L_w(y) = \\ &\frac{f(x)+f(a)}{2}\int_a^x w(s)ds - \frac{f(y)+f(a)}{2}\int_a^y w(s)ds - \int_y^x f(s)w(s)ds\end{aligned} \tag{8}$$

Dividing with "$x-y$" and then letting $x \to y$ we obtain that

$$2L'_{+w}(y) - f(a)w(y) + f(y)w(y) = f'_+(y)\int_a^y w(s)ds. \tag{9}$$

Also from the convexity of f we have

$$f'_+(y) \leq \frac{f(x)-f(y)}{x-y},$$

which, along with the fact that w is nonincreasing, implies that

$$\begin{aligned} f'_+(y)\int_a^y w(s)ds &\leq \frac{f(x)-f(y)}{x-y}\int_a^y w(s)ds \\ &\leq \frac{f(x)+f(a)}{x-y}\int_a^x w(s)ds - \frac{f(y)+f(a)}{x-y}\int_a^y w(s)ds \\ &+ [f(y)-f(a)]w(y) - \frac{f(x)+f(y)}{x-y}\int_y^x w(s)ds. \end{aligned} \tag{10}$$

So from (9) and (10) we get

$$L'_{+w}(y) \leq \tag{11}$$
$$\frac{f(x)+f(a)}{2(x-y)}\int_a^x w(s)ds - \frac{f(y)+f(a)}{2(x-y)}\int_a^y w(s)ds - \frac{f(x)+f(y)}{2(x-y)}\int_y^x w(s)ds.$$

On the other hand from (8) and Theorem 3 (1), we have

$$\frac{L_w(x)-L_w(y)}{x-y} \geq$$
$$\frac{f(x)+f(a)}{2(x-y)}\int_a^x w(s)ds - \frac{f(y)+f(a)}{2(x-y)}\int_a^y w(s)ds - \frac{f(x)+f(y)}{2(x-y)}\int_y^x w(s)ds,$$

and, along with (11), we obtain that

$$\frac{L_w(x)-L_w(y)}{x-y} \geq L'_{+w}(y).$$

This implies the convexity of $L_w(t)$.

For the fact that L is monotonically nondecreasing, from convexity of f on $[a,b]$ we have

$$f'_+(y) \geq \frac{f(y)-f(a)}{y-a},$$

for all $y \in [a,b]$ and so

$$\begin{aligned} &\frac{L_w(x)-L_w(y)}{x-y} \geq L'_{+w}(y) = \frac{f'_+(y)}{2}\int_a^y w(s)ds + \frac{f(a)w(y)}{2} - \frac{f(y)w(y)}{2} \\ &= \frac{1}{2}\Big[f'_+(y)\int_a^y w(s)ds + (f(a)-f(y))w(y)\Big] \geq \\ &\frac{1}{2}\Big[f'_+(y)(y-a) - (f(y)-f(a))\Big]w(y) \geq 0, \end{aligned}$$

for any $x > y$.

(iii) Since L_w is monotonically nondecreasing we have $0 \leq L_w(y) \leq L_w(b)$, for all $y \in [a,b]$ and so

$$\begin{aligned} &\frac{f(y)+f(a)}{2}\int_a^y w(x)dx - \int_a^y f(x)w(x)dx \leq \\ &\frac{f(b)+f(a)}{2}\int_a^b w(x)dx - \int_a^b f(x)w(x)dx, \end{aligned}$$

which implies that

$$\int_y^b f(x)w(x)dx + \frac{f(a)+f(y)}{2}\int_a^y w(x)dx \leq \frac{f(a)+f(b)}{2}\int_a^b w(x)dx. \tag{12}$$

Also, by the use of Theorem 3 (1) we get

$$\int_y^b f(x)w(x)dx + \frac{f(a)+f(y)}{2}\int_a^y w(x)dx \tag{13}$$
$$\geq \int_y^b f(x)w(x)dx + \int_a^y f(x)w(x)dx = \int_a^b f(x)w(x)dx.$$

Now from (12) and (13), we have the result.

(iv) Since L_w is convex, then from the fact that

$$L_w(tu+(1-t)v) \leq tL_w(u) + (1-t)L_w(v),$$

for any $u, v \in [a,b]$ and each $t \in [0,1]$, we have the result.

(v) The following identity was obtained in [8]:

$$\frac{f(a)+f(t)}{2}\int_a^t w(x)dx - \int_a^t f(x)w(x)dx = \frac{(t-a)^2}{2}\int_0^1 p(s)f'(sa+(1-s)t)ds, \tag{14}$$

for any $t \in [a,b]$ where

$$p(s) = \int_s^1 w(ua+(1-u)t)du + \int_s^0 w(ua+(1-u)t)du, \qquad s \in [0,1].$$

Since w is nonincreasing, then we obtain

$$\int_s^1 w(ua+(1-u)t)du \leq w(sa+(1-s)t)(as+(1-s)t-a) =$$
$$w(sa+(1-s)t)(1-s)(t-a),$$

and

$$\int_s^0 w(ua+(1-u)t)du \leq w(sa+(1-s)t)(t-sa-(1-s)t) =$$
$$w(sa+(1-s)t)s(t-a).$$

So

$$|p(s)| \leq w(sa+(1-s)t)(t-a), \qquad s \in [0,1]. \tag{15}$$

Now by the use of (15) in (14) we get

$$|L_w(t)| < \frac{(t-a)^3}{2}\int_0^1 w(sa+(1-s)t)|f'(sa+(1-s)t)|ds, \tag{16}$$

for any $t \in [a,b]$. Using the change of variable $x = sa+(1-s)t$ and some calculations imply that

$$|L_w(t)| \leq \frac{(t-a)^2}{2}\int_a^t w(x)|f'(x)|dx,$$

for any $t \in [a,b]$. Furthermore if $|f'|$ is convex on $[a,b]$, then from (16) and by the use of the change of variable $x = sa+(1-s)t$ we get

$$|L_w(t)| \leq \frac{(t-a)^3}{2}\left[|f'(a)|\int_a^t \frac{t-x}{t-a}w(x)\frac{dx}{t-a} + |f'(t)|\int_a^t \frac{x-a}{t-a}w(x)\frac{dx}{t-a}\right],$$

which implies that

$$|L_w(t)| \leq \frac{(t-a)}{2}\left[|f'(a)|\int_a^t (t-x)w(x)dx + |f'(t)|\int_a^t (x-a)w(x)dx\right],$$

for any $t \in [a,b]$. □

Remark 1. *(i) By the use of Theorem 3 (1), it is not hard to see that if w is nondecreasing on $[a,b]$, then some properties of L_w and corresponding results obtained in Theorem 4 may change. However the argument of proof is similar. The details are omitted.*

(ii) Theorem 4 gives a generalization of Theorem 1, along with some new results.

The following result is including some properties of the mapping P_w in the case that w is nondecreasing.

Theorem 5. *Let $f : I \subset \mathbb{R} \to \mathbb{R}$ be a convex function on the interval I and differentiable on I°. Consider $a, b \in I^\circ$ with $a < b$ such that $w : [a,b] \to \mathbb{R}$ is a nonnegative and continuous function with $w'(x) \geq 0$ for all $a \leq x \leq b$. Then*

(i) P_w is nonnegative, if $f(a) \leq f(\frac{a+t}{2})$ for any $t \in [a,b]$.
(ii) If for any $x < y$ we have $f(x) \leq f(\frac{x+y}{2})$, then P_w is nondecreasing on $[a,b]$.
(iii) If $f' \in L([a,b])$, then for each $t \in [a,b]$ we have

$$|P_w(t)| \leq (t-a)\left[\int_a^{\frac{a+t}{2}} w(x)(x-a)|f'(x)|dx + \int_{\frac{a+t}{2}}^t w(x)(t-x)|f'(x)|dx\right]. \tag{17}$$

Furthermore when $|f'|$ is convex on $[a,b]$, then:

$$|P_w(t)| \leq \left[\int_a^{\frac{a+t}{2}} w(x)(t-x)(x-a)dx + \int_{\frac{a+t}{2}}^t w(x)(t-x)^2dx\right]|f'(a)| + \left[\int_a^{\frac{a+t}{2}} w(x)(x-a)^2dx + \int_{\frac{a+t}{2}}^t w(x)(t-x)(x-a)dx\right]|f'(t)|. \tag{18}$$

(iv) The following inequality holds:

$$P_w(t) - L_w(t) \leq \int_a^t f(x)w(x)dx, \tag{19}$$

provided that $f(a) \leq f(\frac{a+t}{2})$ for all $t \in [a,b]$.
(v) If for any $x < y$ we have $f(x) \leq f(\frac{x+y}{2})$, then the following refinement of (3) holds:

$$f\left(\frac{a+b}{2}\right)\int_a^b w(x)dx \leq \int_a^t f(x)w(x)dx + f\left(\frac{a+b}{2}\right)\int_a^b w(x)dx - f\left(\frac{a+t}{2}\right)\int_a^t w(x)dx \leq \int_a^b f(x)w(x)dx, \tag{20}$$

for all $t \in [a,b]$.

Proof. (i) It follows from Theorem 3 (2).

(ii) Suppose that $a \leq x < y < b$. So from Theorem 3 (2) and the facts that w is nondecreasing and f is convex, we get

$$\begin{aligned}
&P_w(y) - P_w(x) = \\
&\int_a^y f(t)w(t)dt - f\Big(\frac{a+y}{2}\Big)\int_a^y w(t)dt - \int_a^x f(t)w(t)dt + f\Big(\frac{a+x}{2}\Big)\int_a^x w(t)dt = \\
&\int_x^y f(t)w(t)dt + f\Big(\frac{a+x}{2}\Big)\int_a^x w(t)dt - f\Big(\frac{a+y}{2}\Big)\int_a^y w(t)dt \geq \\
&f\Big(\frac{x+y}{2}\Big)\int_x^y w(t)dt + f\Big(\frac{a+x}{2}\Big)\int_a^x w(t)dt - f\Big(\frac{a+y}{2}\Big)\int_a^y w(t)dt \geq \\
&f\Big(\frac{x+y}{2}\Big)(y-x)w(x) + f\Big(\frac{a+x}{2}\Big)(x-a)w(a) - f\Big(\frac{a+y}{2}\Big)(y-a)w(y) \geq \\
&\Big[f\Big(\frac{x+y}{2}\Big)(y-x) + f\Big(\frac{a+x}{2}\Big)(x-a) - f\Big(\frac{a+y}{2}\Big)(y-a)\Big]w(a) \geq 0.
\end{aligned}$$

This completes the proof.

(iii) The following identity is obtained in [8]:

$$\int_a^t f(x)w(x)dx - f\Big(\frac{a+t}{2}\Big)\int_a^t w(x)dx = (t-a)^2\int_0^1 k(s)f'(sa+(1-s)t)ds,$$

for any $t \in [a,b]$, where

$$k(s) = \begin{cases} \displaystyle\int_0^s w(ua+(1-u)t)du, & s \in [0, \frac{1}{2}); \\ \displaystyle -\int_s^1 w(ua+(1-u)t)du, & s \in [\frac{1}{2}, 1]. \end{cases}$$

By similar method used to prove part (v) of Theorem 4, we can obtain the results. We omitted the details here.

(iv) By Theorem 3 (1), for any $t \in (a,b]$ we have

$$\int_a^{\frac{a+t}{2}} f(x)w(x)dx \leq \frac{f(\frac{a+t}{2}) + f(a)}{2}\int_a^{\frac{a+t}{2}} w(x)dx \leq \frac{f(\frac{a+t}{2}) + f(a)}{2}\int_a^t w(x)dx, \tag{21}$$

and

$$\int_{\frac{a+t}{2}}^t f(x)w(x)dx \leq \frac{f(\frac{a+t}{2}) + f(t)}{2}\int_{\frac{a+t}{2}}^t w(x)dx \leq \frac{f(\frac{a+t}{2}) + f(t)}{2}\int_a^t w(x)dx. \tag{22}$$

If we add (21) to (22), we obtain

$$\int_a^t f(x)w(x)dx \leq \Big[f\Big(\frac{a+t}{2}\Big) + \frac{f(a)+f(t)}{2}\Big]\int_a^t w(x)dx,$$

which is equivalent with

$$\int_a^t f(x)w(x)dx \leq -P_w(t) + L_w(t) + 2\int_a^t f(x)w(x)dx.$$

This implies the desired result.

(v) The left side of (20) is a consequence of assertion (i) and the following inequality:

$$\int_a^t f(x)w(x)dx - f\Big(\frac{a+t}{2}\Big)\int_a^t w(x)dx \geq 0,$$

for all $t \in [a,b]$.

Since P_w is nondecreasing we have $P_w(t) \leq P_w(b)$ for all $t \in [a,b]$, i. e.

$$\int_a^t f(x)w(x)dx - f\Big(\frac{a+t}{2}\Big)\int_a^t w(x)dx \leq$$
$$\int_a^b f(x)w(x)dx - f\Big(\frac{a+b}{2}\Big)\int_a^b w(x)dx.$$

Then we have the right side of (20). □

Remark 2. *(i) By the use of Theorem 3 (2) (w is nonincreasing on $[a,b]$) in the proof of Theorem 5, we can obtain some different properties for P_w with new corresponding results. The details are omitted.*

(ii) Theorem 5 gives a generalization of Theorem 2, along with some new results.

3. Applications

The following means for real numbers $a, b \in \mathbb{R}$ are well known:

$$A(a,b) = \frac{a+b}{2} \qquad \textit{arithmetic mean},$$
$$L_n(a,b) = \Big[\frac{b^{n+1}-a^{n+1}}{(n+1)(b-a)}\Big]^{\frac{1}{n}} \qquad \textit{generalized log-mean},\ n \in \mathbb{R},\ a < b.$$

The following result holds between the two above special means:

Theorem 6. *For any $a, b \in \mathbb{R}$ with $0 < a < b$ and $n \in \mathbb{N}$ we have*

$$A^n(a,b) \leq L_n^n(a,b) \leq A(a^n, b^n). \tag{23}$$

In this section as applications of our results in previous section, we give some refinements for the inequalities mentioned in (23).

Consider $a, b \in (0,\infty)$ with $a < b$. Define

$$\begin{cases} f(x) = x^n, & x \in [a,b] \text{ and } n \geq 1; \\ w(x) = x^{-s}, & x \in [a,b] \text{ and } s \in [0,1) \cup (1,\infty). \end{cases}$$

From (4) with some calculations we have

$$\frac{b^{n-s+1}-a^{n-s+1}}{n-s+1} \leq$$
$$\frac{b^{n-s+1}-t^{n-s+1}}{n-s+1} + \frac{a^n+t^n}{2}\Big(\frac{t^{1-s}-a^{1-s}}{1-s}\Big) \leq$$
$$\frac{a^n+b^n}{2}\Big(\frac{b^{1-s}-a^{1-s}}{1-s}\Big),$$

for all $t \in [a,b]$, which implies that

$$(b-a)L_{n-s}^{n-s}(a,b) \leq$$
$$(b-t)L_{n-s}^{n-s}(t,b) + A(a^n,t^n)\Big(\frac{t^{1-s}-a^{1-s}}{1-s}\Big) \leq \tag{24}$$
$$A(a^n,b^n)\Big(\frac{b^{1-s}-a^{1-s}}{1-s}\Big).$$

Inequality (24) gives a refinement for the right part of (23).

In the case that $s = 1$ we have

$$(b-a)L_{n-1}^{n-1}(a,b) \leq (b-t)L_{n-1}^{n-1}(t,b) + ln\frac{t}{a}A(a^n,t^n) \leq ln\frac{t}{a}A(a^n,b^n).$$

In the case that $s = 0$ we get

$$L_n^n(a,b) \leq \Big(\frac{b-t}{b-a}\Big)L_n^n(t,b) + \Big(\frac{t-a}{b-a}\Big)A(a^n,t^n) \leq A(a^n,b^n), \tag{25}$$

for all $t \in [a,b]$. In fact inequality (25) is equivalent with the first inequality obtained in the applications section of [4].

Now with the same assumption for f and w as was used to obtain (24), by the use of (20) we get:

$$\begin{aligned} &A^n(a,b)\Big(\frac{b^{1-s}-a^{1-s}}{1-s}\Big) \leq \\ &A^n(a,b)\Big(\frac{b^{1-s}-a^{1-s}}{1-s}\Big) + (t-a)L_{n-s}^{n-s}(t,a) - A^n(a,t)\Big(\frac{t^{1-s}-a^{1-s}}{1-s}\Big) \leq \\ &(b-a)L_{n-s}^{n-s}(b,a), \end{aligned} \tag{26}$$

for all $t \in [a,b]$ and $s \in [0,1) \cup (1,\infty)$. Inequality (26) gives a refinement for the left part of (23). Also if we consider $s = 1$, then we obtain

$$ln\frac{b}{a}A^n(a,b) \leq ln\frac{b}{a}A^n(a,b) + (t-a)L_{n-1}^{n-1}(t,a) - ln\frac{t}{a}A^n(a,t) \leq (b-a)L_{n-1}^{n-1}(b,a),$$

for all $t \in [a,b]$. In a more special case, if we set $s = 0$, then we get:

$$A^n(a,b) \leq A^n(a,b) + \Big(\frac{t-a}{b-a}\Big)\big[L_n^n(t,a) - A^n(a,t)\big] \leq L_n^n(b,a),$$

for all $t \in [a,b]$.

Finally we encourage interested readers to use inequalities (4)–(7) and inequalities (17)–(20), for appropriate functions f and w to obtain some new special means types and numerical inequalities.

Author Contributions: Conceptualization, M.R.D.; methodology, M.R.D. and M.D.L.S.; investigation, M.R.D. and M.D.L.S.; writing–original draft preparation, M.R.D.; writing–review and editing, M.D.L.S.; project administration, M.R.D.; funding acquisition, M.D.L.S.

Funding: This research was in part supported by a grant from University of Bojnord (No. 97/367/19164). The second author thanks the Basque Government for its support through Grant IT1207-19.

Acknowledgments: The authors are grateful to the referees and the editor for their valuable comments and suggestions.

Conflicts of Interest: The authors declare no conflict of interest.

References

1. Fejér, L. Über die fourierreihen, II. *Math. Naturwise. Anz Ungar. Akad. Wiss.* **1906**, *24*, 369–390.
2. Mitrinović, D.S.; Lacković, I.B. Hermite and convexity. *Aequ. Math.* **1985**, *28*, 229–232. [CrossRef]
3. Dragomir, S.S.; Pearce, C.E.M. Selected Topics on Hermite-Hadamard Inequalities and Applications; RGMIA Monographs, Victoria University. 2000. Available online: http://ajmaa.org/RGMIA/monographs.php/ (accessed on 15 March 2016).
4. Dragomir, S.S.; Agarwal, R.P. Two new mappings associated with Hadamard's inequalities for convex functions. *Appl. Math. Lett.* **1998**, *11*, 33–38. [CrossRef]
5. Niculescu, C.P.; Persson, L.E. *Convex Functions and Their Applications: A Contemporary Approach*; CMS Books in Mathematics; Springer: Berlin, Germany, 2006.
6. Robert, A.W.; Varbeg, D.E. *Convex Functions*; Academic Press: New York, NY, USA; London, UK, 1973.

7. Abramovich, S.; Persson, L.-E. Extensions and Refinements of Fejér and Hermite-Hadamard type inequalities. *Math. Inequal. Appl.* **2018**, *21*, 759–772. [CrossRef]
8. Sarikaya, M.Z. On new Hermite Hadamard Fejér type integral inequalities. *Stud. Univ. Babeş-Bolyai Math.* **2012**, *57*, 377–386.

mathematics

MDPI

Article

Generalized Steffensen's Inequality by Fink's Identity

Asfand Fahad [1], Saad Ihsan Butt [2,*] and Josip Pečarić [3]

[1] Department of Mathematics, COMSATS University Islamabad, Vehari Campus, Vehari 61100, Pakistan; asfandfahad1@yahoo.com
[2] Department of Mathematics, COMSATS University Islamabad, Lahore Campus, Lahore 54000, Pakistan
[3] RUDN University, Miklukho-Maklaya str. 6, 117198 Moscow, Russia; pecaric@element.hr
* Correspondence: saadihsanbutt@gmail.com

Received: 11 February 2019; Accepted: 28 March 2019; Published: 4 April 2019

Abstract: By using Fink's Identity, Green functions, and Montgomery identities we prove some identities related to Steffensen's inequality. Under the assumptions of n-convexity and n-concavity, we give new generalizations of Steffensen's inequality and its reverse. Generalizations of some inequalities (and their reverse), which are related to Hardy-type inequality. New bounds of Grüss and Ostrowski-type inequalities have been proved. Moreover, we formulate generalized Steffensen's-type linear functionals and prove their monotonicity for the generalized class of $(n+1)$-convex functions at a point. At the end, we present some applications of our study to the theory of exponentially convex functions.

Keywords: Steffensen's inequality; higher order convexity; Green functions; Montgomery identity; Fink's identity

1. Introduction

Integral inequalities such as Hardy's inequality, Steffensen's inequality, and Ostrowski's inequality are topics of interest of many Mathematicians since their pronouncement. Several generalizations of these inequalities have been proved for different classes of functions, such as convex functions, n-convex functions, and other types of functions, for example see [1–4]. Moreover, integral inequalities have been proved for different integrals, such as Jensen-steffensen inequality for diamond integral and bounds of related identities have been obtained in [5]. Other than that, Hardy's inequality for fractional integral on general domains have been proved in [6].

Steffensen's inequality was proved in [7]: if $\psi, f : [c,d] \to \mathbb{R}$, with ψ be a decreasing function and function f having range in $[0,1]$, then

$$\int_c^d \psi(z)f(z)\,dt \le \int_c^{c+\theta} \psi(z)\,dz, \qquad \text{where } \theta = \int_c^d f(z)\,dz. \tag{1}$$

A massive literature dealing with several variants and improvements of Steffensen's inequality can be seen in [8,9] and references therein. A well known generalization of Steffensen's inequality has been presented in [4]. Several results of [4] have been recently generalized by using non-bounded Montgomery's identity in [10]. To proceed further, we recall a nice generalization of Steffensen's inequality proved by Pečarić, see [11].

Theorem 1. *Let $\psi : J \to \mathbb{R}$ be a increasing function (J is an interval in $\mathbb{R}$ such that $c, d, f(c), f(d) \in J$) and $f : [c,d] \to \mathbb{R}$ be increasing and differentiable function.*

(*i*) *If* $f(t) \le t$, *then*

$$\int_{f(c)}^{f(d)} \psi(z)\,dz \le \int_c^d \psi(z)f'(z)\,dz. \tag{2}$$

(*ii*) *If* $f(t) \ge t$, *then* (2) *holds in reverse direction.*

Remark 1. *We can consider* f *to be absolute continuous instead of differentiable function and the suppositions of Theorem 1 can also be weakened. In fact for an increasing function* ψ, *the function* $\Psi(x) = \int_c^x \psi(z)\,dz$ *is well defined and satisfies* $\Psi' = \psi$ *at all except the set of points with measure zero. One can substitute* $x = f(z)$ *in* (2) *(see [12] (Corollary 20.5)), provided that* f *is absolutely continuous increasing function, therefore*

$$\Psi(f(d)) - \Psi(f(c)) = \int\limits_{f(c)}^{f(d)} \psi(x)\,dx = \int\limits_c^d \psi(f(z))f'(z)\,dz \le \int\limits_c^d \psi(z)f'(z)\,dz, \tag{3}$$

where the last inequality holds when $f(z) \le z$. *In [1], substitutions presented conclude that (3) yields (2) and generalization of a result proved by Rabier in [4], which gives (1).*

Recently, Fahad et al. introduced new generalization [1] of (1) by extending the results of [4,11]. By using Hermite interpolation, several inequalities related to the results of [1,4,11] have also been proved in [13]. We consider the important conclusions given in [1].

Corollary 1. *Suppose* $\psi : J \to \mathbb{R}$, $f : [c,d] \to \mathbb{R}$ *two differentiable functions with* f *non-decreasing as well, where* J *is an interval containing* $[c,d]$, $f(c)$ *and* $f(d)$. *If* ψ *is convex, then:*

(*i*) *If* f *satisfies condition* (*i*) *given in Theorem 1, then*

$$\psi(f(d)) \le \psi(f(c)) + \int\limits_c^d \psi'(z)f'(z)\,dz. \tag{4}$$

(*ii*) (4) *holds in reverse direction, if* f *satisfies condition* (*ii*) *given in Theorem 1.*

Corollary 1 gives (3) and therefore leads to (1), (2) and generalization of Rabier's result in [4]. Next we narrate some further important results of [1].

Corollary 2. *Consider* $\psi : [0,d] \to \mathbb{R}$ *be differentiable convex function with* $\psi(0) = 0$ *and* $f : [0,d] \to [0,+\infty)$ *be another function.*

(*i*) *If* $\int\limits_0^t f(z)\,dz \le t$ *for every* $t \in [0,b]$, *then*

$$\psi\left(\int\limits_0^d f(z)\,dz\right) \le \int\limits_0^d \psi'(z)f(z)\,dz. \tag{5}$$

(*ii*) (5) *holds reversely if* $t \le \int\limits_0^t f(z)\,dz$ *for every* $t \in [0,d]$.

Corollary 3. *Consider* ψ *and* f *as defined in Corollary 2 and let* $\lambda : [0,d] \to [0,+\infty)$ *and denote* $\Lambda(z) = \int\limits_z^d \lambda(t)\,dt$.

(*i*) *If* $\int\limits_0^t f(z)\,dz \leq t$ *for every* $t \in [0,d]$, *then*

$$\int\limits_0^d \lambda(t)\psi\left(\int\limits_0^t f(z)\,dz\right)dt \leq \int\limits_0^d \Lambda(z)\psi'(z)f(z)\,dz. \tag{6}$$

(*ii*) (6) *holds reversely if* $t \leq \int\limits_0^t f(z)\,dz$ *for every* $t \in [0,d]$.

Following two lemmas will be useful in our construction as well, see [14,15].

Lemma 1. *For a function* $\psi \in C^2([c,d])$, *we have:*

$$\psi(\xi) = \frac{d-\xi}{d-c}\psi(c) + \frac{\xi-c}{d-c}\psi(d) + \int_c^d G_{*,1}(\xi,u)\psi''(u)\,du, \tag{7}$$

$$\psi(\xi) = \psi(c) + (\xi-c)\psi'(d) + \int_c^d G_{*,2}(\xi,u)\psi''(u)\,du, \tag{8}$$

$$\psi(\xi) = \psi(d) + (d-\xi)\psi'(c) + \int_c^d G_{*,3}(\xi,u)\psi''(u)\,du, \tag{9}$$

$$\psi(\xi) = \psi(d) - (d-c)\psi'(d) + (\xi-c)\psi'(c) + \int_c^d G_{*,4}(\xi,u)\psi''(u)\,du, \tag{10}$$

$$\psi(\xi) = \psi(c) + (d-c)\psi'(c) - (d-\xi)\psi'(d) + \int_c^d G_{*,5}(\xi,u)\psi''(u)\,du, \tag{11}$$

where

$$G_{*,1}(\xi,u) = \begin{cases} \frac{(\xi-d)(u-c)}{d-c}, & \text{if } c \leq u \leq \xi, \\ \frac{(u-d)(\xi-c)}{d-c}, & \text{if } \xi < u \leq d. \end{cases} \tag{12}$$

$$G_{*,2}(\xi,u) = \begin{cases} c-u, & \text{if } c \leq u \leq \xi, \\ c-\xi, & \text{if } \xi < u \leq d. \end{cases} \tag{13}$$

$$G_{*,3}(\xi,u) = \begin{cases} \xi-d, & \text{if } c \leq u \leq \xi, \\ u-d, & \text{if } \xi < u \leq d. \end{cases} \tag{14}$$

$$G_{*,4}(\xi,u) = \begin{cases} \xi-c, & \text{if } c \leq u \leq \xi, \\ u-c, & \text{if } \xi < u \leq d. \end{cases} \tag{15}$$

and

$$G_{*,5}(\xi,u) = \begin{cases} d-u, & \text{if } c \leq u \leq \xi, \\ d-\xi, & \text{if } \xi < u \leq d. \end{cases} \tag{16}$$

Lemma 2. *Let* $\psi \in C^1[c,d]$, *then*

$$\psi(\xi) = \frac{1}{d-c}\int_c^d \psi(u)\,du + \int_c^d p_1(\xi,u)\psi'(u)\,du, \tag{17}$$

$$\psi(\xi) = \psi(d) + \int_c^d p_2(\xi,u)\psi'(u)\,du \tag{18}$$

and

$$\psi(\xi) = \psi(c) + \int_c^d p_3(\xi,u)\psi'(u)\,du, \tag{19}$$

where

$$p_1(\xi,u)=\begin{cases}\frac{u-c}{d-c}, & \text{if } c\le u\le\xi,\\ \frac{u-d}{d-c}, & \text{if } \xi<u\le d.\end{cases} \tag{20}$$

$$p_2(\xi,u)=\begin{cases}0, & \text{if } c\le u\le\xi,\\ -1, & \text{if } \xi<u\le d.\end{cases} \tag{21}$$

$$p_3(\xi,u)=\begin{cases}1, & \text{if } c\le u\le\xi,\\ 0, & \text{if } \xi<u\le d.\end{cases} \tag{22}$$

Clearly,

$$p_i(\xi,u)=\frac{\partial G_{*,i}(\xi,u)}{\partial\xi}\quad for\ all\ i=1,2,3,$$
$$p_2(\xi,u)=\frac{\partial G_{*,5}(\xi,u)}{\partial\xi}\quad and\quad p_3(\xi,u)=\frac{\partial G_{*,4}(\xi,u)}{\partial\xi}. \tag{23}$$

Throughout the calculations in the main results, we will use $p_i(\xi,u)$ corresponding to $\frac{\partial G_{*,i}(\xi,u)}{\partial\xi}$ for $i=1,2,3$, and for $\frac{\partial G_{*,4}(\xi,u)}{\partial\xi}$, $\frac{\partial G_{*,5}(\xi,u)}{\partial\xi}$ we use $p_3(\xi,s)$ and $p_2(\xi,s)$, respectively.

We also require the classical Fink's identity given in [16]:

Lemma 3. *Let $c,d\in\mathbb{R}$ and $\psi:[c,d]\to\mathbb{R}$, $n\ge 1$ and $\psi^{(n-1)}$ is absolutely continuous on $[c,d]$.*

$$\begin{aligned}\psi(u) &= \frac{n}{d-c}\int_c^d\psi(s)ds-\sum_{w=1}^{n-1}\left(\frac{n-w}{(d-c)w!}\right)\left(\psi^{(w-1)}(c)(u-c)^w-\psi^{(w-1)}(d)(u-d)^w\right)\\ &\quad+\frac{1}{(n-1)!(d-c)}\int_c^d(u-t)^{n-1}W^{[c,d]}(t,u)\psi^{(n)}(t)dt,\end{aligned} \tag{24}$$

where $W^{[c,d]}(t,u)$ is given by:

$$W^{[c,d]}(t,u)=\begin{cases}t-c, & \text{if } c\le t\le u\le d,\\ t-d, & \text{if } c\le u<t\le d.\end{cases} \tag{25}$$

Divided differences are fairly ascribed to Newton, and the term "divided difference" was used by Augustus de Morgan in 1842. Divided differences are found to be very helpful when we are dealing with functions having different degrees of smoothness. The following definition of divided difference is given in [8] (p. 14).

Definition 1. *The nth-order divided difference of a function $\psi:[c,d]\to\mathbb{R}$ at mutually distinct points $z_0,\dots,z_n\in[c,d]$ is defined recursively by*

$$[z_i;\psi]=\psi(z_i),\quad i=0,\dots,n,$$
$$[z_0,\dots,z_n;\psi]=\frac{[z_1,\dots,z_n;\psi]-[z_0,\dots,z_{n-1};\psi]}{z_n-z_0}. \tag{26}$$

It is easy to see that (26) is equivalent to

$$[z_0,\dots,z_n;\psi]=\sum_{i=0}^{n}\frac{\psi(z_i)}{q'(z_i)},\quad where\ q(z)=\prod_{j=0}^{n}(z-z_j).$$

The following definition of a real valued convex function is characterized by nth-order divided difference (see [8] (p. 15)).

Definition 2. *A function $\psi : [c,d] \to \mathbb{R}$ is said to be n-convex $(n \geq 0)$ if and only if for all choices of $(n+1)$ distinct points $z_0, \ldots, z_n \in [c,d]$, $[z_0, \ldots, z_n; \psi] \geq 0$ holds.*

If this inequality is reversed, then ψ is said to be n-concave. If the inequality is strict, then ψ is said to be a strictly n-convex (n-concave) function.

Remark 2. *Note that 0-convex functions are non-negative functions, 1-convex functions are increasing functions, and 2-convex functions are simply the convex functions.*

The following theorem gives an important criteria to examine the n-convexity of a function ψ (see [8] (p. 16)).

Theorem 2. *If $\psi^{(n)}$ exists, then ψ is n-convex if and only if $\psi^{(n)} \geq 0$.*

In this article, we use Fink's identity, Montgomery identities, and Green functions to prove some identities related to Steffensen's inequality. By using these identities we obtain a generalization of (4). In addition, we construct new identities which enable us to prove generalizations of inequalities (5) and (6) as one can obtain Classical Hardy-type inequalities from them, see [1]. We use Čebyšev functional to construct new bounds of Grüss and Ostrowski-type inequalities. Finally, we give several applications of our work.

2. Main Results

For our convenience, we use the following notations and assumptions:

$$\mathbb{S}_1(\psi, f, c, d) = \psi(f(c)) + \int_c^d \psi'(z) f'(z)\, dz - \psi(f(d)).$$

$$\mathbb{S}_2(\psi, f, d) = \int_0^d \psi'(z) f(z)\, dz - \psi\left(\int_0^d f(z)\, dz\right).$$

$$\mathbb{S}_3(\psi, f, w, d) = \int_0^d \Lambda(z)\psi'(z) f(z)\, dz - \int_0^d \lambda(t)\psi\left(\int_0^t f(z)\, dz\right) dt.$$

(A_1) For $n \in \mathbb{N}$, $n \geq 3$, let $\psi : [c,d] \to \mathbb{R}$ be n times differentiable function with $\psi^{(n-1)}$ absolutely continuous on $[c,d]$.

(A_2) For $n \in \mathbb{N}$, $n \geq 3$, let $\psi : [0,d] \to \mathbb{R}$ be n times differentiable function with $\psi(0) = 0$ and $\psi^{(n-1)}$ absolutely continuous on $[0,d]$.

The first part of this section is the generalization of (4). For this, we start with the following theorem:

Theorem 3. *Consider (A_1) with f be as in Corollary 1 (i) then:*

(a) For $j = 1, 2, 4, 5$, we have:

$$\mathbb{S}_1(\psi, f, c, d) = \frac{(n-2)(\psi'(d)-\psi'(c))}{d-c} \int\limits_c^d \mathbb{S}_1(G_{*,j}(.,u), f, c, d)du + \sum_{w=1}^{n-3} \left(\frac{n-w-2}{(d-c)w!}\right) \times$$
$$\left(\psi^{(w+1)}(d) \int\limits_c^d \mathbb{S}_1(G_{*,j}(.,u), f, c, d)(u-d)^w du - \psi^{(w+1)}(c) \int\limits_c^d \mathbb{S}_1(G_{*,j}(.,u), f, c, d)(u-c)^w du\right) \tag{27}$$
$$+ \frac{1}{(n-3)!(d-c)} \int\limits_c^d \psi^{(n)}(t) \left(\int\limits_c^d \mathbb{S}_1(G_{*,j}(.,u), f, c, d)(u-t)^{n-3} \mathcal{W}^{[c,d]}(t,u)du\right) dt.$$

(b) If $\psi'(c) = 0$, then

$$\mathbb{S}_1(\psi, f, c, d) = \frac{(n-2)(\psi'(d)-\psi'(c))}{d-c} \int\limits_c^d \mathbb{S}_1(G_{*,3}(.,u), f, c, d)du + \sum_{w=1}^{n-3} \left(\frac{n-w-2}{(d-c)w!}\right) \times$$
$$\left(\psi^{(w+1)}(d) \int\limits_c^d \mathbb{S}_1(G_{*,3}(.,u), f, c, d)(u-d)^w du - \psi^{(w+1)}(c) \int\limits_c^d \mathbb{S}_1(G_{*,3}(.,u), f, c, d)(u-c)^w du\right)$$
$$+ \frac{1}{(n-3)!(d-c)} \int\limits_c^d f^{(n)}(t) \left(\int\limits_c^d \mathbb{S}_1(G_{*,3}(.,u), f, c, d)(u-t)^{n-3} \mathcal{W}^{[c,d]}(t,u)du\right) dt.$$

Proof. (a) We first prove by fixing $j = 1$, other cases for $j = 2, 4, 5$ can be treated analogously. Utilizing (7) and (17) for ψ and ψ' respectively, we get

$$\mathbb{S}_1(\psi, f, c, d) = \psi(f(c)) - \psi(f(d)) + \int\limits_c^d \psi'(t) f'(t)\, dt = \frac{d-f(c)}{d-c}\psi(c) + \frac{f(c)-c}{d-c}\psi(d) +$$
$$\int\limits_c^d G_{*,1}(f(c), u)\psi''(u)\, du - \frac{d-f(d)}{d-c}\psi(c) - \frac{f(d)-c}{d-c}\psi(d) - \int\limits_c^d G_{*,1}(f(d), u)\psi''(u)\, du$$
$$+ \int\limits_c^d \left[\frac{\psi(d)-\psi(c)}{d-c} + \int\limits_c^d p_1(t,u)\psi''(u)\, du\right] f'(t)\, dt.$$

Simplifying and employing Fubini's theorem, we get

$$\begin{aligned}
\mathbb{S}_1(\psi, f, c, d) &= \frac{f(d)-f(c)}{d-c}\psi(c) - \frac{f(d)-f(c)}{d-c}\psi(d) \\
&+ \int\limits_c^d \left[G_{*,1}(f(c), u) - G_{*,1}(f(d), u)\right] \psi''(u)\, du \\
&+ \frac{\psi(d)-\psi(c)}{d-c}(f(d) - f(c)) + \int\limits_c^d \int\limits_c^d p_1(t,u) f'(t)\psi''(u)\, dt\, du \\
&= \int\limits_c^d \mathbb{S}_1(G_{*,1}(.,u), f, c, d)\psi''(u)\, du.
\end{aligned}$$

Now by replacing n with $n-2$ in (24) for ψ'', we have:

$$\mathbb{S}_1(\psi, f, c, d) = \int\limits_c^d \mathbb{S}_1(G_{*,1}(.,u), f, c, d)\Big(\frac{(n-2)(\psi'(d)-\psi'(c))}{d-c} + \sum_{w=1}^{n-3}\left(\frac{n-w-2}{(d-c)w!}\right) \times$$
$$\left(\psi^{(w+1)}(d)(u-d)^w - \psi^{(w+1)}(c)(u-c)^w\right) + \frac{1}{(n-3)!(d-c)} \int\limits_c^d (u-t)^{n-3} \mathcal{W}^{[c,d]}(t,u)\psi^{(n)}(t)dt\Big)\, du.$$

Rest follows from simplification and Fubini's theorem.

(b) Using assumption $\psi'(c) = 0$ and employing a similar method as in (a).

□

From the next two theorems we get a generalization of Steffensen's inequality and its reverse by generalizing (4) and its reverse.

Theorem 4. *Consider* (A_1) *with* f *be as in Corollary 1* (i) *and let*

$$(u-t)^{n-3}W^{[c,d]}(t,u) \geq 0. \tag{28}$$

(a) *If* ψ *is n-convex, then for each* $j \in \{1,2,3,4,5\}$ *(where* $\psi'(0) = 0$ *for* $j = 3$*), we have:*

$$\begin{gathered}\mathbb{S}_1(\psi, f, c, d) \geq \\ \frac{(n-2)(\psi'(d)-\psi'(c))}{d-c}\int_c^d \mathbb{S}_1(G_{*,1}(.,u), f, c, d)du + \sum_{w=1}^{n-3}\left(\frac{n-w-2}{(d-c)w!}\right) \times \\ \left(\psi^{(w+1)}(d)\int_c^d \mathbb{S}_1(G_{*,1}(.,u), f, c, d)(u-d)^w du - \psi^{(w+1)}(c)\int_c^d \mathbb{S}_1(G_{*,1}(.,u), f, c, d)(u-c)^w du\right).\end{gathered} \tag{29}$$

(b) *If* $-\psi$ *is n-convex, then for each* j, *(29) holds in the reverse direction.*

Proof. For each j, the function $G_{*,j}(.,u)$ is convex and differentiable. Since f is non-decreasing with $f(z) \leq z$, therefore Corollary 1 (i) gives $\mathbb{S}_1(G_{*,1}(.,u), f, c, d) \geq 0$. On the other hand, if ψ is n-convex ($-\psi$ is n-convex), then $\psi^{(n)}(z) \geq (\leq) 0$. Therefore, given assumption together with n-convexity of ψ ($-\psi$) implies $\int_c^d \psi^{(n)}(t)\left(\int_c^d \mathbb{S}_1(G_{*,j}(.,u), f, c, d)(u-t)^{n-3}W^{[c,d]}(t,u)du\right)dt \geq (\leq) 0$. The rest follows from (27). □

Theorem 5. *Consider* (A_1) *for even n and* f *as in Corollary 1* (i)*. Then*

(a) *If* ψ *is n-convex, then (29) holds.*
(b) *If* $-\psi$ *is n-convex, then the reverse of (29) holds.*
(c) *Let (29) (reverse of (29)) holds and*

$$\sum_{w=0}^{n-3}\left(\frac{n-w-2}{(d-c)w!}\right)\left(\psi^{(w+1)}(d)(u-d)^w du - \psi^{(w+1)}(c)(u-c)^w du\right) \geq (\leq) 0. \tag{30}$$

Then $\mathbb{S}_1(\psi, f, c, d) \geq (\leq)\ 0.$

Proof.
$(a),(b)$ We define

$$H(u,t) = (u-t)^{n-3}W^{[c,d]}(t,u) = \begin{cases} (u-t)^{n-3}(t-c), & \text{if } c \leq t \leq u \leq d, \\ (u-t)^{n-3}(t-d), & \text{if } c \leq u < t \leq d. \end{cases}$$

Clearly $H(u,t) \geq 0$ for even n. Consequently, we get (28), n-convexity of ψ ($-\psi$), and Theorem 4 (a) (Theorem 4 (b)) yields (29) (and its reverse).

(c) By definition of $G_{*,j}(.,u)$ and assumption on f, Corollary 1 (i) gives $\mathbb{S}_1(G_{*,j}(.,u), f, c, d) \geq 0$. Therefore, by using (30) and $\mathbb{S}_1(G_{*,j}(.,u), f, c, d) \geq 0$ in (29) (and its reverse), we get $\mathbb{S}_1(\psi, f, c, d) \geq (\leq)$ $(\leq)0$, which completes the proof. □

Now, we prove the following theorem which enables us to prove a generalization of (5).

Theorem 6. *Consider* (A_2) *and let* f *be as in Corollary 2* (i) *then:*

(a)

$$\mathbb{S}_2(\psi,f,d) = \tfrac{(n-2)(\psi'(d)-\psi'(0))}{d}\int_0^d \mathbb{S}_2(G_{*,j}(.,u),f,d)du + \textstyle\sum_{w=1}^{n-3}\left(\tfrac{n-w-2}{dw!}\right)$$
$$\times\left(\psi^{(w+1)}(d)\int_0^d \mathbb{S}_2(G_{*,j}(.,u),f,d)(u-d)^w du - \psi^{(w+1)}(0)\int_0^d \mathbb{S}_2(G_{*,j}(.,u),f,d)u^w du\right)$$
$$+\tfrac{1}{d(n-3)!}\int_0^d \psi^{(n)}(t)\Big(\int_0^d \mathbb{S}_2(G_{*,j}(.,u),f,d)(u-t)^{n-3}W^{[0,d]}(t,u)du\Big)dt$$

for $j = 1,2$.

(b) *If* $\psi'(0) = 0$*, then*

$$\mathbb{S}_2(\psi,f,d) + \psi(d) = \tfrac{(n-2)(\psi'(d)-\psi'(0))}{d}\int_0^d \mathbb{S}_2(G_{*,3}(.,u),f,d)du + \textstyle\sum_{w=1}^{n-3}\left(\tfrac{n-w-2}{dw!}\right)$$
$$\times\left(\psi^{(w+1)}(d)\int_0^d \mathbb{S}_2(G_{*,3}(.,u),f,d)(u-d)^w du - f^{(w+1)}(0)\int_0^d \mathbb{S}_2(G_{*,3}(.,u),f,d)u^w du\right)$$
$$+\tfrac{1}{d(n-3)!}\int_0^d \psi^{(n)}(t)\Big(\int_0^d \mathbb{S}_2(G_{*,3}(.,u),f,d)(u-t)^{n-3}W^{[0,d]}(t,u)du\Big)dt.$$

(c)

$$\mathbb{S}_2(\psi,f,d) + \psi(d) - d\psi'(d) =$$
$$\tfrac{(n-2)(\psi'(d)-\psi'(0))}{d}\int_0^d \mathbb{S}_2(G_{*,4}(.,u),f,d)du + \textstyle\sum_{w=1}^{n-3}\left(\tfrac{n-w-2}{dw!}\right)$$
$$\times\left(\psi^{(w+1)}(d)\int_0^d \mathbb{S}_2(G_{*,4}(.,u),f,d)(u-d)^w du - f^{(w+1)}(0)\int_0^d \mathbb{S}_2(G_{*,4}(.,u),f,d)u^w du\right)$$
$$+\tfrac{1}{d(n-3)!}\int_0^d \psi^{(n)}(t)\Big(\int_0^d \mathbb{S}_2(G_{*,4}(.,u),f,d)(u-t)^{n-3}W^{[0,d]}(t,u)du\Big)dt.$$

(d) *If* $\psi'(0) = 0$*, then*

$$\mathbb{S}_2(\psi,f,d) - d\psi'(d) = \tfrac{(n-2)(\psi'(d)-\psi'(0))}{d}\int_0^d \mathbb{S}_2(G_{*,5}(.,u),f,d)du + \textstyle\sum_{w=1}^{n-3}\left(\tfrac{n-w-2}{dw!}\right)$$
$$\times\left(\psi^{(w+1)}(d)\int_0^d \mathbb{S}_2(G_{*,5}(.,u),f,d)(u-d)^w du - f^{(w+1)}(0)\int_0^d \mathbb{S}_2(G_{*,5}(.,u),f,d)u^w du\right)$$
$$+\tfrac{1}{d(n-3)!}\int_0^d \psi^{(n)}(t)\Big(\int_0^d \mathbb{S}_2(G_{*,5}(.,u),f,d)(u-t)^{n-3}W^{[0,d]}(t,u)du\Big)dt.$$

Proof. We give proof of our results by fixing $j = 1$, and other cases can be proved in the similar way. By using (7) and (17) for ψ and ψ' respectively and applying assumption $\psi(0) = 0$, we get

$$\mathbb{S}_2(\psi,f,d) = \int_0^d \psi'(t)f(t)\,dt - \psi\left(\int_0^d f(t)\,dt\right) =$$
$$\int_0^d \tfrac{1}{d}\psi(d)f(t)\,dt + \int_0^d\left[\int_0^d \tfrac{\partial G_{*,1}(t,u)}{\partial t}\psi''(u)\,du\right]f(t)\,dt - \frac{\int_0^d f(t)\,dt}{d}\psi(d)$$
$$-\int_0^d G_{*,1}\left(\int_0^d f(t)\,dt,u\right)\psi''(u)\,du$$
$$= \int_0^d \mathbb{S}_2(G_{*,1}(.,u),f,d)\psi''(u)\,du.$$

Now replacing n with $n-2$ in (24) for ψ'' and simplifying we get the required identities. □

Our next result gives a generalization of (5).

Theorem 7. *Consider* (A_2), *f as in Corollary 2 (i) and let*

$$(u-t)^{n-3}W^{[0,d]}(t,u) \geq 0, \tag{31}$$

then the following hold:

(a) *If ψ is n-convex, then*
(i)

$$\mathbb{S}_2(\psi,f,d) \geq \frac{(n-2)(\psi'(d)-\psi'(0))}{d}\int_0^d \mathbb{S}_2(G_{*,j}(.,u),f,d)du + \sum_{w=1}^{n-3}\left(\frac{n-w-2}{dw!}\right) \times \left(\psi^{(w+1)}(d)\int_0^d \mathbb{S}_2(G_{*,j}(.,u),f,d)(u-d)^w du - f^{(w+1)}(0)\int_0^d \mathbb{S}_2(G_{*,j}(.,u),f,d)u^w du\right) \tag{32}$$

for $j=1,2$.

(ii) *If $\psi'(0)=0$, then*

$$\mathbb{S}_2(\psi,f,d)+\psi(d) \geq \frac{(n-2)(\psi'(d)-\psi'(0))}{d}\int_0^d \mathbb{S}_2(G_{*,3}(.,u),f,d)du + \sum_{w=1}^{n-3}\left(\frac{n-w-2}{dw!}\right) \times \left(\psi^{(w+1)}(d)\int_0^d \mathbb{S}_2(G_{*,3}(.,u),f,d)(u-d)^w du - f^{(w+1)}(0)\int_0^d \mathbb{S}_2(G_{*,3}(.,u),f,d)u^w du\right). \tag{33}$$

(iii)

$$\mathbb{S}_2(\psi,f,d)+\psi(d)-d\psi'(d) \geq \frac{(n-2)(\psi'(d)-\psi'(0))}{d}\int_0^d \mathbb{S}_2(G_{*,4}(.,u),f,d)du + \sum_{w=1}^{n-3}\left(\frac{n-w-2}{dw!}\right) \times \left(\psi^{(w+1)}(d)\int_0^d \mathbb{S}_2(G_{*,4}(.,u),f,d)(u-d)^w du - f^{(w+1)}(0)\int_0^d \mathbb{S}_2(G_{*,4}(.,u),f,d)u^w du\right). \tag{34}$$

(iv) *If $\psi'(0)=0$, then*

$$\mathbb{S}_2(\psi,f,d)-d\psi'(d) \geq \frac{(n-2)(\psi'(d)-\psi'(0))}{d}\int_0^d \mathbb{S}_2(G_{*,5}(.,u),f,d)du + \sum_{w=1}^{n-3}\left(\frac{n-w-2}{dw!}\right) \times \left(\psi^{(w+1)}(d)\int_0^d \mathbb{S}_2(G_{*,5}(.,u),f,d)(u-d)^w du - f^{(w+1)}(0)\int_0^d \mathbb{S}_2(G_{*,5}(.,u),f,d)u^w du\right). \tag{35}$$

(b) *Inequalities (32)–(35) are reversed provided that $-\psi$ is n-convex.*

Proof. The proof is similar to that of Theorem 4 except using Theorem 6 and Corollary 2 (i). □

Theorem 8. *Consider* (A_2) *for even n and f be as in Corollary 2 (i). Then*

(a) *If ψ is n-convex, then (32)–(35) hold.*
(b) *If $-\psi$ is n-convex, then the reverse of (32)–(35) holds.*
(c) *If any of (32)–(35) (reverse of (32)–(35)) hold and*

$$\sum_{w=0}^{n-3}\left(\frac{n-w-2}{dw!}\right)\left(\psi^{(w+1)}(d)(u-d)^w du - \psi^{(w+1)}(0)u^w du\right) \geq (\leq) 0. \tag{36}$$

Then $\mathbb{S}_2(\psi, f, d) \geq (\leq)\ 0$.

Proof. The proof is similar to that of Theorem 5 except using Theorem 7 and Corollary 2 (i). □

Next we give some generalized identities considering (6).

Theorem 9. *Consider* (A_2) *and let* f, λ *and* Λ *be as in Corollary 3* (i) *then:*

(a) *For* $j = 1, 2$, *we have*

$$\mathbb{S}_3(\psi, f, \lambda, d) - \tfrac{(n-2)(\psi'(d)-\psi'(0))}{d} \int_0^d \mathbb{S}_3(G_{*,j}(.,u), f, \lambda, d)du + \textstyle\sum_{w=1}^{n-3} \left(\tfrac{n-w-2}{dw!}\right)$$
$$\times \left(\psi^{(w+1)}(d) \int_0^d \mathbb{S}_3(G_{*,j}(.,u), f, \lambda, d)(u-d)^w du - \psi^{(w+1)}(0) \int_0^d \mathbb{S}_3(G_{*,j}(.,u), f, \lambda, d)u^w du\right)$$
$$+ \tfrac{1}{d(n-3)!} \int_0^d \psi^{(n)}(t) \left(\int_0^d \mathbb{S}_3(G_{*,j}(.,u), f, \lambda, d)(u-t)^{n-3} W^{[0,d]}(t,u)du\right)dt.$$

(b) *If* $\psi'(0) = 0$, *then*

$$\mathbb{S}_3(\psi, f, \lambda, d) + \psi(d) \int_0^d \lambda(x)\, dx =$$
$$\tfrac{(n-2)(\psi'(d)-\psi'(0))}{d} \int_0^d \mathbb{S}_3(G_{*,3}(.,u), f, \lambda, d)du + \textstyle\sum_{w=1}^{n-3} \left(\tfrac{n-w-2}{dw!}\right)$$
$$\times \left(\psi^{(w+1)}(d) \int_0^d \mathbb{S}_3(G_{*,3}(.,u), f, \lambda, d)(u-d)^w du - \psi^{(w+1)}(0) \int_0^d \mathbb{S}_3(G_{*,3}(.,u), f, \lambda, d)u^w du\right)$$
$$+ \tfrac{1}{d(n-3)!} \int_0^d \psi^{(n)}(t) \left(\int_0^d \mathbb{S}_3(G_{*,3}(.,u), f, \lambda, d)(u-t)^{n-3} W^{[0,d]}(t,u)du\right)dt.$$

(c)

$$\mathbb{S}_3(\psi, f, \lambda, d) + (\psi(d) - d\psi'(d)) \int_0^d \lambda(x)\, dx =$$
$$\tfrac{(n-2)(\psi'(d)-\psi'(0))}{d} \int_0^d \mathbb{S}_3(G_{*,4}(.,u), f, \lambda, d)du + \textstyle\sum_{w=1}^{n-3} \left(\tfrac{n-w-2}{dw!}\right)$$
$$\times \left(\psi^{(w+1)}(d) \int_0^d \mathbb{S}_3(G_{*,4}(.,u), f, \lambda, d)(u-d)^w du - \psi^{(w+1)}(0) \int_0^d \mathbb{S}_3(G_{*,4}(.,u), f, \lambda, d)u^w du\right)$$
$$+ \tfrac{1}{d(n-3)!} \int_0^d \psi^{(n)}(t) \left(\int_0^d \mathbb{S}_3(G_{*,4}(.,u), f, \lambda, d)(u-t)^{n-3} W^{[0,d]}(t,u)du\right)dt.$$

(d) *If* $\psi'(0) = 0$, *then*

$$\mathbb{S}_3(\psi, f, \lambda, d) - d\psi'(d) \int_0^d \lambda(x)\, dx =$$
$$\tfrac{(n-2)(\psi'(d)-\psi'(0))}{d} \int_0^d \mathbb{S}_3(G_{*,5}(.,u), f, \lambda, d)du + \textstyle\sum_{w=1}^{n-3} \left(\tfrac{n-w-2}{dw!}\right)$$
$$\times \left(\psi^{(w+1)}(d) \int_0^d \mathbb{S}_3(G_{*,5}(.,u), f, \lambda, d)(u-d)^w du - \psi^{(w+1)}(0) \int_0^d \mathbb{S}_3(G_{*,5}(.,u), f, \lambda, d)u^w du\right)$$
$$+ \tfrac{1}{d(n-3)!} \int_0^d \psi^{(n)}(t) \left(\int_0^d \mathbb{S}_3(G_{*,5}(.,u), f, \lambda, d)(u-t)^{n-3} W^{[0,d]}(t,u)du\right)dt.$$

Proof. We give a proof of our results by fixing $j = 1$, and other cases can be proved in a similar way. By using (7) and (17) for ψ and ψ' respectively and applying assumption $\psi(0) = 0$, we get:

$$\begin{aligned}
&\mathbb{S}_3(\psi, f, \lambda, d) = \int_0^d \Lambda(t)\psi'(t)f(t)\,dt - \int_0^d \lambda(x)\psi\left(\int_0^x f(t)\,dt\right)dx = \\
&\int_0^d \Lambda(t)f(t)\left[\tfrac{1}{d}\psi(d) + \int_0^d \tfrac{\partial G_{*,1}(t,u)}{\partial t}\psi''(u)\,du\right]dt - \int_0^d \lambda(x)\left[\tfrac{1}{d}\psi(d)\int_0^x f(t)\,dt\right. \\
&\left.+\int_0^d G_{*,1}\left(\int_0^x f(t)\,dt, u\right)\psi''(u)\,du\right]dx = \tfrac{1}{d}\psi(d)\left[\int_0^d \Lambda(t)f(t)\,dt - \int_0^d \lambda(x)\int_0^x f(t)\,dt\,dx\right] \\
&+\int_0^d \Lambda(t)f(t)\int_0^d \tfrac{\partial G_{*,1}(t,u)}{\partial t}\psi''(u)\,du\,dt - \int_0^d \lambda(x)\int_0^d G_{*,1}\left(\int_0^x f(t)\,dt, u\right)\psi''(u)\,du\,dx.
\end{aligned}$$

Since $\int_0^d \lambda(x)\int_0^x f(t)\,dt\,dx = \int_0^d f(t)\left(\int_t^d \lambda(x)\,dx\right)dt = \int_0^d \Lambda(t)f(t)dt$, therefore

$$\begin{aligned}
&\mathbb{S}_3(\psi, f, \lambda, d) \\
&= \int_0^d\left[\int_0^d \Lambda(t)f(t)\frac{\partial G_{*,1}(t,u)}{\partial t}\,dt - \int_0^d \lambda(x)G_{*,1}\left(\int_0^x f(t)\,dt, u\right)dx\right]\psi''(u)\,du \\
&= \int_0^d \mathbb{S}_3(G_{*,1}(.,u), f, \lambda, d)\psi''(u)\,du.
\end{aligned}$$

The rest follows from (24). □

Next, we present a generalization of (6).

Theorem 10. *Consider* (A_2) *and let* f, λ, Λ *be as in Corollary 3 (i) and (31) holds, then:*

(a) *If* ψ *is n-convex, then*

(i) *For* $j = 1, 2$, *we have*

$$\begin{aligned}
&\mathbb{S}_3(\psi, f, \lambda, d) \geq \tfrac{(n-2)(\psi'(d)-\psi'(0))}{d}\int_0^d \mathbb{S}_3(G_{*,j}(.,u), f, \lambda, d)du + \textstyle\sum_{w=1}^{n-3}\left(\tfrac{n-w-2}{dw!}\right) \\
&\times\left(\psi^{(w+1)}(d)\int_0^d \mathbb{S}_3(G_{*,j}(.,u), f, \lambda, d)(u-d)^w du - \psi^{(w+1)}(0)\int_0^d \mathbb{S}_3(G_{*,j}(.,u), f, \lambda, d)u^w du\right).
\end{aligned} \tag{37}$$

(ii) *If* $\psi'(0) = 0$, *then*

$$\begin{aligned}
&\mathbb{S}_3(\psi, f, \lambda, d) + \psi(d)\int_0^d \lambda(x)\,dx \geq \\
&\tfrac{(n-2)(\psi'(d)-\psi'(0))}{d}\int_0^d \mathbb{S}_3(G_{*,3}(.,u), f, \lambda, d)du + \textstyle\sum_{w=1}^{n-3}\left(\tfrac{n-w-2}{dw!}\right) \\
&\times\left(\psi^{(w+1)}(d)\int_0^d \mathbb{S}_3(G_{*,3}(.,u), f, \lambda, d)(u-d)^w du - \psi^{(w+1)}(0)\int_0^d \mathbb{S}_3(G_{*,3}(.,u), f, \lambda, d)u^w du\right).
\end{aligned} \tag{38}$$

(iii)

$$\begin{aligned}&\mathbb{S}_3(\psi,f,\lambda,d)+(\psi(d)-d\psi'(d))\int_0^d\lambda(x)\,dx\geq\\&\frac{(n-2)(\psi'(d)-\psi'(0))}{d}\int_0^d\mathbb{S}_3(G_{*,4}(.,u),f,\lambda,d)du+\sum_{w=1}^{n-3}\left(\frac{n-w-2}{dw!}\right)\\&\times\left(\psi^{(w+1)}(d)\int_0^d\mathbb{S}_3(G_{*,4}(.,u),f,\lambda,d)(u-d)^w du-\psi^{(w+1)}(0)\int_0^d\mathbb{S}_3(G_{*,4}(.,u),f,\lambda,d)u^w du\right).\end{aligned}\tag{39}$$

(iv) If $\psi'(0)=0$, then

$$\begin{aligned}&\mathbb{S}_3(\psi,f,\lambda,d)-d\psi'(d)\int_0^d\lambda(x)\,dx\geq\\&\frac{(n-2)(\psi'(d)-\psi'(0))}{d}\int_0^d\mathbb{S}_3(G_{*,5}(.,u),f,\lambda,d)du+\sum_{w=1}^{n-3}\left(\frac{n-w-2}{dw!}\right)\\&\times\left(\psi^{(w+1)}(d)\int_0^d\mathbb{S}_3(G_{*,5}(.,u),f,\lambda,d)(u-d)^w du-\psi^{(w+1)}(0)\int_0^d\mathbb{S}_3(G_{*,5}(.,u),f,\lambda,d)u^w du\right).\end{aligned}\tag{40}$$

(b) *Inequalities (37)–(40) are reversed provided that $-\psi$ is n-convex.*

Proof. The proof is similar to that of Theorem 4 except using Theorem 9 and Corollary 3 (i). □

Theorem 11. *Consider (A_2) for even n and let f, λ, and Λ be as in Corollary 3 (i). Then*

(a) *If ψ is n-convex, then (37)–(40) hold.*
(b) *If $-\psi$ is n-convex, then the reverses of (37)–(40) hold.*
(c) *If any of (37)–(40) (reverse of (37)–(40)) hold and (36) is valid. Then $\mathbb{S}_3(\psi,f,\lambda,d)\geq(\leq)\,0$.*

Proof. The proof is similar to that of Theorem 5 except using Theorem 10 and Corollary 3 (i). □

3. New Upper Bounds Via Čebyšev Functional

Consider the Čebyšev functional for two Lebesgue integrable functions $\mathbb{F}_1,\mathbb{F}_2:[c,d]\to\mathbb{R}$ given as:

$$T(\mathbb{F}_1,\mathbb{F}_2)=\frac{1}{d-c}\int_c^d\mathbb{F}_1(\xi)\mathbb{F}_2(\xi)d\xi-\frac{1}{d-c}\int_c^d\mathbb{F}_1(\xi)d\xi.\frac{1}{d-c}\int_c^d\mathbb{F}_2(\xi)d\xi.$$

Cerone and Dragomir in [17] proposed new bounds utilizing Čebyšev functional given as:

Theorem 12. *For $\mathbb{F}_1\in L[c,d]$ and $\mathbb{F}_2:[c,d]\to\mathbb{R}$ be an absolutely continuous function along with $(.-c)(d-.)[\mathbb{F}_2']^2\in L[c,d]$. The following inequality holds*

$$|T(\mathbb{F}_1,\mathbb{F}_2)|\leq\frac{1}{\sqrt{2}}\left[\frac{T(\mathbb{F}_1,\mathbb{F}_1)}{(d-c)}\right]^{\frac{1}{2}}\left(\int_c^d(\xi-c)(d-\xi)[\mathbb{F}_2'(\xi)]^2d\xi\right)^{\frac{1}{2}}.\tag{41}$$

Theorem 13. *For $\mathbb{F}_1:[c,d]\to\mathbb{R}$ be an absolutely continuous with $\mathbb{F}_1'\in L_\infty[c,d]$ and $\mathbb{F}_2:[c,d]\to\mathbb{R}$ is an increasing function. The following inequality holds*

$$|T(\mathbb{F}_1,\mathbb{F}_2)|\leq\frac{||\mathbb{F}_1'||_\infty}{2(d-c)}\int_c^d(\xi-c)(d-\xi)d\mathbb{F}_2(\xi).\tag{42}$$

The constants $\frac{1}{\sqrt{2}}$ and $\frac{1}{2}$ are the optimal constants.

Now we utilize the above theorems to construct new upper bounds for our obtained generalized identities. For our convenience we denote

$$\mathfrak{O}_j(t) = \int_c^d \mathbb{S}_1(G_{*,j}(.,u), f, c, d)(u-t)^{n-3} W^{[c,d]}(t,u)du, \ \ t \in [c,d], \tag{43}$$

for $\{j = 1, \ldots, 5\}$. Consider the Čebyšev functional $T_j(\mathfrak{O}_j, \mathfrak{O}_j)$ $\{j = 1, \ldots, 5\}$ given as:

$$T_j(\mathfrak{O}_j, \mathfrak{O}_j) = \frac{1}{d-c}\int_c^d \mathfrak{O}_j^2(\xi)d\xi - \left(\frac{1}{d-c}\int_c^d \mathfrak{O}_j(\xi)d\xi\right)^2. \tag{44}$$

Grüss type inequalities associated with Theorems 12 and 13 can be given as:

Theorem 14. *Under the assumptions of Theorem 3, let $\psi : [c,d] \to \mathbb{R}$ be absolutely continuous along with $(.-c)(d-.)[\psi^{(n+1)}]^2 \in L[c,d]$ and $\mathfrak{O}_j$ $\{j = 1,2,3,4,5\}$ be defined as in (43). Then*

$$\begin{aligned}&\mathbb{S}_1(\psi, f, c, d) - \sum_{w=0}^{n-3}\left(\frac{n-w-2}{(d-c)w!}\right) \times \\ &\left(\psi^{(w+1)}(d)\int_c^d \mathbb{S}_1(G_{*,1}(.,u), f, c, d)(u-d)^w du - \psi^{(w+1)}(c)\int_c^d \mathbb{S}_1(G_{*,1}(.,u), f, c, d)(u-c)^w du\right) \\ &- \frac{\psi^{(n-1)}(d) - \psi^{(n-1)}(c)}{(d-c)^2(n-3)!}\int_c^d \mathfrak{O}_j(t)dt = Rem(c, d, \mathfrak{O}_j, \psi^{(n)})\end{aligned} \tag{45}$$

where

$$|Rem(c,d,\mathfrak{O}_j,\psi^{(n)})| \leq \frac{1}{\sqrt{2}\,(n-3)!}\left[\frac{T_j(\mathfrak{O}_j,\mathfrak{O}_j)}{(d-c)}\right]^{\frac{1}{2}} \left|\int_c^d (t-c)(d-t)[\psi^{(n+1)}(t)]^2 dt\right|^{\frac{1}{2}}.$$

Proof. Fix $\{j = 1, \ldots, 5\}$. Using Čebyšev functional for $\mathbb{F}_1 = \mathfrak{O}_j$, $\mathbb{F}_2 = \psi^{(n)}$ and by comparing (45) with (27), we have

$$Rem(c,d,\mathfrak{O}_j,\psi^{(n)}) = \frac{1}{(n-3)!}T_j(\mathfrak{O}_j,\psi^{(n)}).$$

Employing Theorem 12 for the new functions, we get the required bound. □

Theorem 15. *Under the assumptions of Theorem 3, let $\psi : [c,d] \to \mathbb{R}$ be absolutely continuous along with $\psi^{(n+1)} \geq 0$ and $\mathfrak{O}_j$ $\{j = 1,2,3,4,5\}$ be defined as in (44). Then $Rem(c,d,\mathfrak{O}_j,\psi^{(n)})$ in (45) satisfies a bound*

$$|Rem(c,d,\mathfrak{O}_j,\psi^{(n)})| \leq \frac{||\mathfrak{O}_j'||_\infty}{(n-3)!}\left[\frac{\psi^{(n-1)}(d) + \psi^{(n-1)}(c)}{2} - \frac{\psi^{(n-2)}(d) - \psi^{(n-2)}(c)}{d \quad c}\right]. \tag{46}$$

Proof. In the proof of Theorem 14, we have established that

$$Rem(c,d,\mathfrak{O}_j,\psi^{(n)}) = \frac{1}{(n-3)!}T_j(\mathfrak{O}_j,\psi^{(n)}).$$

Now applying Theorem 13 for $\mathbb{F}_1 = \mathfrak{O}_j$, $\mathbb{F}_2 = \psi^{(n)}$, we have

$$|Rem(c,d,\mathfrak{O}_j,\psi^{(n)})| = \frac{1}{(n-3)!}|T_j(\mathfrak{O}_j,\psi^{(n)})|$$

$$\leq \frac{||\mathfrak{O}_j'||_\infty}{2(d-c)\,(n-3)!}\int_c^d (t-c)(d-t)\psi^{(n+1)}(t)dt$$

Now since

$$\int_c^d (t-c)(d-t)\psi^{(n+1)}(t)dt = \int_c^d [2t-(c+d)]\psi^{(n)}(t)dt$$
$$= (d-c)\left[\psi^{(n-1)}(d)+\psi^{(n-1)}(c)\right] - 2(\psi^{(n-2)}(d) - f^{(n-2)}(c))$$

therefore the required bound in (46) follows. □

Ostrowski-type inequalities associated with generalized Steffensen's inequality can be given as:

Theorem 16. *Under the assumptions of Theorem 3, let $|\psi^{(n)}|^s : [c,d] \to \mathbb{R}$ be a R-integrable function and consider (s,s') pair of conjugate exponents from $[1,\infty]$ such that $\frac{1}{s}+\frac{1}{s'}=1$. Then, we have*

$$\left| \begin{array}{c} \mathbb{S}_1(\psi,f,c,d) - \sum_{w=0}^{n-3}\left(\frac{n-w-2}{(d-c)w!}\right)\times \\ \left(\psi^{(w+1)}(d)\int_c^d \mathbb{S}_1(G_{*,1}(.,u),f,c,d)(u-d)^w du - \psi^{(w+1)}(c)\int_c^d \mathbb{S}_1(G_{*,1}(.,u),f,c,d)(u-c)^w du\right) \end{array} \right|$$
$$\leq \frac{||\psi^{(n)}||_s}{(n-3)!(d-c)}\left(\int_c^d \left|\int_c^d \mathbb{S}_1(G_{*,j}(.,u),f,c,d)(u-t)^{n-3}W^{[c,d]}(t,u)du\right|^{s'} dt\right)^{1/s'}. \quad (47)$$

The constant on the R.H.S. of (47) is sharp for $1 < s \leq \infty$ and the best possible for $s=1$.

Proof. Fix $\{j=1,\ldots,5\}$. Let us denote by

$$\Im_j = \frac{1}{(n-3)!(d-c)}\left(\int_c^d \mathbb{S}_1(G_{*,j}(.,u),f,c,d)(u-t)^{n-3}W^{[c,d]}(t,u)du\right), \quad t\in[c,d].$$

Using identity (27), we find

$$\left| \begin{array}{c} \mathbb{S}_1(\psi,f,c,d) - \sum_{w=0}^{n-3}\left(\frac{n-w-2}{(d-c)w!}\right)\times \\ \left(\psi^{(w+1)}(d)\int_c^d \mathbb{S}_1(G_{*,1}(.,u),f,c,d)(u-d)^w du - \psi^{(w+1)}(c)\int_c^d \mathbb{S}_1(G_{*,1}(.,u),f,c,d)(u-c)^w du\right) \end{array} \right|$$
$$= \left|\int_c^d \Im_j(t)\psi^{(n)}(t)dt\right|. \quad (48)$$

Applying Hölder's inequality for integrals on the R. H. S. of (48), we obtain

$$\left|\int_c^d \Im_j(t)\psi^{(n)}(t)dt\right| \leq \left(\int_c^d \left|\psi^{(n)}(t)\right|^s dt\right)^{\frac{1}{s}} \left(\int_c^d |\Im_j(t)|^{s'} dt\right)^{\frac{1}{s'}},$$

which combined together with (48) gives (47).

For sharpness of the constant $\left(\int_c^d |\Im_j(t)|^{s'} dt\right)^{1/s'}$ let us define the function ψ for which the equality in (47) holds.

For $1 < s \leq \infty$ let ψ be such that

$$\psi^{(n)}(t) = \operatorname{sgn}\Im_j(t)|\Im_j(t)|^{\frac{1}{s-1}}$$

and for $s=\infty$ let $\psi^{(n)}(t) = \operatorname{sgn}\Im_j(t)$.

For $s = 1$, we shall show that

$$\left| \int_c^d \mathfrak{I}_j(t)\psi^{(n)}(t)dt \right| \leq \max_{t\in[c,d]} |\mathfrak{I}_j(t)| \left(\int_c^d \psi^{(n)}(t)dt \right) \tag{49}$$

is the best possible inequality. Suppose that $|\mathfrak{I}_j(t)|$ attains its maximum at $t_0 \in [c,d]$. To start with first we assume that $\mathfrak{I}_j(t_0) > 0$. For Θ small enough we define $\psi_\Theta(t)$ by

$$\psi_\Theta(t) = \begin{cases} 0, & c \leq t \leq t_0, \\ \frac{1}{\Theta n!}(t-t_0)^n, & t_o \leq t \leq t_0+\Theta, \\ \frac{1}{n!}(t-t_0)^{n-1}, & t_0+\Theta \leq t \leq d. \end{cases}$$

Then for Θ small enough

$$\left| \int_c^d \mathfrak{I}_j(t)\psi^{(n)}(t)dt \right| = \left| \int_{t_0}^{t_0+\Theta} \mathfrak{I}_j(t)\frac{1}{\Theta}dt \right| = \frac{1}{\Theta}\int_{t_0}^{t_0+\Theta} \mathfrak{I}_j(t)dt.$$

Now from inequality (49), we have

$$\frac{1}{\Theta}\int_{t_0}^{t_0+\Theta} \mathfrak{I}_j(t)dt \leq \mathfrak{I}_j(t_0)\int_{t_0}^{t_0+\Theta}\frac{1}{\Theta}dt = \mathfrak{I}_j(t_0).$$

Since

$$\lim_{\Theta\to 0}\frac{1}{\Theta}\int_{t_0}^{t_0+\Theta} \mathfrak{I}_j(t)dt = \mathfrak{I}_j(t_0),$$

the statement follows. In the case when $\mathfrak{I}_j(t_0) < 0$, we define $f_\Theta(t)$ by

$$\psi_\Theta(t) = \begin{cases} \frac{1}{n!}(t-t_0-\Theta)^{n-1}, & c \leq t \leq t_0, \\ \frac{-1}{\Theta n!}(t-t_0-\Theta)^n, & t_o \leq t \leq t_0+\Theta, \\ 0, & t_0+\Theta \leq t \leq d, \end{cases}$$

then the rest of the proof is the same as above. □

Remark 3. *Similar bounds of Grüss and Ostrowski-type inequalities can be obtained by using Theorems 6 and 9.*

4. Monotonic Steffensen's-Type Functionals

The notion of $(n+1)$-convex function at a point was introduced in [18]. In the current section, we define some linear functionals from the differences of the generalized Steffensen's-type inequalities. By proving monotonicity of these functionals, we obtain new inequalities which contribute to the theory of more generalized class of functions, i.e., $(n+1)$-convex functions at a point. Below is the definition of $(n+1)$-convex function at point, see [18].

Definition 3. *Let $I \subseteq \mathbb{R}$ be an interval, $\xi \in I^0$ and $n \in \mathbb{N}$. A function $f : I \to \mathbb{R}$ is said to be $(n+1)$-convex at point ξ if there exists a constant K_ξ such that the function*

$$F(x) = f(x) - K_\xi \frac{x^n}{n!}$$

is n-concave on $I \cap (-\infty, \xi]$ and n-convex on $I \cap [\xi, \infty)$.

Pečarić et al. in [18] studied necessary and sufficient conditions on two linear functionals $\Omega : C([\delta_1, \xi]) \to \mathbb{R}$ and $\Gamma : C([\xi, \delta_2] \to \mathbb{R}$ so that the inequality $\Omega(f) \leq \Gamma(f)$ holds for every function f that is $(n+1)$-convex at point ξ. In this section, we define some linear functionals and obtained certain inequalities associated with these linear functionals. Let $n \in \mathbb{N}$ be even, $\psi : [c,d] \to \mathbb{R}$ be n times differentiable function with $\psi^{(n-1)}$ absolutely continuous on $[c,d]$. Let c_1, $c_2 \in [c,d]$ and $\xi \in (c,d)$, where $c_1 < \xi < c_2$. Let $f_1 : [c_1, \xi] \to \mathbb{R}$ and $f_2 : [\xi, c_2] \to \mathbb{R}$ be increasing with $f_i(t) \leq t$ for $i = 1,2$. For $j = 1,2,\ldots,5$, we construct:

$$\begin{aligned}&\Omega_{1,j}(\psi) = \mathbb{S}_1(\psi, f_1, c_1, \xi) - \textstyle\sum_{w=0}^{n-3}\left(\frac{n-w-2}{(\xi-c_1)w!}\right) \times \\ &\left(\psi^{(w+1)}(\xi)\int\limits_{c_1}^{\xi}\mathbb{S}_1(G_{*,1}(.,u), f_1, c_1, \xi)(u-\xi)^w du - \psi^{(w+1)}(c_1)\int\limits_{c_1}^{\xi}\mathbb{S}_1(G_{*,1}(.,u), f_1, c_1, \xi)(u-c_1)^w du\right)\end{aligned} \tag{50}$$

and

$$\begin{aligned}&\Gamma_{1,j}(\psi) = \mathbb{S}_1(\psi, f_2, \xi, c_2) - \textstyle\sum_{w=0}^{n-3}\left(\frac{n-w-2}{(c_2-\xi)w!}\right) \times \\ &\left(\psi^{(w+1)}(c_2)\int\limits_{\xi}^{c_2}\mathbb{S}_1(G_{*,1}(.,u), f_2, \xi, c_2)(u-c_2)^w du - \psi^{(w+1)}(\xi)\int\limits_{\xi}^{c_2}\mathbb{S}_1(G_{*,1}(.,u), f_2, \xi, c_2)(u-\xi)^w du\right).\end{aligned} \tag{51}$$

Theorem 5 (a) enables $\Gamma_{1,j}(\psi) \geq 0$ for $j = 1,2,\ldots,5$ (and $\psi'(0) = 0$ for $j = 3$), provided that ψ is n-convex. Furthermore, Theorem 5 (b) enables $\Omega_{1,j}(\psi) \leq 0$ for $j = 1,2,\ldots,5$ (and $f'(0) = 0$ for $j = 3$), provided that $-\psi$ is n-convex.

Theorem 17. *Let ψ, f_1, f_2 be as defined above and $\psi : [c,d] \to \mathbb{R}$ be $(n+1)$-convex at a point ξ for even $n > 3$. If $\Omega_{1,j}(P_n) = \Gamma_{1,j}(P_n)$, for all $j = 1,2,\ldots,5$ and $\psi'(0) = 0$ for $j = 3$, where $P_n(u) = u^n$ then:*

$$\Omega_{1,j}(\psi) \leq \Gamma_{1,j}(\psi),$$

for $j = 1,2,\ldots,5$.

Proof. Since ψ is $(n+1)$-convex, it follows from Definition 3 that there exist K_ξ such that $\Psi(u) = \psi(u) - \frac{K_\xi u^n}{n!}$ is n-concave on $[c_1, \xi]$ and n-convex on $[\xi, c_2]$. Therefore, for each $j = 1,2,\ldots,5$, we have

$$\Omega_{1,j}(\psi) - \frac{K_\xi}{n!}\Omega_{1,j}(P_n) = \Omega_{1,j}(\Psi) \leq 0 \leq \Gamma_{1,j}(\Psi) = \Gamma_{1,j}(\psi) - \frac{K_\xi}{n!}\Gamma_{1,j}(P_n).$$

Since $\Omega_{1,j}(P_n) = \Gamma_{1,j}(P_n)$, therefore $\Omega_{1,j}(\psi) \leq \Gamma_{1,j}(\psi)$, which completes the proof. □

Remark 4. *We may proceed further by defining linear functionals with the inequalities proved in Theorems 8 and 11. Moreover, by proving monotonicity of new functionals we extend the inequalities in Theorems 8 and 11.*

5. Application to Exponentially Convex Functions

We start this section by an important Remark given as:

Remark 5. *By the virtue of Theorem 4 (a), for $j = 1,2,\ldots,5$, we define the positive linear functionals with respect to n-convex function ψ as follows*

$$\begin{aligned}&\Delta_{1,j}(\psi) := \mathbb{S}_1(\psi, f, c, d) - \textstyle\sum_{w=0}^{n-3}\left(\frac{n-w-2}{(d-c)w!}\right) \times \\ &\left(\psi^{(w+1)}(d)\int\limits_{c}^{d}\mathbb{S}_1(G_{*,1}(.,u), f, c, d)(u-d)^w du - \psi^{(w+1)}(c)\int\limits_{c}^{d}\mathbb{S}_1(G_{*,1}(.,u), f, c, d)(u-c)^w du\right) \geq 0.\end{aligned} \tag{52}$$

Next we construct the non trivial examples of exponentially convex functions (see [19]) from positive linear functionals $\Delta_{1,j}(\psi)$ for $(j = 1,2,\ldots,5)$.

For this consider the family of real valued functions on $[0,\infty)$ given as

$$\psi_s(u) = \begin{cases} \frac{u^s}{s(s-1)\cdots(s-n+1)}, & s \notin \{0,1,\ldots,n-1\}; \\ \frac{u^t \ln u}{(-1)^{n-1-t} t!(n-1-t)!}, & s = t \in \{0,1,\ldots,n-1\}. \end{cases} \tag{53}$$

It is interesting to note that this is a family of n-convex functions as

$$\frac{d^n}{du^n}\psi_s(u) = u^{s-n} \geq 0.$$

Since $s \mapsto u^{s-n} = e^{(s-n)\ln u}$ is exponentially convex function, therefore the mapping $s \mapsto \Delta_{1,j}(\psi_s)$ is exponential convex and as a special case, it is also log-convex mapping. The log-convexity of this mapping enables us to construct the known Lyapunov inequality given as

$$\left(\Delta_{1,j}(\psi_s)\right)^{t-r} \leq \left(\Delta_{1,j}(\psi_r)\right)^{t-s}\left(\Delta_{1,j}(\psi_t)\right)^{s-r} \tag{54}$$

for $r,s,t \in \mathbb{R}$ such that $r < s < t$ where $j = 1,2,\ldots,5$.

Remark 6. *We have not given the proof of the above mentioned results (see [19] for details). The Lyapunov inequality empowered us to refine lower (upper) bound for action of the functional on the class of functions given in (53) because if exponentially convex mapping attains zero value at some point it is zero everywhere (see [19]).*

One can also consider some other classes of n-convex functions given in the paper [19,20] and can get similar estimations. A similar technique can also be employed by considering the results of Theorems 7 and 10.

6. Conclusions and Outlooks

In this article, we extended the pool of inequalities by proving generalizations of well-known Steffensen's inequalities and their reverses. The inequalities proved in the main results provide generalizations of the results from [1,4,7,11]. Moreover, Hardy's inequality is also one of the well-known inequalities. In this article, we also proved generalizations of inequalities, from [1], which are closely related to Hardy's inequality. We also developed new bounds of Grüss and Ostrowski-type inequalities. Further, the contribution of these inequalities to the theory of $(n+1)$-convex functions has been presented by defining functionals from new inequalities and describing their properties. Lastly, new inequalities related to exponentially convex functions and log-convex functions, such as the Lyapunov inequality, have been developed. In the future, it can be investigated whether we can use other interpolations, such as Hermite interpolation, to prove new generalizations of Steffensen's inequality and related results.

Author Contributions: Supervision, J.P.; Writing original draft, S.I.B.; Writing - review and editing, A.F. and S.I.B.

Funding: The research of S.I.B. was supported by the H.E.C. Pakistan under NRPU project 5327. The research of J.P. was supported by the Ministry of Education and Science of the Russian Federation (the Agreement number No. 02.a03.21.0008).

Conflicts of Interest: The authors declare no conflict of interest.

References

1. Fahad, A.; Pečarić, J.; Praljak, M. Generalized Steffensen's Inequaliy. *J. Math. Inequal.* **2015**, *9*, 481–487.
2. Iqbal, S.; Pečarić, J.; Samraiz, M. Hardy-type Inequalities in Quotients Involving Fractional Calculus Operators. *J. Math. Anal.* **2017**, *8*, 47–70.
3. Irshad, N.; Khan, A.R. Generalization of Ostrowski Inequality for Differentiable Functions and its Applications to Numerical Quadrature Rules. *J. Math. Anal.* **2017**, *8*, 79–102.

4. Rabier, P. Steffensen's inequality and $L^1 - L^\infty$ estimates of weighted integrals. *Proc. Am. Math. Soc.* **2012**, *140*, 665–675.
5. Nosheen, A.; Bibi, R.; Pečarić, J. Jensen-Steffensen inequality for diamond integrals, its converse and improvements via Green function and Taylor's formula. *Aequat. Math.* **2018**, *92*, 289–309.
6. Loss, M.; Sloane, C. Hardy inequalities for fractional integrals on general domains. *J. Funct. Anal.* **2010**, *259*, 1369–1379.
7. Steffensen, J.F. On certain inequalities between mean values, and their application to actuarial problems. *Scand. Actuar. J.* **1918**, *1*, 82–97.
8. Pečarić, J.; Proschan, F.; Tong, Y.L. *Convex Functions, Partial Orderings and Statistical Applications*; Academic Press: New York, NY, USA, 1992.
9. Pečarić, J.; Smoljak Kalamir, K.; Varošanec, S. *Steffensen's and Related Inequalities (A Comprehensive Survey and Recent Advances)*; Element: Zagreb, Croatia, 2014.
10. Fahad. A.; Jakšetić, J.; Pečarić, J. On Rabier's result and nonbounded Montgomery's identity. *Math. Inequal. Appl.* **2019**, *22*, 175–180.
11. Pečarić, J. Connections among some inequalities of Gauss, Steffensen and Ostrowski. *Southeast Asian Bull. Math.* **1989**, *13*, 89–91.
12. Hewitt, E.; Stromberg, K. *Real and Abstract Analysis*, 3rd ed.; Springer: New York, NY, USA, 1975.
13. Fahad, A.; Pečarić, J.; Praljak, M. Hermite Interpolation of composition function and Steffensen-type Inequalities. *J. Math. Inequal.* **2016**, *10*, 1051–1062.
14. Fahad, A.; Pečarić, J.; Qureshi, M.I. Generalized Steffensen's Inequaliy by Lidstone Interpolation and Montgomery Identity. *J. Inequal. Appl.* **2018**, *2018*, 237.
15. Mehmood, N.; Agrwal, R.P.; Butt, S.I.; Pečarić, J. New Generalizations of Popoviciu-type inequalities via new Green's functions and Montgomery identity. *J. Inequal. Appl.* **2017**, *2017*, 108.
16. Fink, A.M. Bounds of the deviation of a function from its averages. *Czechoslovak Math.* **1992**, *42*, 289–310.
17. Cerone, P.; Dragomir, S.S. Some new Ostrowski-type bounds for the Čebyšev functional and applications. *J. Math. Inequal.* **2014**, *8*, 159–170.
18. Pečarić, J.; Praljak, M.; Witkowski, A. Linear operator inequality for n-convex functions at a point. *Math. Inequal. Appl.* **2015**, *18*, 1201–1217.
19. Jakšetić, J.; Pečarić, J. Exponential Convexity method. *J. Convex Anal.* **2013**, *20*, 181–187.
20. Butt, S.I.; Khan, K.A.; Pečarić, J. Popoviciu Type Inequalities via Green Function and Generalized Montgomery Identity. *Math. Inequal. Appl.* **2015**, *18*, 1519–1538.

mathematics

MDPI

Article

On a Reverse Half-Discrete Hardy-Hilbert's Inequality with Parameters

Bicheng Yang [1], Shanhe Wu [2,*] and Aizhen Wang [3]

[1] Institute of Applied Mathematics, Longyan University, Longyan 364012, China; bcyang@gdei.edu.cn
[2] Department of Mathematics, Longyan University, Longyan 364012, China
[3] Department of Mathematics, Guangdong University of Education, Guangzhou 510303, China; ershimath@163.com
* Correspondence: shanhewu@163.com

Received: 22 August 2019; Accepted: 29 October 2019; Published: 4 November 2019

Abstract: By means of the weight functions, the idea of introduced parameters, and the Euler-Maclaurin summation formula, a reverse half-discrete Hardy-Hilbert's inequality and the reverse equivalent forms are given. The equivalent statements of the best possible constant factor involving several parameters are considered. As applications, two results related to the case of the non-homogeneous kernel and some particular cases are obtained.

Keywords: weight function; half-discrete Hardy-Hilbert's inequality; parameter; Euler-Maclaurin summation formula; reverse inequality

MSC: 26D15; 26D10; 26A42

1. Introduction

If $0 < \sum_{m=1}^{\infty} a_m^2 < \infty$ and $0 < \sum_{n=1}^{\infty} b_n^2 < \infty$, then we have the following discrete Hilbert's inequality with the best possible constant factor π (cf., [1], Theorem 315):

$$\sum_{m=1}^{\infty}\sum_{n=1}^{\infty}\frac{a_m b_n}{m+n} < \pi(\sum_{m=1}^{\infty} a_m^2 \sum_{n=1}^{\infty} b_n^2)^{1/2} \tag{1}$$

Correspondingly, if $0 < \int_0^\infty f^2(x)dx < \infty$ and $0 < \int_0^\infty g^2(y)dy < \infty$, we still have the following Hilbert's integral inequality (cf., [1], Theorem 316):

$$\int_0^\infty \int_0^\infty \frac{f(x)g(y)}{x+y}dxdy < \pi(\int_0^\infty f^2(x)dx \int_0^\infty g^2(y)dy)^{1/2} \tag{2}$$

where the constant factor π is the best possible.

As is known to us, Inequalities (1) and (2) and their extensions with conjugate exponents as well as independent parameters play an important role in analysis and their applications (cf., [2–13]).

Concerning with Inequalities (1) and (2), we have the following half-discrete Hilbert-type inequality (cf., [1], Theorem 351):

If $K(x)(x > 0)$ is a decreasing function and $p > 1, \frac{1}{p} + \frac{1}{q} = 1, 0 < \phi(s) = \int_0^\infty K(x)x^{s-1}dx < \infty$, $f(x) \geq 0, 0 < \int_0^\infty f^p(x)dx < \infty$, then

$$\sum_{n=1}^{\infty} n^{p-2}(\int_0^\infty K(nx)f(x)dx)^p < \phi^p(\frac{1}{q})\int_0^\infty f^p(x)dx. \tag{3}$$

In recent years, some new extensions of the Inequality (3) were provided in [14–19].

In 2006, with the help of the Euler-Maclaurin summation formula, Krnic et al. [20] gave an extension of (1) with the kernel $\frac{1}{(m+n)^{\lambda}}(0 < \lambda \leq 14)$. In 2019, Adiyasuren et al. [21] considered an extension of (1) with $p, q > 1(\frac{1}{p} + \frac{1}{q} = 1)$ involving the partial sums. In 2016–2017, by using the weight functions, Hong [22,23] considered some equivalent statements of the extensions of (1) and (2) with several parameters. Some related works can be found in [24–26].

In this paper, following the way of [20,22], by using the weight functions, the idea of introduced parameters, and the Euler-Maclaurin summation formula, a reverse half-discrete Hardy-Hilbert's inequality with the homogeneous kernel $\frac{1}{(x+n)^{\lambda}}(0 < \lambda \leq 5)$ and the reverse equivalent forms are established. The equivalent statements of the best possible constant factor related to several parameters are presented. As applications, two corollaries related to the case of the non-homogeneous kernel and some particular cases are obtained.

2. Some Lemmas

In what follows, we assume that

$$0 < p < 1(q < 0), \frac{1}{p} + \frac{1}{q} = 1, \lambda \in (0, 5], \sigma \in (0, 2] \cap (0, \lambda), \mu \in (0, \lambda),$$

$f(x) \geq 0\ (x \in R_{+} = (0, \infty)), a_n \geq 0\ (n \in \mathrm{N} = \{1, 2, \cdots\})$ satisfying

$$0 < \int_0^{\infty} x^{p[1-(\frac{\lambda-\sigma}{p}+\frac{\mu}{q})]-1} f^p(x)dx < \infty \ and \ 0 < \sum_{n=1}^{\infty} n^{q[1-(\frac{\sigma}{p}+\frac{\lambda-\mu}{q})]-1} a_n^q < \infty.$$

Lemma 1. *Define a weight function by*

$$\varpi(\sigma, x) := x^{\lambda-\sigma} \sum_{n=1}^{\infty} \frac{n^{\sigma-1}}{(x+n)^{\lambda}} (x \in \mathrm{R}_{+}). \tag{4}$$

Then, we have

$$B(\sigma, \lambda - \sigma)(1 - \rho_{\sigma}(x)) < \varpi(\sigma, x) < B(\sigma, \lambda - \sigma)(x \in \mathrm{R}_{+}), \tag{5}$$

where, $\rho_{\sigma}(x) := \frac{(1+\theta_x)^{-\lambda}}{\sigma B(\sigma,\lambda-\sigma)} \frac{1}{x^{\sigma}} = O(\frac{1}{x^{\sigma}}) \in (0,1)(\theta_x \in (0, \frac{1}{x}); x > 0)$. $B(u,v) := \int_0^{\infty} \frac{t^{u-1}}{(1+t)^{u+v}} dt\ (u, v > 0)$ *is the beta function.*

Proof. For fixed $x > 0$, we set function $g_x(t) := \frac{t^{\sigma-1}}{(x+t)^{\lambda}} (t > 0)$. Using the Euler-Maclaurin summation formula (cf., [20]), for $\rho(t) := t - [t] - \frac{1}{2}$, we have

$$\sum_{n=1}^{\infty} g_x(n) = \int_1^{\infty} g_x(t)dt + \tfrac{1}{2} g_x(1) + \int_1^{\infty} \rho(t) g_x'(t)dt = \int_0^{\infty} g_x(t)dt - h(x),$$
$$h(x) := \int_0^1 g_x(t)dt - \tfrac{1}{2} g_x(1) - \int_1^{\infty} \rho(t) g_x'(t)dt.$$

Thus, we obtain $-\frac{1}{2} g_x(1) = \frac{-1}{2(x+1)^{\lambda}}$,

$$\begin{aligned}\int_0^1 g_x(t)dt &= \int_0^1 \frac{t^{\sigma-1}}{(x+t)^{\lambda}} dt = \frac{1}{\sigma} \int_0^1 \frac{dt^{\sigma}}{(x+t)^{\lambda}} = \frac{1}{\sigma} \frac{t^{\sigma}}{(x+t)^{\lambda}} \Big|_0^1 + \frac{\lambda}{\sigma} \int_0^1 \frac{t^{\sigma} dt}{(x+t)^{\lambda+1}} \\ &= \frac{1}{\sigma} \frac{1}{(x+1)^{\lambda}} + \frac{\lambda}{\sigma(\sigma+1)} \int_0^1 \frac{dt^{\sigma+1}}{(x+t)^{\lambda+1}}\end{aligned}$$

$$\begin{aligned}&> \frac{1}{\sigma} \frac{1}{(x+1)^{\lambda}} + \frac{\lambda}{\sigma(\sigma+1)} \left[\frac{t^{\sigma+1}}{(x+t)^{\lambda+1}}\right]_0^1 + \frac{\lambda(\lambda+1)}{\sigma(\sigma+1)(x+1)^{\lambda+2}} \int_0^1 t^{\sigma+1} dt \\ &= \frac{1}{\sigma} \frac{1}{(x+1)^{\lambda}} + \frac{\lambda}{\sigma(\sigma+1)} \frac{1}{(x+1)^{\lambda+1}} + \frac{\lambda(\lambda+1)}{\sigma(\sigma+1)(\sigma+2)} \frac{1}{(x+1)^{\lambda+2}},\end{aligned}$$

$$\begin{aligned}-g_x'(t) &= -\tfrac{(\sigma-1)t^{\sigma-2}}{(x+t)^{\lambda}} + \tfrac{\lambda t^{\sigma-1}}{(x+t)^{\lambda+1}} = \tfrac{(1-\sigma)t^{\sigma-2}}{(x+t)^{\lambda}} + \tfrac{\lambda t^{\sigma-2}}{(x+t)^{\lambda}} - \tfrac{\lambda x t^{\sigma-2}}{(x+t)^{\lambda+1}} \\ &= \tfrac{(\lambda+1-\sigma)t^{\sigma-2}}{(x+t)^{\lambda}} - \tfrac{\lambda x t^{\sigma-2}}{(x+t)^{\lambda+1}}.\end{aligned}$$

For $0 < \sigma \leq 2, \sigma < \lambda \leq 5$, we find

$$(-1)^i \frac{d^i}{dt^i}[\frac{t^{\sigma-2}}{(x+t)^{\lambda}}] > 0, (-1)^i \frac{d^i}{dt^i}[\frac{t^{\sigma-2}}{(x+t)^{\lambda+1}}] > 0 (t > 0; i = 0,1,2,3),$$

and then by using the Euler-Maclaurin summation formula (cf., [20]), we find

$$(\lambda+1-\sigma)\int_1^{\infty} \rho(t)\frac{t^{\sigma-2}}{(x+t)^{\lambda}}dt > -\frac{\lambda+1-\sigma}{12(x+1)^{\lambda}},$$

$$\begin{aligned}-x\lambda\int_1^{\infty}\rho(t)\tfrac{t^{\sigma-2}}{(x+t)^{\lambda+1}}dt &> \tfrac{x\lambda}{12(x+1)^{\lambda+1}} - \tfrac{x\lambda}{720}[\tfrac{t^{\sigma-2}}{(x+t)^{\lambda+1}}]''_{t=1} \\ &> \tfrac{(x+1)\lambda-\lambda}{12(x+1)^{\lambda+1}} - \tfrac{(x+1)\lambda}{720}[\tfrac{(\lambda+1)(\lambda+2)}{(x+1)^{\lambda+3}} + \tfrac{2(\lambda+1)(2-\sigma)}{(x+1)^{\lambda+2}} + \tfrac{(2-\sigma)(3-\sigma)}{(x+1)^{\lambda+1}}] \\ &= \tfrac{\lambda}{12(x+1)^{\lambda}} - \tfrac{\lambda}{12(x+1)^{\lambda+1}} - \tfrac{\lambda}{720}[\tfrac{(\lambda+1)(\lambda+2)}{(x+1)^{\lambda+2}} + \tfrac{2(\lambda+1)(2-\sigma)}{(x+1)^{\lambda+1}} + \tfrac{(2-\sigma)(3-\sigma)}{(x+1)^{\lambda}}].\end{aligned}$$

Hence, we have

$$h(x) > \frac{h_1}{(x+1)^{\lambda}} + \frac{\lambda h_2}{(x+1)^{\lambda+1}} + \frac{\lambda(\lambda+1)h_3}{(x+1)^{\lambda+2}},$$

where, $h_1 := \frac{1}{\sigma} - \frac{1}{2} - \frac{1-\sigma}{12} - \frac{\lambda(2-\sigma)(3-\sigma)}{720}, h_2 := \frac{1}{\sigma(\sigma+1)} - \frac{1}{12} - \frac{(\lambda+1)(2-\sigma)}{720}$, and

$$h_3 := \frac{1}{\sigma(\sigma+1)(\sigma+2)} - \frac{\lambda+2}{720}.$$

For $\lambda \in (0,5], \frac{\lambda}{720} < \frac{1}{24}, \sigma \in (0,2]$, it follows that

$$h_1 > \frac{1}{\sigma} - \frac{1}{2} - \frac{1-\sigma}{12} - \frac{(2-\sigma)(3-\sigma)}{24} = \frac{24-20\sigma+7\sigma^2-\sigma^3}{24\sigma} > 0.$$

In fact, setting $g(\sigma) := 24 - 20\sigma + 7\sigma^2 - \sigma^3$ $(\sigma \in (0,2])$, we obtain

$$g'(\sigma) = -20 + 14\sigma^2 - 3\sigma^2 = -3(\sigma - \frac{7}{3})^2 - \frac{11}{3} < 0,$$

and then we obtain $h_1 > \frac{g(\sigma)}{24\sigma} \geq \frac{g(2)}{24\sigma} = \frac{4}{24\sigma} > 0$ $(\sigma \in (0,2])$.

We observe that $h_2 > \frac{1}{6} - \frac{1}{12} - \frac{12}{360} = \frac{1}{20} > 0$, and $h_3 \geq \frac{1}{24} - \frac{7}{720} = \frac{23}{720} > 0$. Hence, we deduce that $h(x) > 0$, and thus we have

$$\begin{aligned}\varpi(\sigma,x) &= x^{\lambda-\sigma}\sum_{n=1}^{\infty} g_x(n) < x^{\lambda-\sigma}\int_0^{\infty} g_x(t)dt \\ &= x^{\lambda-\sigma}\int_0^{\infty}\tfrac{t^{\sigma-1}dt}{(x+t)^{\lambda}} = \int_0^{\infty}\tfrac{u^{\sigma-1}du}{(1+u)^{\lambda}} = B(\sigma,\lambda-\sigma).\end{aligned}$$

On the other-hand, we also have

$$\begin{aligned}\sum_{n=1}^{\infty} g_x(n) &= \int_1^{\infty} g_x(t)dt + \tfrac{1}{2}g_x(1) + \int_1^{\infty}\rho(t)g_x'(t)dt \\ &= \int_1^{\infty} g_x(t)dt + H(x), \\ H(x) &:= \tfrac{1}{2}g_x(1) + \int_1^{\infty}\rho(t)g_x'(t)dt.\end{aligned}$$

We obtain $\frac{1}{2}g_x(1) = \frac{1}{2(x+1)^\lambda}$ and

$$g_x'(t) = \frac{-(\lambda+1-\sigma)t^{\sigma-2}}{(x+t)^\lambda} + \frac{\lambda x t^{\sigma-2}}{(x+t)^{\lambda+1}}.$$

For $\sigma \in (0,2] \cap (0,\lambda), 0 < \lambda \le 5$, by the Euler-Maclaurin summation formula, we obtain

$$-(\lambda+1-\sigma)\int_1^\infty \rho(t)\frac{t^{\sigma-2}}{(x+t)^\lambda}dt > 0,$$

$$\begin{aligned} &x\lambda\int_1^\infty \rho(t)\frac{t^{\sigma-2}}{(x+t)^{\lambda+1}}dt \\ &> -\frac{x\lambda}{12(x+1)^{\lambda+1}} = -\frac{(x+1)\lambda-\lambda}{12(x+1)^{\lambda+1}} = \frac{-\lambda}{12(x+1)^\lambda} + \frac{\lambda}{12(x+1)^{\lambda+1}} > \frac{-\lambda}{12(x+1)^\lambda}. \end{aligned}$$

Hence, we have

$$H(x) > \frac{1}{2(x+1)^\lambda} - \frac{\lambda}{12(x+1)^\lambda} = \frac{6-\lambda}{12(x+1)^\lambda} > 0,$$

and then

$$\begin{aligned} \varpi(\sigma,x) &= x^{\lambda-\sigma}\sum_{n=1}^\infty g_x(n) > x^{\lambda-\sigma}\int_1^\infty g_x(t)dt \\ &= x^{\lambda-\sigma}\int_0^\infty g_x(t)dt - x^{\lambda-\sigma}\int_0^1 g_x(t)dt \\ &= B(\sigma,\lambda-\sigma)[1-\frac{1}{B(\sigma,\lambda-\sigma)}\int_0^{\frac{1}{x}}\frac{u^{\sigma-1}}{(1+u)^\lambda}du] > 0. \end{aligned}$$

By the integral mid-value theorem, we find

$$\int_0^{\frac{1}{x}}\frac{u^{\sigma-1}}{(1+u)^\lambda}du = \frac{1}{(1+\theta_x)^\lambda}\int_0^{\frac{1}{x}}u^{\sigma-1}du = \frac{1}{\sigma(1+\theta_x)^\lambda}\frac{1}{x^\sigma}(\theta_x \in (0,\frac{1}{x})).$$

This proves Inequality (5). □

Lemma 2. *The following reverse inequality is valid:*

$$\begin{aligned} I &= \int_0^\infty\sum_{n=1}^\infty\frac{f(x)a_n}{(x+n)^\lambda}dx > B^{\frac{1}{p}}(\sigma,\lambda-\sigma)B^{\frac{1}{q}}(\mu,\lambda-\mu) \\ &\times\{\int_0^\infty(1-\rho_\sigma(x))x^{p[1-(\frac{\lambda-\sigma}{p}+\frac{\mu}{q})]-1}f^p(x)dx\}^{\frac{1}{p}}\{\sum_{n=1}^\infty n^{q[1-(\frac{\sigma}{p}+\frac{\lambda-\mu}{q})]-1}a_n^q\}^{\frac{1}{q}}. \end{aligned} \tag{6}$$

Proof. For $n \in \mathbf{N}$, setting $x = nu$, we obtain the following weight function:

$$\omega(\mu,n) := n^{\lambda-\mu}\int_0^\infty\frac{x^{\mu-1}dx}{(x+n)^\lambda} = \int_0^\infty\frac{u^{\mu-1}du}{(u+1)^\lambda} = B(\mu,\lambda-\mu). \tag{7}$$

For $0 < p < 1, q < 0$, by the reverse Hölder's inequality (cf., [27]) and the Lebesgue term by term integration theorem (cf., [28]), we obtain

$$\begin{aligned} &\int_0^\infty\sum_{n=1}^\infty\frac{f(x)a_n}{(x+n)^\lambda}dx = \int_0^\infty\sum_{n=1}^\infty\frac{1}{(x+n)^\lambda}[\frac{n^{(\sigma-1)/p}}{x^{(\mu-1)/q}}f(x)][\frac{x^{(\mu-1)/q}}{n^{(\sigma-1)/p}}a_n]dx \\ &\ge\{\int_0^\infty[\sum_{n=1}^\infty\frac{1}{(x+n)^\lambda}\frac{n^{\sigma-1}}{x^{(\mu-1)(p-1)}}]f^p(x)dx\}^{\frac{1}{p}}\{\sum_{n=1}^\infty[\int_0^\infty\frac{1}{(x+n)^\lambda}\frac{x^{\mu-1}}{n^{(\sigma-1)(q-1)}}dx]a_n^q\}^{\frac{1}{q}} \\ &= \{\int_0^\infty\varpi(\sigma,x)x^{p[1-(\frac{\lambda-\sigma}{p}+\frac{\mu}{q})]-1}f^p(x)dx\}^{\frac{1}{p}}\{\sum_{n=1}^\infty\omega(\mu,n)n^{q[1-(\frac{\sigma}{p}+\frac{\lambda-\mu}{q})]-1}a_n^q\}^{\frac{1}{q}}. \end{aligned}$$

Then by (5) and (7), we obtain Inequality (6). □

Remark 1. *For $\mu+\sigma=\lambda$, we find*

$$\varpi(\sigma,x)=x^{\mu}\sum_{n=1}^{\infty}\frac{n^{\sigma-1}}{(x+n)^{\lambda}}(x\in R_{+}),$$

$$0<\int_{0}^{\infty}x^{p(1-\mu)-1}f^{p}(x)dx<\infty \text{ and } 0<\sum_{n=1}^{\infty}n^{q(1-\sigma)-1}a_{n}^{q}<\infty,$$

and then we reduce (6) as follows:

$$\int_{0}^{\infty}\sum_{n=1}^{\infty}\frac{f(x)a_{n}}{(x+n)^{\lambda}}dx>B(\mu,\sigma)[\int_{0}^{\infty}(1-\rho_{\sigma}(x))x^{p(1-\mu)-1}f^{p}(x)dx]^{\frac{1}{p}}[\sum_{n=1}^{\infty}n^{q(1-\sigma)-1}a_{n}^{q}]^{\frac{1}{q}}. \quad (8)$$

Lemma 3. *The constant factor $B(\mu,\sigma)$ in (8) is the best possible.*

Proof. For $0<\varepsilon<p\mu$, we set

$$\widetilde{f}(x):=\{\begin{matrix}0,0<x<1,\\ x^{\mu-\frac{\varepsilon}{p}-1},x\geq 1\end{matrix},\widetilde{a}_{n}:=n^{\sigma-\frac{\varepsilon}{q}-1}(n\in N).$$

If there exists a positive constant $M(M\geq B(\mu,\sigma))$ such that (8) is valid when replacing $B(\mu,\sigma)$ by M, then by a substitution of $f(x)=\widetilde{f}(x),a_{n}=\widetilde{a}_{n}$, we get

$$\widetilde{I}:=\int_{0}^{\infty}\sum_{n=1}^{\infty}\frac{\widetilde{f}(x)\widetilde{a}_{n}}{(x+n)^{\lambda}}dx>M$$

$$\times[\int_{0}^{\infty}(1-\rho_{\sigma}(x))x^{p(1-\mu)-1}\widetilde{f}^{p}(x)dx]^{\frac{1}{p}}[\sum_{n=1}^{\infty}n^{q(1-\sigma)-1}\widetilde{a}_{n}^{q}]^{\frac{1}{q}}$$
$$=M(\int_{1}^{\infty}(1-O(\tfrac{1}{x^{\sigma}}))x^{-\varepsilon-1}dx)^{\frac{1}{p}}(\sum_{n=1}^{\infty}n^{-\varepsilon-1})^{\frac{1}{q}}$$
$$\geq M(\int_{1}^{\infty}x^{-\varepsilon-1}dx-\int_{1}^{\infty}O(\tfrac{1}{x^{\sigma+\varepsilon+1}})dx)^{\frac{1}{p}}(\int_{1}^{\infty}x^{-\varepsilon-1}dx)^{\frac{1}{q}}$$
$$=\frac{M}{\varepsilon}(1-\varepsilon O(1))^{\frac{1}{p}}.$$

For $\mu-\frac{\varepsilon}{p}>0(0<p<1)$, by (7), we obtain

$$\widetilde{I}=\sum_{n=1}^{\infty}n^{-\varepsilon-1}[n^{(\sigma+\frac{\varepsilon}{p})}\int_{1}^{\infty}\frac{x^{(\mu-\frac{\varepsilon}{p})-1}}{(x+n)^{\lambda}}dx]\leq\sum_{n=1}^{\infty}n^{-\varepsilon-1}[n^{(\sigma+\frac{\varepsilon}{p})}\int_{0}^{\infty}\frac{x^{(\mu-\frac{\varepsilon}{p})-1}}{(x+n)^{\lambda}}dx]$$
$$=\sum_{n=1}^{\infty}n^{-\varepsilon-1}\omega(\mu-\tfrac{\varepsilon}{p},n)=B(\mu-\tfrac{\varepsilon}{p},\sigma+\tfrac{\varepsilon}{p})(1+\sum_{n=2}^{\infty}n^{-\varepsilon-1})$$
$$\leq B(\mu-\tfrac{\varepsilon}{p},\sigma+\tfrac{\varepsilon}{p})(1+\int_{1}^{\infty}x^{-\varepsilon-1}dx)=\tfrac{\varepsilon+1}{\varepsilon}B(\mu-\tfrac{\varepsilon}{p},\sigma+\tfrac{\varepsilon}{p}).$$

Then we have

$$(\varepsilon+1)B(\mu-\frac{\varepsilon}{p},\sigma+\frac{\varepsilon}{p})\geq\varepsilon\widetilde{I}\geq M(1-\varepsilon O(1))^{\frac{1}{p}}.$$

For $\varepsilon\to 0^{+}$, in view of the continuity of the beta function, it follows that $B(\mu,\sigma)\geq M$. Therefore, $M=B(\mu,\sigma)$ is the best possible constant factor of (8). Lemma 3 is proved. □

Remark 2. *Setting* $\hat{\mu} := \frac{\lambda-\sigma}{p} + \frac{\mu}{q}, \hat{\sigma} := \frac{\sigma}{p} + \frac{\lambda-\mu}{q}$, *we have*

$$\hat{\mu} + \hat{\sigma} = \frac{\lambda-\sigma}{p} + \frac{\mu}{q} + \frac{\sigma}{p} + \frac{\lambda-\mu}{q} = \frac{\lambda}{p} + \frac{\lambda}{q} = \lambda,$$

and for $\lambda - \mu - \sigma \in (-p\mu, p(\lambda-\mu))$, *we find*

$$\begin{aligned} &\hat{\mu} > \tfrac{(1-p)\mu}{p} + \tfrac{\mu}{q} = 0, \hat{\mu} < \tfrac{\mu+p(\lambda-\mu)}{p} + \tfrac{\mu}{q} = \lambda, \\ &0 < \hat{\sigma} = \lambda - \hat{\mu} < \lambda, B(\hat{\mu},\hat{\sigma}) \in \mathrm{R}_+. \end{aligned}$$

We can reduce (6) to the following

$$\begin{aligned} &\int_0^\infty \sum_{n=1}^\infty \tfrac{f(x)a_n}{(x+n)^\lambda} dx > B^{\frac{1}{p}}(\sigma,\lambda-\sigma) B^{\frac{1}{q}}(\mu,\lambda-\mu) \\ &\quad \times [\int_0^\infty (1-\rho_\sigma(x)) x^{p(1-\hat{\mu})-1} f^p(x) dx]^{\frac{1}{p}} [\sum_{n=1}^\infty n^{q(1-\hat{\sigma})-1} a_n^q]^{\frac{1}{q}}. \end{aligned} \tag{9}$$

Lemma 4. *If* $\lambda - \mu - \sigma \in (-p\mu, p(\lambda-\mu))$, *the constant factor* $B^{\frac{1}{p}}(\sigma,\lambda-\sigma)B^{\frac{1}{q}}(\mu,\lambda-\mu)$ *in (9) is the best possible, then we have* $\mu + \sigma = \lambda$.

Proof. If the constant factor $B^{\frac{1}{p}}(\sigma,\lambda-\sigma)B^{\frac{1}{q}}(\mu,\lambda-\mu)$ in (9) is the best possible, then by (8), the unique best possible constant factor must be $B(\hat{\mu},\hat{\sigma}) (\in \mathrm{R}_+)$, namely,

$$B(\hat{\mu},\hat{\sigma}) = B^{\frac{1}{p}}(\sigma,\lambda-\sigma)B^{\frac{1}{q}}(\mu,\lambda-\mu).$$

By the reverse Hölder's inequality (cf., [27]), we find

$$\begin{aligned} B(\hat{\mu},\hat{\sigma}) &= \int_0^\infty \tfrac{t^{\hat{\mu}-1}}{(1+t)^\lambda} dt = \int_0^\infty \tfrac{t^{\frac{\lambda-\sigma}{p}+\frac{\mu}{q}-1}}{(1+t)^\lambda} dt = \int_0^\infty \tfrac{1}{(1+t)^\lambda} (t^{\frac{\lambda-\sigma-1}{p}})(t^{\frac{\mu-1}{q}}) dt \\ &\geq [\int_0^\infty \tfrac{1}{(1+t)^\lambda} t^{\lambda-\sigma-1} dt]^{\frac{1}{p}} [\int_0^\infty \tfrac{1}{(1+t)^\lambda} t^{\mu-1} dt]^{\frac{1}{q}} \\ &= B^{\frac{1}{p}}(\sigma,\lambda-\sigma)B^{\frac{1}{q}}(\mu,\lambda-\mu). \end{aligned} \tag{10}$$

We observe that (10) keeps the form of equality if and only if there exist constants A, B such that they are not all zero and

$$At^{\lambda-\sigma-1} = Bt^{\mu-1} \text{ a.e. in } \mathrm{R}_+.$$

Suppose that $A \neq 0$. We find that $t^{\lambda-\mu-\sigma} = \frac{B}{A}$ *a.e. in* R_+, and thus we conclude that $\lambda - \mu - \sigma = 0$, i.e., $\mu + \sigma = \lambda$. Lemma 4 is proved. □

3. Main Results

Theorem 1. *Inequality (6) is equivalent to the following inequalities:*

$$\begin{aligned} J_1 &:= \{\sum_{n=1}^\infty n^{p(\frac{\sigma}{p}+\frac{\lambda-\mu}{q})-1} [\int_0^\infty \tfrac{f(x)}{(x+n)^\lambda} dx]^p\}^{\frac{1}{p}} \\ &> B^{\frac{1}{p}}(\sigma,\lambda-\sigma)B^{\frac{1}{q}}(\mu,\lambda-\mu)\{\int_0^\infty (1-\rho_\sigma(x)) x^{p[1-(\frac{\lambda-\sigma}{p}+\frac{\mu}{q})]-1} f^p(x) dx\}^{\frac{1}{p}}, \end{aligned} \tag{11}$$

$$J_2 := \{\int_0^\infty \frac{x^{q(\frac{\lambda-\sigma}{p}+\frac{\mu}{q})-1}}{(1-\rho_\sigma(x))^{q-1}}[\sum_{n=1}^\infty \frac{a_n}{(x+n)^\lambda}]^q dx\}^{\frac{1}{q}} > B^{\frac{1}{p}}(\sigma,\lambda-\sigma)B^{\frac{1}{q}}(\mu,\lambda-\mu)\{\sum_{n=1}^\infty n^{q[1-(\frac{\sigma}{p}+\frac{\lambda-\mu}{q})]-1}a_n^q\}^{\frac{1}{q}}. \tag{12}$$

If the constant factor $B^{\frac{1}{p}}(\sigma,\lambda-\sigma)B^{\frac{1}{q}}(\mu,\lambda-\mu)$ in (6) is the best possible, then so is the constant factor in (11) and (12).

In particular, for $\mu+\sigma=\lambda$ in (6), (11) and (12), we have Inequality (8) and the following equivalent versions of reverse inequalities with the best possible constant factor $B(\mu,\sigma)$:

$$\{\sum_{n=1}^\infty n^{p\sigma-1}[\int_0^\infty \frac{f(x)}{(x+n)^\lambda}dx]^p\}^{\frac{1}{p}} > B(\mu,\sigma)[\int_0^\infty (1-\rho_\sigma(x))x^{p(1-\mu)-1}f^p(x)dx]^{\frac{1}{p}}, \tag{13}$$

$$\{\int_0^\infty \frac{x^{q\mu-1}}{(1-\rho_\sigma(x))^{q-1}}[\sum_{n=1}^\infty \frac{a_n}{(x+n)^\lambda}]^q dx\}^{\frac{1}{q}} > B(\mu,\sigma)[\sum_{n=1}^\infty n^{q(1-\sigma)-1}a_n^q]^{\frac{1}{q}}. \tag{14}$$

Proof. Suppose that (11) is valid. By the Lebesgue term by term integration theorem and the reverse Hölder's inequality (cf., [27,28]), we have

$$I = \sum_{n=1}^\infty \int_0^\infty \frac{f(x)a_n}{(x+n)^\lambda}dx = \sum_{n=1}^\infty [n^{\frac{-1}{p}+(\frac{\sigma}{p}+\frac{\lambda-\mu}{q})}\int_0^\infty \frac{f(x)}{(x+n)^\lambda}dx][n^{\frac{1}{p}-(\frac{\sigma}{p}+\frac{\lambda-\mu}{q})}a_n] \geq J_1\{\sum_{n=1}^\infty n^{q[1-(\frac{\sigma}{p}+\frac{\lambda-\mu}{q})]-1}a_n^q\}^{\frac{1}{q}}. \tag{15}$$

Then by (11), we have Inequality (6). On the other-hand, assuming that Inequality (6) is valid, we set

$$a_n := n^{p(\frac{\sigma}{p}+\frac{\lambda-\mu}{q})-1}[\int_0^\infty \frac{f(x)}{(x+n)^\lambda}dx]^{p-1}, n \in \mathrm{N}.$$

If $J_1=\infty$, then Inequality (11) is naturally valid; if $J_1=0$, so it is impossible to make Inequality (11) valid, namely $J_1>0$. Suppose that $0<J_1<\infty$. By (6), we have

$$\sum_{n=1}^\infty n^{q[1-(\frac{\sigma}{p}+\frac{\lambda-\mu}{q})]-1}a_n^q = J_1^p = I > B^{\frac{1}{p}}(\sigma,\lambda-\sigma)B^{\frac{1}{q}}(\mu,\lambda-\mu) \times\{\int_0^\infty (1-\rho_\sigma(x))x^{p[1-(\frac{\lambda-\sigma}{p}+\frac{\mu}{q})]-1}f^p(x)dx\}^{\frac{1}{p}}\{\sum_{n=1}^\infty n^{q[1-(\frac{\sigma}{p}+\frac{\lambda-\mu}{q})]-1}a_n^q\}^{\frac{1}{q}},$$

$$\{\sum_{n=1}^\infty n^{q[1-(\frac{\sigma}{p}+\frac{\lambda-\mu}{q})]-1}a_n^q\}^{\frac{1}{p}} = J_1 > B^{\frac{1}{p}}(\sigma,\lambda-\sigma)B^{\frac{1}{q}}(\mu,\lambda-\mu) \times\{\int_0^\infty (1-\rho_\sigma(x))x^{p[1-(\frac{\lambda-\sigma}{p}+\frac{\mu}{q})]-1}f^p(x)dx\}^{\frac{1}{p}},$$

namely, Inequality (11) follows, which is equivalent to Inequality (6).

Suppose that Inequality (12) is valid. By the reverse Hölder's inequality, we have

$$I = \int_0^\infty [(1-\rho_\sigma(x))^{\frac{1}{p}}x^{\frac{1}{q}-(\frac{\lambda-\sigma}{p}+\frac{\mu}{q})}f(x)][\frac{x^{\frac{-1}{q}+(\frac{\lambda-\sigma}{p}+\frac{\mu}{q})}}{(1-\rho_\sigma(x))^{1/p}}\sum_{n=1}^\infty \frac{a_n}{(x+n)^\lambda}]dx \geq \{\int_0^\infty (1-\rho_\sigma(x))x^{p[1-(\frac{\lambda-\sigma}{p}+\frac{\mu}{q})]-1}f^p(x)dx\}^{\frac{1}{p}}J_2. \tag{16}$$

Then by (12), we obtain Inequality (6). On the other-hand, assuming that Inequality (6) is valid, we set

$$f(x) := x^{q(\frac{\lambda-\sigma}{p}+\frac{\mu}{q})-1}[\sum_{n=1}^{\infty}\frac{a_n}{(x+n)^{\lambda}}]^{q-1}, x \in \mathrm{R}_+$$

If $J_2 = \infty$, then Inequality (12) is naturally valid; if $J_2 = 0$, then it is impossible to make Inequality (12) valid, namely $J_2 > 0$. Suppose that $0 < J_2 < \infty$. By (6), we have

$$\int_0^{\infty}(1-\rho_\sigma(x))x^{p[1-(\frac{\lambda-\sigma}{p}+\frac{\mu}{q})]-1}f^p(x)dx = J_2^q = I > B^{\frac{1}{p}}(\sigma,\lambda-\sigma)B^{\frac{1}{q}}(\mu,\lambda-\mu)$$
$$\times\{\int_0^{\infty}(1-\rho_\sigma(x))x^{p[1-(\frac{\lambda-\sigma}{p}+\frac{\mu}{q})]-1}f^p(x)dx\}^{\frac{1}{p}}\{\sum_{n=1}^{\infty}n^{q[1-(\frac{\sigma}{p}+\frac{\lambda-\mu}{q})]-1}a_n^q\}^{\frac{1}{q}},$$

$$\{\int_0^{\infty}(1-\rho_\sigma(x))x^{p[1-(\frac{\lambda-\sigma}{p}+\frac{\mu}{q})]-1}f^p(x)dx\}^{\frac{1}{q}} = J_2 > B^{\frac{1}{p}}(\sigma,\lambda-\sigma)B^{\frac{1}{q}}(\mu,\lambda-\mu)$$
$$\times\{\sum_{n=1}^{\infty}n^{q[1-(\frac{\sigma}{p}+\frac{\lambda-\mu}{q})]-1}a_n^q\}^{\frac{1}{q}},$$

namely, Inequality (12) follows, which is equivalent to Inequality (6).

Hence, Inequalities (6), (11) and (12) are equivalent.

If the constant factor $B^{\frac{1}{p}}(\sigma,\lambda-\sigma)B^{\frac{1}{q}}(\mu,\lambda-\mu)$ in (6) is the best possible, then so is the constant factor in (11) and (12). Otherwise, by (15) (or (16)), we would reach a contradiction that the constant factor in (6) is not the best possible. This completes the proof of Theorem 1. □

Theorem 2. *The following statements (i), (ii), (iii) and (iv) are equivalent.*

(i) $B^{\frac{1}{p}}(\sigma,\lambda-\sigma)B^{\frac{1}{q}}(\mu,\lambda-\mu)$ *is independent of* p,q*;*
(ii) $B^{\frac{1}{p}}(\sigma,\lambda-\sigma)B^{\frac{1}{q}}(\mu,\lambda-\mu)$ *is expressible as a single integral;*
(iii) $B^{\frac{1}{p}}(\sigma,\lambda-\sigma)B^{\frac{1}{q}}(\mu,\lambda-\mu)$ *is the best possible of (6);*
(iv) *If* $\lambda-\mu-\sigma \in (-p\mu, p(\lambda-\mu))$*, then* $\mu+\sigma=\lambda$.

Proof. (i) ⇒ (ii). In view of $B^{\frac{1}{p}}(\sigma,\lambda-\sigma)B^{\frac{1}{q}}(\mu,\lambda-\mu)$ is independent of p,q, we find

$$B^{\frac{1}{p}}(\sigma,\lambda-\sigma)B^{\frac{1}{q}}(\mu,\lambda-\mu) = \lim_{\substack{p\to 1^-,\\ q\to-\infty}} B^{\frac{1}{p}}(\sigma,\lambda-\sigma)B^{\frac{1}{q}}(\mu,\lambda-\mu) = B(\sigma,\lambda-\sigma),$$

which is a single integral $\int_0^{\infty}\frac{t^{\sigma-1}}{(1+t)^{\lambda}}dt$.

(ii) ⇒ (iv). Suppose that $B^{\frac{1}{p}}(\sigma,\lambda-\sigma)B^{\frac{1}{q}}(\mu,\lambda-\mu)$ is expressible as a single integral $\int_0^{\infty}\frac{t^{\frac{\lambda-\sigma}{p}+\frac{\mu}{q}-1}}{(1+t)^{\lambda}}dt$. Then (10) keeps the form of equality. By the proof of Lemma 4, for $\lambda-\mu-\sigma \in (-p\mu, p(\lambda-\mu))$, we have $\mu+\sigma=\lambda$.

(iv) ⇒ (i). If $\mu+\sigma=\lambda$, then

$$B^{\frac{1}{p}}(\sigma,\lambda-\sigma)B^{\frac{1}{q}}(\mu,\lambda-\mu) = B(\mu,\sigma),$$

which is independent of p,q.

Hence, (i) ⇔ (ii) ⇔ (iv).

(vi) ⇒ (iii). By Lemma 3, for $\mu+\sigma=\lambda$, $B^{\frac{1}{p}}(\sigma,\lambda-\sigma)B^{\frac{1}{q}}(\mu,\lambda-\mu)$ is the best possible of (6).

(iii) ⇒ (iv). By Lemma 4, we have $\mu+\sigma=\lambda$.

Therefore, we show that (iv)⇔ (iii), and then the statements (i), (ii), (iii) and (iv) are equivalent. The proof Theorem 2 is complete. □

4. Two Corollaries and Some Particular Inequalities

Replacing x by $\frac{1}{x}$, and then setting $F(x) = x^{\lambda-2}f(\frac{1}{x})$ in Theorems 1 and 2, we find

$$\rho_\sigma(x^{-1}) = \frac{(1+\theta_{x^{-1}})^{-\lambda}}{\sigma B(\sigma,\lambda-\sigma)}x^\sigma = O(x^\sigma) \in (0,1)(\theta_{x^{-1}} \in (0,x); x>0),$$

and obtain the following corollaries:

Corollary 1. *If $F(x)$, $a_n \geq 0$ such that*

$$0 < \int_0^\infty x^{p[1-(\frac{\sigma}{p}+\frac{\lambda-\mu}{q})]-1}F^p(x)dx < \infty \text{ and } 0 < \sum_{n=1}^\infty n^{q[1-(\frac{\sigma}{p}+\frac{\lambda-\mu}{q})]-1}a_n^q < \infty,$$

then the following inequalities are equivalent:

$$\int_0^\infty \sum_{n=1}^\infty \frac{F(x)a_n}{(1+xn)^\lambda}dx > B^{\frac{1}{p}}(\sigma,\lambda-\sigma)B^{\frac{1}{q}}(\mu,\lambda-\mu) \times\{\int_0^\infty (1-\rho_\sigma(x^{-1}))x^{p[1-(\frac{\sigma}{p}+\frac{\lambda-\mu}{q})]-1}F^p(x)dx\}^{\frac{1}{p}}\{\sum_{n=1}^\infty n^{q[1-(\frac{\sigma}{p}+\frac{\lambda-\mu}{q})]-1}a_n^q\}^{\frac{1}{q}}, \tag{17}$$

$$\{\sum_{n=1}^\infty n^{p(\frac{\sigma}{p}+\frac{\lambda-\mu}{q})-1}[\int_0^\infty \frac{F(x)}{(1+xn)^\lambda}dx]^p\}^{\frac{1}{p}} > B^{\frac{1}{p}}(\sigma,\lambda-\sigma)B^{\frac{1}{q}}(\mu,\lambda-\mu)\{\int_0^\infty (1-\rho_\sigma(x^{-1}))x^{p[1-(\frac{\sigma}{p}+\frac{\lambda-\mu}{q})]-1}F^p(x)dx\}^{\frac{1}{p}}, \tag{18}$$

$$\{\int_0^\infty \frac{x^{q(\frac{\sigma}{p}+\frac{\lambda-\mu}{q})-1}}{(1-\rho_\sigma(x^{-1}))^{q-1}}[\sum_{n=1}^\infty \frac{a_n}{(1+xn)^\lambda}]^q dx\}^{\frac{1}{q}} > B^{\frac{1}{p}}(\sigma,\lambda-\sigma)B^{\frac{1}{q}}(\mu,\lambda-\mu)\{\sum_{n=1}^\infty n^{q[1-(\frac{\sigma}{p}+\frac{\lambda-\mu}{q})]-1}a_n^q\}^{\frac{1}{q}}. \tag{19}$$

If the constant factor $B^{\frac{1}{p}}(\sigma,\lambda-\sigma)B^{\frac{1}{q}}(\mu,\lambda-\mu)$ in (17) is the best possible, then so is the constant factor in (18) and (19).

In particular, for $\mu = \lambda - \sigma$ in (17), (18) and (19), we have the following equivalent inequalities with the best possible constant factor $B(\lambda-\sigma,\sigma)$:

$$\int_0^\infty \sum_{n=1}^\infty \frac{F(x)a_n}{(1+xn)^\lambda}dx > B(\lambda-\sigma,\sigma) \times[\int_0^\infty (1-\rho_\sigma(x^{-1}))x^{p(1-\sigma)-1}F^p(x)dx]^{\frac{1}{p}}[\sum_{n=1}^\infty n^{q(1-\sigma)-1}a_n^q]^{\frac{1}{q}}, \tag{20}$$

$$\{\sum_{n=1}^\infty n^{p\sigma-1}[\int_0^\infty \frac{F(x)}{(1+xn)^\lambda}dx]^p\}^{\frac{1}{p}} > B(\lambda-\sigma,\sigma)[\int_0^\infty (1-\rho_\sigma(x^{-1}))x^{p(1-\sigma)-1}F^p(x)dx]^{\frac{1}{p}}, \tag{21}$$

$$\{\int_0^\infty \frac{x^{q\sigma-1}}{(1-\rho(x^{-1}))^{q-1}}[\sum_{n=1}^\infty \frac{a_n}{(1+xn)^\lambda}]^q dx\}^{\frac{1}{q}} > B(\lambda-\sigma,\sigma)[\sum_{n=1}^\infty n^{q(1-\sigma)-1}a_n^q]^{\frac{1}{q}}. \tag{22}$$

Corollary 2. *The following statements (i), (ii), (iii) and (iv) are equivalent:*

(i) $B^{\frac{1}{p}}(\sigma,\lambda-\sigma)B^{\frac{1}{q}}(\mu,\lambda-\mu)$ *is independent of* p,q*;*
(ii) $B^{\frac{1}{p}}(\sigma,\lambda-\sigma)B^{\frac{1}{q}}(\mu,\lambda-\mu)$ *is expressible as a single integral;*
(iii) $B^{\frac{1}{p}}(\sigma,\lambda-\sigma)B^{\frac{1}{q}}(\mu,\lambda-\mu)$ *is the best possible of (17);*
(iv) *If* $\lambda-\mu-\sigma\in(-q\sigma,q(\lambda-\sigma))$*, then we have* $\mu=\lambda-\sigma$.

Remark 3. *(i) For* $\sigma=2<\lambda(\le 5),\mu=\lambda-2$ *in (8), (13) and (14), since*

$$B(\lambda-2,2)=\frac{\Gamma(\lambda-2)\Gamma(2)}{\Gamma(\lambda)}=\frac{\Gamma(\lambda-2)}{(\lambda-1)(\lambda-2)\Gamma(\lambda-2)}=\frac{1}{(\lambda-1)(\lambda-2)},$$

$$\rho_2(x)=\frac{(\lambda-1)(\lambda-2)}{2(1+\theta_x)^{\lambda}}\frac{1}{x^2}=O(\frac{1}{x^2})\in(0,1)(\theta_x\in(0,\frac{1}{x});x>0),$$

we have the following equivalent versions of reverse inequalities with the best possible constant factor $\frac{1}{(\lambda-1)(\lambda-2)}$:

$$\int_0^{\infty}\sum_{n=1}^{\infty}\frac{f(x)a_n}{(x+n)^{\lambda}}dx>\frac{1}{(\lambda-1)(\lambda-2)}[\int_0^{\infty}(1-\rho_2(x))x^{p(3-\lambda)-1}f^p(x)dx]^{\frac{1}{p}}(\sum_{n=1}^{\infty}n^{-q-1}a_n^q)^{\frac{1}{q}}, \tag{23}$$

$$\{\sum_{n=1}^{\infty}n^{2p-1}[\int_0^{\infty}\frac{f(x)}{(x+n)^{\lambda}}dx]^p\}^{\frac{1}{p}}>\frac{1}{(\lambda-1)(\lambda-2)}[\int_0^{\infty}(1-\rho_2(x))x^{p(3-\lambda)-1}f^p(x)dx]^{\frac{1}{p}}, \tag{24}$$

$$\{\int_0^{\infty}\frac{x^{q(\lambda-2)-1}}{(1-\rho_2(x))^{q-1}}[\sum_{n=1}^{\infty}\frac{a_n}{(x+n)^{\lambda}}]^q dx\}^{\frac{1}{q}}>\frac{1}{(\lambda-1)(\lambda-2)}(\sum_{n=1}^{\infty}n^{-q-1}a_n^q)^{\frac{1}{q}}. \tag{25}$$

(ii) For $\sigma=2<\lambda(\le 5),\mu=\lambda-2$ *in (20), (21) and (22), we have*

$$\rho_2(x^{-1})=\frac{(\lambda-1)(\lambda-2)}{2(1+\theta_{x^{-1}})^{\lambda}}x^2=O(x^2)\in(0,1)(\theta_{x^{-1}}\in(0,x);x>0),$$

and the following equivalent versions of reverse inequalities with the best possible constant factor $\frac{1}{(\lambda-1)(\lambda-2)}$:

$$\int_0^{\infty}\sum_{n=1}^{\infty}\frac{F(x)a_n}{(1+xn)^{\lambda}}dx>\frac{1}{(\lambda-1)(\lambda-2)}(\int_0^{\infty}(1-\rho_2(x^{-1}))x^{-p-1}F^p(x)dx)^{\frac{1}{p}}(\sum_{n=1}^{\infty}n^{-q-1}a_n^q)^{\frac{1}{q}}, \tag{26}$$

$$\{\sum_{n=1}^{\infty}n^{2p-1}[\int_0^{\infty}\frac{F(x)}{(1+xn)^{\lambda}}dx]^p\}^{\frac{1}{p}}>\frac{1}{(\lambda-1)(\lambda-2)}(\int_0^{\infty}(1-\rho_2(x^{-1}))x^{-p-1}F^p(x)dx)^{\frac{1}{p}}, \tag{27}$$

$$\{\int_0^{\infty}\frac{x^{2q-1}}{(1-\rho_2(x^{-1}))^{q-1}}[\sum_{n=1}^{\infty}\frac{a_n}{(1+xn)^{\lambda}}]^q dx\}^{\frac{1}{q}} \tag{28}$$

5. Conclusions

Let us give a brief summary of this paper, by the way of [20,22] and the use of the weight functions, the idea of introducing parameters and the Euler-Maclaurin summation formula, a reverse half-discrete Hardy-Hilbert's inequality and the reverse equivalent forms are given in Lemma 2 and Theorem 1. The equivalent statements of the best possible constant factor related to some parameters are proved in Theorem 2. As applications, two corollaries about the reverse cases of the non-homogeneous kernel and some particular cases are considered in Corollaries 1, 2 and Remark 3. The above-mentioned lemmas and theorems reveal some essential characters of this type of Hardy-Hilbert inequality.

Author Contributions: B.Y. carried out the mathematical studies, participated in the sequence alignment and drafted the manuscript. S.W. and A.W. participated in the design of the study and performed the numerical analysis. All authors read and approved the final manuscript.

Funding: This work is supported by the National Natural Science Foundation (No. 61772140), and Science and Technology Planning Project Item of Guangzhou City (No. 201707010229). We are grateful for their help.

Acknowledgments: The authors thanks the referees for their useful proposes for reforming the paper.

Conflicts of Interest: The authors declare that they have no competing interest.

References

1. Hardy, G.H.; Littlewood, J.E.; Polya, G. *Inequalities*; Cambridge University Press: Cambridge, UK, 1934.
2. Yang, B.C. *The Norm of Operator and Hilbert-Type Inequalities*; Science Press: Beijing, China, 2009.
3. Yang, B.C. *Hilbert-Type Integral Inequalities*; Bentham Science Publishers Ltd.: Sharjah, UAE, 2009.
4. Yang, B.C. On the norm of an integral operator and applications. *J. Math. Anal. Appl.* **2006**, *321*, 182–192. [CrossRef]
5. Xu, J.S. Hardy-Hilbert's inequalities with two parameters. *Adv. Math.* **2007**, *36*, 63–76.
6. Yang, B.C. On the norm of a Hilbert's type linear operator and applications. *J. Math. Anal. Appl.* **2007**, *325*, 529–541. [CrossRef]
7. Xie, Z.T.; Zeng, Z.; Sun, Y.F. A new Hilbert-type inequality with the homogeneous kernel of degree −2. *Adv. Appl. Math. Sci.* **2013**, *12*, 391–401.
8. Zhen, Z.; Raja Rama Gandhi, K.; Xie, Z.T. A new Hilbert-type inequality with the homogeneous kernel of degree −2 and with the integral. *Bull. Math. Sci. Appl.* **2014**, *3*, 11–20.
9. Xin, D.M. A Hilbert-type integral inequality with the homogeneous kernel of zero degree. *Math. Theory Appl.* **2010**, *30*, 70–74.
10. Azar, L.E. The connection between Hilbert and Hardy inequalities. *J. Inequal. Appl.* **2013**, *452*, 2013. [CrossRef]
11. Batbold, T.; Sawano, Y. Sharp bounds for m-linear Hilbert-type operators on the weighted Morrey spaces. *Math. Inequal. Appl.* **2017**, *20*, 263–283. [CrossRef]
12. Adiyasuren, V.; Batbold, T.; Krnic, M. Multiple Hilbert-type inequalities involving some differential operators. *Banach. J. Math. Anal.* **2016**, *10*, 320–337. [CrossRef]
13. Adiyasuren, V.; Batbold, T.; Krnic, M. Hilbert–type inequalities involving differential operators, the best constants and applications. *Math. Inequal. Appl.* **2015**, *18*, 111–124. [CrossRef]
14. Rassias, M.T.H.; Yang, B.C. On half-discrete Hilbert's inequality. *Appl. Math. Comput.* **2013**, *220*, 75–93. [CrossRef]
15. Yang, B.C.; Krnic, M. A half-discrete Hilbert-type inequality with a general homogeneous kernel of degree 0. *J. Math. Inequal.* **2012**, *6*, 401–417.
16. Rassias, M.T.H.; Yang, B.C. A multidimensional half – discrete Hilbert–type inequality and the Riemann zeta function. *Appl. Math. Comput.* **2013**, *225*, 263–277. [CrossRef]
17. Rassias, M.T.H.; Yang, B.C. On a multidimensional half–discrete Hilbert–type inequality related to the hyperbolic cotangent function. *Appl. Math. Comput.* **2013**, *242*, 800–813. [CrossRef]
18. Huang, Z.X.; Yang, B.C. On a half-discrete Hilbert-Type inequality similar to Mulholland's inequality. *J. Inequal. Appl.* **2013**, *290*, 2013. [CrossRef]
19. Yang, B.C.; Lebnath, L. *Half-Discrete Hilbert-Type Inequalities*; World Scientific Publishing: Singapore, 2014.
20. Krnic, M.; Pecaric, J. Extension of Hilbert's inequality. *J. Math. Anal. Appl.* **2006**, *324*, 150–160. [CrossRef]
21. Adiyasuren, V.; Batbold, T.; Azar, L.E. A new discrete Hilbert-type inequality involving partial sums. *J. Inequal. Appl.* **2019**, *127*, 2019. [CrossRef]
22. Hong, Y.; Wen, Y. A necessary and Sufficient condition of that Hilbert type series inequality with homogeneous kernel has the best constant factor. *Ann. Math.* **2016**, *37*, 329–336.
23. Hong, Y. On the structure character of Hilbert's type integral inequality with homogeneous kernel and applications. *J. Jilin Univ. Sci. Ed.* **2017**, *55*, 189–194.
24. Hong, Y.; Huang, Q.L.; Yang, B.C.; Liao, J.L. The necessary and sufficient conditions for the existence of a kind of Hilbert-type multiple integral inequality with the non-homogeneous kernel and its applications. *J. Inequal. Appl.* **2017**, *2017*, 316. [CrossRef]

25. Xin, D.M.; Yang, B.C.; Wang, A.Z. Equivalent property of a Hilbert-type integral inequality related to the beta function in the whole plane. *J. Funct. Spaces* **2018**. [CrossRef]
26. Hong, Y.; He, B.; Yang, B.C. Necessary and Sufficient Conditions for the Validity of Hilbert Type Integral Inequalities with a Class of Quasi-Homogeneous Kernels and Its Application in Operator Theory. *J. Math. Inequal.* **2018**, *12*, 777–788. [CrossRef]
27. Kuang, J.C. *Applied Inequalities*; Shangdong Science and Technology Press: Jinan, China, 2004.
28. Kuang, J.C. *Real and Functional Analysis (Continuation)*; Higher Education Press: Beijing, China, 2015.

mathematics

MDPI

Article

Variation Inequalities for One-Sided Singular Integrals and Related Commutators

Feng Liu [1], Seongtae Jhang [2], Sung-Kwun Oh [3] and Zunwei Fu [2,4,*]

1 College of Mathematics and System Science, Shandong University of Science and Technology, Qingdao 266590, China; FLiu@sdust.edu.cn
2 Department of Computer Science, The University of Suwon, Wau-ri, Bongdam-eup, Hwaseong-si, Gyeonggi-do 445-743, Korea; stjhang@suwon.ac.kr
3 Department of Electrical Engineering, The University of Suwon, Wau-ri, Bongdam-eup, Hwaseong-si, Gyeonggi-do 445-743, Korea; ohsk@suwon.ac.kr
4 School of Mathematical Sciences, Qufu Normal University, Qufu 273000, China
* Correspondence: fuzunwei@eyou.com

Received: 26 August 2019; Accepted: 18 September 2019; Published: 20 September 2019

Abstract: We establish one-sided weighted endpoint estimates for the ϱ-variation ($\varrho > 2$) operators of one-sided singular integrals under certain priori assumption by applying one-sided Calderón–Zygmund argument. Using one-sided sharp maximal estimates, we further prove that the ϱ-variation operators of related commutators are bounded on one-sided weighted Lebesgue and Morrey spaces. In addition, we also show that these operators are bounded from one-sided weighted Morrey spaces to one-sided weighted Campanato spaces. As applications, we obtain some results for the λ-jump operators and the numbers of up-crossings. Our main results represent one-sided extensions of many previously known ones.

Keywords: ϱ-variation; one-sided singular integral; commutator; one-sided weighted Morrey space; one-sided weighted Campanato space

JEL Classification: 42B20; 42B25

---

1. Introduction

Given a family of bounded operators $\mathcal{T} = \{T_\epsilon\}_{\epsilon>0}$ acting between spaces of functions, one of the most significative problems in harmonic analysis is the existence of limits $\lim_{\epsilon\to 0^+} T_\epsilon f$ and $\lim_{\epsilon\to\infty} T_\epsilon f$, when f belongs to a certain space of functions. The question that arises naturally is how to measure the speed of convergence of the above limits. A classic method is to investigate square functions of the type $(\sum_{i=1}^{\infty} |T_{\epsilon_i} f - T_{\epsilon_{i+1}} f|^2)^{1/2}$. Along this line, there is a more general way to study the following oscillation operator

$$\mathcal{O}(\mathcal{T})f(x) = \Big(\sum_{i=1}^{\infty} \sup_{t_{i+1} \le \epsilon_{i+1} < \epsilon_i \le t_i} |T_{\epsilon_{i+1}} f(x) - T_{\epsilon_i} f(x)|^2 \Big)^{1/2},$$

with $\{t_i\}$ being a fixed sequence decreasing to zero. However, beyond that, another typical method is to consider the ϱ-variation operator defined by

$$\mathcal{V}_\varrho(\mathcal{T})f(x) = \sup_{\{\epsilon_i\}\searrow 0} \Big(\sum_{i=1}^{\infty} |T_{\epsilon_i} f(x) - T_{\epsilon_{i+1}} f(x)|^\varrho \Big)^{1/\varrho},$$

where $\varrho > 2$ and the supremum runs over all sequences $\{\epsilon_i\}$ of positive numbers decreasing to zero.

The investigation on variation inequalities is an active research topic in probability, ergodic theory and harmonic analysis. The first variation inequality was proved by Lépingle [15] for martingales (also see [25] for a simple proof). Bourgain [2] proved the similar variation estimates for the ergodic averages of a dynamic system later. Bourgain's work has inspired a number of authors to investigate oscillation and variation inequalities for several families of operators from ergodic theory (see [12,13,24] for examples) and harmonic analysis (cf. [3,4,6,11,14]). Recently, the variation inequalities and their weighed case for singular integrals and related operators have also been studied by many authors. The first work in this direction is due to Campbell et al. [3] who proved that $\mathcal{O}(\mathcal{H})$ and $\mathcal{V}_\varrho(\mathcal{H})$ with $\varrho > 2$ are of type (p, p) for $1 < p < \infty$ and of weak type $(1, 1)$, where $\mathcal{H} = \{H_\epsilon\}_{\epsilon>0}$ is the family of the truncated Hilbert transforms, i.e., $H_\epsilon f(x) = \int_{|x-y|>\epsilon} \frac{f(y)}{x-y}dy$. Subsequently, the aforementioned authors [4] also studied the variation operators related to the classical Riesz transform in $\mathbb{R}^d$ for $d \geq 2$. In 2004, Gillespie and Torrea [9] established the $L^p(\mathbb{R}, w(x)dx)$ bounds for $\mathcal{O}(\mathcal{H})$ and $\mathcal{V}_\varrho(\mathcal{H})$ with $\varrho > 2$, $1 < p < \infty$ and $w \in A_p$ (the Muckenhoupt weights class) (also see [10,14] for the related investigations). Later on, Crescimbeni et al. [5] proved that $\mathcal{O}(\mathcal{H})$ and $\mathcal{V}_\varrho(\mathcal{H})$ with $\rho > 2$ map $L^1(\mathbb{R}, w(x)dx)$ into $L^{1,\infty}(\mathbb{R}, w(x)dx)$ for $w \in A_1$. In particular, Ma et al. [21,22] presented the weighted oscillation and variation inequalities for differential operators and Calderón–Zygmund singular integrals. Recently, Liu and Wu [19] established the weighted oscillation and variational inequalities for the commutator of one-dimensional Calderón–Zygmund singular integrals.

The primary purpose of this paper is to study weighted boundedness of oscillation and variational operators for one-sided singular integrals and their commutators. We say a function K belongs to one-sided Calderón–Zygmund kernel OCZK(B_1, B_2, B_3) if $K \in L^1_{\text{loc}}(\mathbb{R}\backslash\{0\})$ satisfies the following conditions: there exist constants $B_1, B_2, B_3 > 0$ such that

$$\left| \int_{\{\epsilon<|x|<N\}} K(x)dx \right| \leq B_1 \quad \text{for all } \epsilon \text{ and all } N \text{ with } 0 < \epsilon < N,$$

and furthermore $\lim_{\epsilon\to 0^+} \int_{\epsilon<|x|<N} K(x)dx$ exists,

$$|K(x)| \leq B_2|x|^{-1} \quad \text{for all } \ x \neq 0,$$

$$|K(x-y) - K(x)| \leq B_3|y||x|^{-2} \text{ for all } x \text{ and } y \text{ with } |x| > 2|y|.$$

An example of a one-sided Calderón–Zygmund kernel is $K(x) = \frac{\sin(\log x)}{x \log x}\chi_{(0,\infty)}$; see [1]. We mention here that the kernel of one-sided truncated Hilbert Transform, $K_0(x) = \frac{1}{x}\chi_{(0,\infty)}$, is not a OCZK for there does not exist a $B_1 > 0$ such that the first condition above holds.

Let $K \in$ OCZK(B_1, B_2, B_3) with support in $(-\infty, 0)$ and $b \in$ BMO$(\mathbb{R})$. For $m \in \mathbb{N}$, we consider the one-sided operator

$$T_b^{+,m} f(x) = \lim_{\epsilon\to 0^+} T_\epsilon^{+,b,m} f(x) = \text{p.v.} \int_x^\infty (b(x) - b(y))^m K(x-y) f(y) dy,$$

where

$$T_\epsilon^{+,b,m} f(x) := \int_{x+\epsilon}^\infty (b(x) - b(y))^m K(x-y) f(y) dy. \tag{1}$$

For $m \geq 1$, the operator $T_b^{+,m}$ is the m-th order commutator of one-sided singular integral. When $m = 0$, we denote by $T_\epsilon^{+,b,0} = T_\epsilon^+$, and then the operator $T_b^{+,m}$ reduces to the one-sided Calderón–Zygmund singular integral operator T^+, which is defined by

$$T^+ f(x) = \lim_{\epsilon\to 0^+} T_\epsilon^+ f(x) = \text{p.v.} \int_x^\infty K(x-y) f(y) dy. \tag{2}$$

In 1997, Aimar et al. [1] observed that the operator T^+ maps $L^p(\mathbb{R}, w(x)dx)$ into $L^p(\mathbb{R}, w(x)dx)$ for $1 < p < \infty$ and $w \in A_p^+$, and maps $L^1(\mathbb{R}, w(x)dx)$ into $L^{1,\infty}(\mathbb{R}, w(x)dx)$ for $w \in A_1^+$. Subsequently, Lorente and Riveros [20] proved that there exist constants $C > 0$ such that

$$\|T_b^{+,m} f\|_{L^p(\mathbb{R},w(x)dx)} \leq C\|b\|_{\mathrm{BMO}(\mathbb{R})}^m \|f\|_{L^p(\mathbb{R},w(x)dx)}$$

for $w \in A_p^+$ and $1 < p < \infty$, and

$$w(\{x : |T_b^{+,m} f(x)| > \lambda\}) \leq C\phi_m(\|b\|_{\mathrm{BMO}(\mathbb{R})}^m) \int_{\mathbb{R}} \frac{|f(x)|}{\lambda} \left(1 + \log^+ \left(\frac{|f(x)|}{\lambda}\right)\right)^m w(x)dx$$

for $w \in A_1^+$ and $\lambda > 0$, where $\phi_m(t) = t(1 + \log^+ t)^m$ and $z^+ = \max\{z, 0\}$. Other interesting related results for the one-sided operators we may refer to [7,8,16–18], among others.

At first, we shall establish the one-sided weighted endpoint and strong estimates for the ϱ-variation ($\varrho > 2$) operators of one-sided singular integral and its commutator. Let us recall the one-sided weighted BMO spaces.

Definition 1. (One-sided weighted BMO spaces.) *For a weight w, the one-sided weighted BMO spaces $\mathrm{BMO}^+(\mathbb{R}, w(x)dx)$ is defined by*

$$\mathrm{BMO}^+(\mathbb{R}, w(x)dx) := \{f \in L^1_{\mathrm{loc}}(\mathbb{R}, dx) : \|f\|_{\mathrm{BMO}^+(\mathbb{R},w(x)dx)} := \|M^{+,\sharp} f\|_{L^\infty(\mathbb{R},w(x)dx)} < \infty\}.$$

Here, $M^{+,\sharp}$ is one-sided sharp maximal operator defined by

$$M^{+,\sharp} f(x) = \sup_{h>0} \frac{1}{h} \int_x^{x+h} \left(f(y) - \frac{1}{h} \int_{x+h}^{x+2h} f(z)dz\right)^+ dy.$$

Remark 1. *When $w(x) \equiv 1$, the space $\mathrm{BMO}^+(\mathbb{R}, w(x)dx)$ reduces to the one-sided BMO space $\mathrm{BMO}^+(\mathbb{R})$, which was introduced by Martín-Reyes and de la Torre [23]. It was proved in [23] that*

$$M^{+,\sharp} f(x) \leq \sup_{h>0} \inf_{a\in\mathbb{R}} \left(\frac{1}{h} \int_x^{x+h} (f(y) - a)^+ dy + \frac{1}{h} \int_{x+h}^{x+2h} (a - f(y))^+ dy\right) \leq \|f\|_{\mathrm{BMO}(\mathbb{R})} \quad (3)$$

for any $x \in \mathbb{R}$. This yields that $\mathrm{BMO}(\mathbb{R}) \subset \mathrm{BMO}^+(\mathbb{R})$.

We now list our first main result as follows:

Theorem 1. *Let $m \in \mathbb{N}$, $\varrho > 2$, $b \in \mathrm{BMO}(\mathbb{R})$ and $K \in \mathrm{OCZK}(B_1, B_2, B_3)$ with supported in $(-\infty, 0)$. Let $\mathcal{T}_b^m = \{T_\epsilon^{+,b,m}\}_{\epsilon>0}$ and $\mathcal{T} = \{T_\epsilon^+\}_{\epsilon>0}$ be given as in Equation (1) and (2), respectively. Assume that $\|\mathcal{V}_\varrho(\mathcal{T})\|_{L^q(\mathbb{R},dx)\to L^q(\mathbb{R},dx)} < \infty$ for some $q \in (1, \infty)$. Then,*

(i) *for any $w \in A_1^+$ and $f \in L^1(\mathbb{R}, w(x)dx)$, it holds that*

$$\|\mathcal{V}_\varrho(\mathcal{T})f\|_{L^{1,\infty}(\mathbb{R},w(x)dx)} \leq C\|f\|_{L^1(\mathbb{R},w(x)dx)};$$

(ii) *for any $1 < p < \infty$, $w \in A_p^+$ and $f \in L^p(\mathbb{R}, w(x)dx)$, it holds that*

$$\|\mathcal{V}_\varrho(\mathcal{T}_b^m)f\|_{L^p(\mathbb{R},w(x)dx)} \leq C\|b\|_{\mathrm{BMO}(\mathbb{R})}^m \|f\|_{L^p(\mathbb{R},w(x)dx)};$$

(iii) *for a weight w satisfying $w^{-1} \in A_1^-$ and $f \in L^\infty(\mathbb{R}, w(x)dx)$, it holds that*

$$\|\mathcal{V}_\varrho(\mathcal{T})f\|_{\mathrm{BMO}^+(\mathbb{R},w(x)dx)} \leq C\|f\|_{L^\infty(\mathbb{R},w(x)dx)}.$$

In addition, we also investigate the boundedness behavior of the ρ-variation operators of one-sided singular integral and its commutator on one-sided weighted Morrey spaces and Companato spaces. In order to study the boundedness of one-sided singular integral operator on weighted Morrey spaces and Campanato spaces, Shi and Fu [27] introduced the one-sided weighted Morrey spaces and one-sided weighted Campanato spaces, which are defined as follows:

Definition 2. (One-sided weighted Morrey spaces and Campanato spaces.) *Let $1 \le p < \infty$, $-1/p \le \beta < 0$ and w be a weight on $\mathbb{R}$.*

(i) *One-sided weighted Morrey spaces $L^{p,\beta,+}(w)$ are defined by*

$$L^{p,\beta,+}(w) := \{f \in L^p_{\text{loc}}(\mathbb{R}, dx) : \|f\|_{L^{p,\beta,+}(w)} < +\infty\},$$

where

$$\|f\|_{L^{p,\beta,+}(w)} := \sup_{x_0 \in \mathbb{R}} \sup_{h>0} \frac{1}{h^\beta}\Big(\frac{1}{w((x_0-h,x_0))}\int_{x_0}^{x_0+h} |f(x)|^p dx\Big)^{1/p}.$$

(ii) *One-sided weighted Campanato spaces $\mathfrak{L}^{p,\beta,+}(w)$ are given by*

$$\mathfrak{L}^{p,\beta,+}(w) := \{f \in L^p_{\text{loc}}(\mathbb{R}, dx) : \|f\|_{\mathfrak{L}^{p,\beta,+}(w)} < +\infty\},$$

where

$$\|f\|_{\mathfrak{L}^{p,\beta,+}(w)} := \sup_{x_0 \in \mathbb{R}} \sup_{h>0} \frac{1}{h^\beta}\Big(\frac{1}{w((x_0-h,x_0))}\int_{x_0}^{x_0+h} |f(x) - f_{(x_0,x_0+h)}|^p dx\Big)^{1/p}.$$

Remark 2. *It is well known that the following are valid:*

$$\|f\|_{\mathfrak{L}^{p,\beta,+}(w)} \sim \sup_{x_0 \in \mathbb{R}} \sup_{h>0} \inf_{a\in\mathbb{R}} \frac{1}{h^\beta}\Big(\frac{1}{w((x_0-h,x_0))}\int_{x_0}^{x_0+h} |f(x) - a|^p dx\Big)^{1/p}; \tag{4}$$

$$L^{p,\beta,+}(w) \subsetneq \mathfrak{L}^{p,\beta,+}(w).$$

The rest of the main results can be listed as follows.

Theorem 2. *Let $m \in \mathbb{N}$, $\varrho > 2$, $b \in \text{BMO}(\mathbb{R})$ and $K \in \text{OCZK}(B_1, B_2, B_3)$ with support in $(-\infty, 0)$. Let $\mathcal{T}_b^m = \{T_\epsilon^{+,b,m}\}_{\epsilon>0}$ and $\mathcal{T} = \{T_\epsilon^+\}_{\epsilon>0}$ be given as in Equation (1) and (2), respectively. Assume that $\|\mathcal{V}_\varrho(\mathcal{T})\|_{L^q(\mathbb{R},dx)\to L^q(\mathbb{R},dx)} < \infty$ for some $q \in (1,\infty)$. Then,*

(i) *for any $1 < p < 1/(\beta+1)$, $-1/p \le \beta < 0$, $w \in A_p^+$ and $f \in L^{p,\beta,+}(w)$,*

$$\|\mathcal{V}_\varrho(\mathcal{T}_b^m)f\|_{L^{p,\beta,+}(w)} \lesssim \|b\|^m_{\text{BMO}(\mathbb{R})}\|f\|_{L^{p,\beta,+}(w)};$$

(ii) *for any $1 < p < \infty$, $-1/p \le \beta < 0$, $w \in A_p^+$ and $f \in L^{p,\beta,+}(w)$,*

$$\|\mathcal{V}_\varrho(\mathcal{T})f\|_{\mathfrak{L}^{p,\beta,+}(w)} \lesssim \|f\|_{L^{p,\beta,+}(w)}.$$

Remark 3. *We remark that we deal only with $\varrho > 2$ for the variation operators in our main theorems, since it was pointed out in [2] that the variation is often not bounded in the case $\varrho \le 2$. In addition, it is unknown what are the endpoint estimates of the variation operators for the commutators of one-sided singular integrals and whether the above operators are bounded from one-sided weighted Morrey spaces to one-sided weighted Campanato spaces, which are interesting.*

This paper is organized as follows. In Section 2, we shall present some basic definitions and necessary lemmas. In Section 3, we give the proofs of Theorems 1 and 2. As applications, we present the corresponding estimates for the λ-jump operators and the number of up-crossing for these operators in Section 4. Finally, some further comments will be given in Section 5. We would like to remark that our works and ideas are taken from [9,19]. It should also be pointed out that all results in this paper are valid for oscillation operator with similar arguments.

Throughout this paper, for any $p \in (1,\infty)$, we denote by p' the dual exponent to p, i.e., $1/p + 1/p' = 1$. The letter C will represent a positive constant that may vary at each occurrence but is independent of the essential variables. For a weight w, an interval I and a function $f : \mathbb{R} \to \mathbb{R}$, we denote by $w(I) = \int_I w(x)dx$ and $f_I = \frac{1}{|I|}\int_I f(x)dx$. We also use the convention $\sum_{i\in\emptyset} a_i = 0$.

2. Preliminaries

We start with the definitions of one-sided Hardy–Littlewood maximal functions

$$M^+f(x) = \sup_{h>0}\frac{1}{h}\int_x^{x+h}|f(y)|dy \quad \text{and} \quad M^-f(x) = \sup_{h>0}\frac{1}{h}\int_{x-h}^{x}|f(y)|dy.$$

For $r > 0$, we set $M_r^+ f(x) := (M^+|f|^r(x))^{1/r}$.

By a weight, we mean a nonnegative measurable function.

Definition 3. *[26] Let $1 < p < \infty$. A weight w belongs to the class A_p^+ (resp., A_p^-), if $[w]_{A_p^+} < \infty$ (resp., $[w]_{A_p^-} < \infty$), where*

$$[w]_{A_p^+} := \sup_{a<b<c}\frac{1}{(c-a)^p}\Big(\int_a^b w(x)dx\Big)\Big(\int_b^c w(x)^{1-p'}dx\Big)^{p-1},$$

$$[w]_{A_p^-} := \sup_{a<b<c}\frac{1}{(c-a)^p}\Big(\int_b^c w(x)dx\Big)\Big(\int_a^b w(x)^{1-p'}dx\Big)^{p-1}.$$

A weight w belongs to the class A_1^+ (resp., A_1^-), if $[w]_{A_1^+} < \infty$ (resp., $[w]_{A_1^-} < \infty$), where

$$[w]_{A_1^+} := \sup_{x\in\mathbb{R}} w(x)^{-1}M^-w(x) \quad \text{and} \quad [w]_{A_1^-} := \sup_{x\in\mathbb{R}} w(x)^{-1}M^+w(x).$$

Since the A_p^+ and A_p^- classes are increasing with respect to p, the A_∞^+ (resp., A_∞^-) class of weights is defined in a natural way by $A_\infty^+ = \bigcup_{1<p<\infty} A_p^+$ (resp., $A_\infty^- = \bigcup_{1<p<\infty} A_p^-$) with

$$[w]_{A_\infty^+} := \inf_{1<p<\infty}\inf_{w\in A_p^+}[w]_{A_p^+}, \quad [w]_{A_\infty^-} := \inf_{1<p<\infty}\inf_{w\in A_p^-}[w]_{A_p^-}.$$

It is easy to see that $A_p \subsetneq A_p^+$, $A_p \subsetneq A_p^-$ and $A_p = A_p^+ \bigcap A_p^-$. Take e^x for example, $e^x \notin A_1$, but $e^x \in A_1^+$. Here, A_p denotes the usual Muckenhoupt weight.

It was shown in [26] that, for any $1 < p < \infty$, $M^+ : L^p(\mathbb{R}, w(x)dx) \to L^p(\mathbb{R}, w(x)dx)$ is bounded if and only if $w \in A_p^+$; moreover, $M^+ : L^1(\mathbb{R}, w(x)dx) \to L^{1,\infty}(\mathbb{R}, w(x)dx)$ is bounded if and only if $w \in A_1^+$. The same results hold for M^- if $w \in A_p^+$ replaced by $w \in A_p^-$ for $1 \le p < \infty$.

The following lemma will play key roles in our main proofs.

Lemma 1.

(i) *Let $1 \le p \le \infty$ and $w \in A_p^+$. Then, for all $x_0 \in \mathbb{R}$ and $h > 0$,*

$$w(x_0 - h, x_0 + h) \le (1 + 2^p[w]_{A_p^+})w(x_0, x_0 + h). \tag{5}$$

(ii) Let $1 \leq p \leq \infty$ and $w \in A_p^+$. Then, for all $x_0 \in \mathbb{R}$, $h > 0$ and $\lambda \geq 1$,

$$w(x_0 - \lambda h, x_0) \leq \lambda^p (2^p [w]_{A_p^+} + (2^p [w]_{A_p^+})^2) w(x_0, x_0 + h). \tag{6}$$

Proof. Fix $h > 0$ and $x_0 \in \mathbb{R}$ and we set $I = (x_0 - h, x_0 + h)$. Given two functions f, g defined on $\mathbb{R}$, by Hölder's inequality, we get

$$\begin{aligned} &\Big(\frac{1}{|I|}\int_I |f(x)g(x)|dx\Big)^p \\ &\leq \frac{1}{|I|^p}\Big(\int_I |f(x)|^p w(x)dx\Big)\Big(\int_I |g(x)|^{p'} w(x)^{1-p'}dx\Big)^{p/p'} \\ &\leq \Big(\frac{1}{|I|}\int_{I^-} w(x)dx\Big)\Big(\frac{1}{|I|}\int_I |g(x)|^{p'} w(x)^{1-p'}dx\Big)^{p-1}\Big(\frac{1}{w(I^-)}\int_I |f(x)|^p w(x)dx\Big). \end{aligned} \tag{7}$$

Applying Equation (7) to the functions $f = \chi_{I^+}$ and $g = \chi_{I^+}$, we get

$$w(I^-) \leq 2^p [w]_{A_p^+} w(I^+). \tag{8}$$

Then, (5) follows easily from (8).

On the other hand, we get from (7) that

$$\begin{aligned} &\Big(\frac{1}{|\lambda I|}\int_{\lambda I} |f(x)g(x)|dx\Big)^p \\ &\leq \Big(\frac{1}{|\lambda I|}\int_{(\lambda I)^-} w(x)dx\Big)\Big(\frac{1}{|\lambda I|}\int_{\lambda I} |g(x)|^{p'} w(x)^{1-p'}dx\Big)^{p-1} \\ &\times\Big(\frac{1}{w((\lambda I)^-)}\int_{\lambda I} |f(x)|^p w(x)dx\Big). \end{aligned} \tag{9}$$

Applying (9) to the functions $f = \chi_I$ and $g = \chi_{(\lambda I)^+}$, we have

$$w((\lambda I)^-) \leq (2\lambda)^p [w]_{A_p^+} w(I), \tag{10}$$

which together with (5) yields (6). □

By Lemma 2.1 in [26] and the similar argument as in classical Calderón–Zygmund decomposition for the usual Hardy–Littlewood maximal function, one can get the following Calderón–Zygmund decomposition for M^+, which will be crucial for the proof of Lemma 3.

Lemma 2. *Let $f \in L^1(\mathbb{R}, dx)$ and $\alpha > 0$. Let $\Omega = \{x : M^+ f(x) > \alpha\}$. Then, Ω can be decomposed into finitely many disjoint intervals of integers: $\Omega = \bigcup_i I_i$ with the following properties:*

(i) $f = g + \varphi$, *where* $g = f\chi_{\mathbb{R}\setminus\Omega}$ *and* $g = f_{I_i}$ *on* I_i *for each* i;
(ii) $\varphi = \sum_i \varphi_i$, *where* $\varphi_i = (f - f_{I_i})\chi_{I_i}$;
(iii) $\|g\|_{L^\infty(\mathbb{R},dx)} \leq 2\alpha$ *and* $\|g\|_{L^1(\mathbb{R},dx)} \leq \|f\|_{L^1(\mathbb{R},dx)}$;
(iv) *for each* i, $\int_{I_i} \varphi_i(y)dy = 0$ *and* $\frac{1}{|I_i|}\int_{I_i} |\varphi_i(y)|dy \leq 4\alpha$;
(v) $\sum_i |I_i| \leq \alpha^{-1}\|f\|_{L^1(\mathbb{R},dx)}$.

3. Proofs of Main Results

Following [9], let $\Theta = \{\beta : \beta = \{\epsilon_i\}, \epsilon_i \in \mathbb{R}, \epsilon_i \searrow 0\}$ and F_ϱ be the mixed norm Banach space of two variables function h defined on $\mathbb{N} \times \Theta$ such that

$$\|h\|_{F_\varrho} \equiv \sup_\beta \Big(\sum_i |h(i,\beta)|^\varrho\Big)^{1/\varrho} < \infty.$$

Given a family of operators $\mathcal{T} = \{T_t\}_{t>0}$ defined on $L^p(\mathbb{R}, dx)$, we consider the F_ϱ-valued operator $V(\mathcal{T}) : f \longrightarrow V(\mathcal{T})f$ on $L^p(\mathbb{R}, dx)$ given by

$$V(\mathcal{T})f(x) := \left\{T_{[\epsilon_{i+1},\epsilon_i]}f(x)\right\}_{\beta=\{\epsilon_i\}\in\Theta},$$

where the expression $\{T_{[\epsilon_{i+1},\epsilon_i]}f(x)\}_{\beta=\{\epsilon_i\}\in\Theta}$ is an abbreviation for the element of F_ϱ given by

$$(i, \beta) = (i, \{\epsilon_i\}) \longrightarrow T_{[\epsilon_{i+1},\epsilon_i]}f(x) := T_{\epsilon_{i+1}}f(x) - T_{\epsilon_i}f(x).$$

Observe that

$$\mathcal{V}_\varrho(\mathcal{T})f(x) = \|V(\mathcal{T})f(x)\|_{F_\varrho}, \ \forall x \in \mathbb{R}. \tag{11}$$

In order to prove Theorem 1, we shall establish the following key result.

Lemma 3. *Let $\varrho > 2$ and $K \in \mathrm{OCZK}(B_1, B_2, B_3)$ with support in $(-\infty, 0)$. Let $\mathcal{T} = \{T_\epsilon^+\}_{\epsilon>0}$ be given as in Equation (2). Assume that $\|\mathcal{V}_\varrho(\mathcal{T})\|_{L^q(\mathbb{R},w(x)dx)\to L^q(\mathbb{R},w(x)dx)} < \infty$ for some $q \in (1, \infty)$ and $w \in A_q^+$. Then,*

$$\|\mathcal{V}_\varrho(\mathcal{T})f\|_{L^{1,\infty}(\mathbb{R},w(x)dx)} \le C\|f\|_{L^1(\mathbb{R},w(x)dx)}, \ \forall f \in L^1(\mathbb{R}, w(x)dx) \text{ and } w \in A_1^+.$$

Proof. We shall adopt the classical Calderón–Zygmund argument to prove Lemma 3. Let $\Omega = \{x : M^+f(x) > 1\}$. Invoking Lemma 2, we can decompose Ω as $\Omega = \bigcup_j I_j$ and decompose f as $f = g + \varphi$, where all I_j are disjoint intervals, $g = f\chi_{\mathbb{R}\setminus\Omega} + \sum_j f_{I_j}\chi_{I_j}$, $\varphi = \sum_j \varphi_j$, $\varphi_j = (f - f_{I_j})\chi_{I_j}$, $\|g\|_{L^\infty(\mathbb{R},dx)} \le 2$, $\|g\|_{L^1(\mathbb{R},dx)} \le \|f\|_{L^1(\mathbb{R},dx)}$, and for each j, $\int_{I_j} \varphi_j(y)dy = 0$ and $\frac{1}{|I_j|}\int_{I_j} |\varphi_j(y)|dy \le 4$. It suffices to show that

$$w(\{x : \mathcal{V}_\varrho(\mathcal{T})f(x) > 1\}) \le C\|f\|_{L^1(\mathbb{R},w(x)dx)}. \tag{12}$$

It is clear that

$$w(\{x : \mathcal{V}_\varrho(\mathcal{T})f(x) > 1\}) \le w(\{x : \mathcal{V}_\varrho(\mathcal{T})g(x) > 1/2\}) + w(\{x : \mathcal{V}_\varrho(\mathcal{T})\varphi(x) > 1/2\}). \tag{13}$$

By our assumption,

$$\begin{aligned} w(\{x : \mathcal{V}_\varrho(\mathcal{T})g(x) > 1/2\}) &\le 2^q \int_{\mathbb{R}} |\mathcal{V}_\varrho(\mathcal{T})g(x)|^q w(x)dx \\ &\le C \int_{\mathbb{R}} |g(x)|^q w(x)dx \le C\|f\|_{L^1(\mathbb{R},w(x)dx)}. \end{aligned} \tag{14}$$

We set $I_j = (c_j, c_j + |I_j|)$ and $\Omega^* = \bigcup_j (c_j - 2|I_j|, c_j + 2|I_j|)$, then

$$w(\{x : \mathcal{V}_\varrho(\mathcal{T})\varphi(x) > 1/2\}) \le w(\Omega^*) + w(\{x \in \mathbb{R} \setminus \Omega^* : \mathcal{V}_\varrho(\mathcal{T})\varphi(x) > 1/2\}). \tag{15}$$

Using Lemma 1 (i) and the $L^1(\mathbb{R}, w(x)dx) \to L^{1,\infty}(\mathbb{R}, w(x)dx)$ bounds for M^+, one has

$$w(\Omega^*) \le C\sum_j w(I_j) = Cw(\Omega) \le C\|f\|_{L^1(\mathbb{R},w(x)dx)}. \tag{16}$$

We now turn to prove

$$w(\{x \in \mathbb{R} \setminus \Omega^* : \mathcal{V}_\varrho(\mathcal{T})\varphi(x) > 1/2\}) \le C\|f\|_{L^1(\mathbb{R},w(x)dx)}. \tag{17}$$

For every $x \in \mathbb{R} \setminus \Omega^*$, we can choose a decreasing sequence $\{\epsilon_i\}$ (that depends on x) such that

$$\mathcal{V}_\varrho(\mathcal{T})\varphi(x) \le 2\Big(\sum_i |T^+_{[\epsilon_{i+1},\epsilon_i]}\varphi(x)|^\varrho\Big)^{1/\varrho}.$$

For each i and $x \in \mathbb{R} \setminus \Omega^*$, we set $B_i(x) = (x + \epsilon_{i+1}, x + \epsilon_i]$ and

$$N_{i,1} = \{j : I_j \subset B_i(x)\} \quad \text{and} \quad N_{i,2} = \{j : I_j \cap B_i(x) \neq \emptyset,\ I_j \not\subset B_i(x)\}.$$

We notice that the cardinal of the $N_{i,2}$ is at most two. Thus, it holds that

$$\begin{aligned}\mathcal{V}_\varrho(\mathcal{T})\varphi(x) &\leq 2\Big(\sum_i \Big|\sum_{j\in N_{i,1}} T^+_{[\epsilon_{i+1},\epsilon_i]}\varphi_j(x)\Big|^\varrho\Big)^{1/\varrho} + 2\Big(\sum_i \Big|\sum_{j\in N_{i,2}} T^+_{[\epsilon_{i+1},\epsilon_i]}\varphi_j(x)\Big|^\varrho\Big)^{1/\varrho} \\ &\leq 2\sum_i \sum_{j\in N_{i,1}} |T^+_{[\epsilon_{i+1},\epsilon_i]}\varphi_j(x)| + 4\Big(\sum_i \sum_{j\in N_{i,2}} |T^+_{[\epsilon_{i+1},\epsilon_i]}\varphi_j(x)|^\varrho\Big)^{1/\varrho}.\end{aligned}$$

It follows that

$$\begin{aligned}& w(\{x \in \mathbb{R} \setminus \Omega^* : \mathcal{V}_\varrho(\mathcal{T})\varphi(x) > 1/2\}) \\ &\leq w\Big(\Big\{x \in \mathbb{R} \setminus \Omega^* : \sum_i \sum_{j\in N_{i,1}} |T^+_{[\epsilon_{i+1},\epsilon_i]}\varphi_j(x)| > \frac{1}{8}\Big\}\Big) \\ &\quad + w\Big(\Big\{x \in \mathbb{R} \setminus \Omega^* : \Big(\sum_i \sum_{j\in N_{i,2}} |T^+_{[\epsilon_{i+1},\epsilon_i]}\varphi_j(x)|^\varrho\Big)^{1/\varrho} > \frac{1}{16}\Big\}\Big).\end{aligned} \tag{18}$$

Fix $x \in \mathbb{R} \setminus \Omega^*$. Note that $|x - c_j| \geq 2|I_j| > 2|y - c_j|$ for any $y \in I_j$. Then, $|K(x-y) - K(x-c_j)| \leq B_3|x-c_j|^{-2}|y-c_j|$. This together with the properties of φ_j yield that

$$|T^+_{[\epsilon_{i+1},\epsilon_i]}\varphi_j(x)| = \Big|\int_{\mathbb{R}} (K(x-y) - K(x-c_j))\varphi_j(y)dy\Big| \leq 2B_3|I_j||x-c_j|^{-2}\int_{I_j} |f(y)|dy.$$

Observing that $T^+_{[\epsilon_{i+1},\epsilon_i]}\varphi_j(x) = 0$ if $x > c_j + |I_j|$, we thus have

$$\begin{aligned}& w\Big(\Big\{x \in \mathbb{R} \setminus \Omega^* : \sum_i \sum_{j\in N_{i,1}} |T^+_{[\epsilon_{i+1},\epsilon_i]}\varphi_j(x)| > \frac{1}{8}\Big\}\Big) \\ &\leq 8\int_{\mathbb{R}\setminus\Omega^*} \sum_i \sum_{j\in N_{i,1}} |T^+_{[\epsilon_{i+1},\epsilon_i]}\varphi_j(x)|w(x)dx \\ &\leq 16B_3 \sum_j |I_j| \int_{(-\infty,c_j-2|I_j|]} \frac{w(x)}{|x-c_j|^2}dx \int_{I_j} |f(y)|dy.\end{aligned} \tag{19}$$

Fix $y \in I_j$. One can easily check that $c_j - x \geq 2(y-x)/3$ for any $x \leq c_j - 2|I_j|$. Then,

$$\begin{aligned}\int_{(-\infty,c_j-2|I_j|]} \frac{w(x)}{|x-c_j|^\delta}dx &\leq \sum_{k=1}^\infty \int_{[c_j-2^{k+1}|I_j|,c_j-2^k|I_j|]} \frac{w(x)}{|x-c_j|^\delta}dx \\ &\leq \sum_{k=1}^\infty (2^k|I_j|)^{-\delta} 2^{k+3}|I_j| \frac{1}{2^{k+3}|I_j|}\int_{y-2^{k+3}|I_j|}^y w(x)dx \\ &\leq C(\delta)|I_j|^{1-\delta} M^- w(y)\end{aligned} \tag{20}$$

for any $\delta > 1$. By (19) and (20) (with $\delta = 2$) and $w \in A_1^+$, we have

$$\begin{aligned}& w\Big(\Big\{x \in \mathbb{R} \setminus \Omega^* : \sum_i \sum_{j\in N_{i,1}} |T^+_{[\epsilon_{i+1},\epsilon_i]}\varphi_j(x)| > \frac{1}{8}\Big\}\Big) \\ &\leq C\sum_j \int_{I_j} |f(y)|M^- w(y)dy \leq C([w]_{A_1^+})\|f\|_{L^1(\mathbb{R},w(x)dx)}.\end{aligned} \tag{21}$$

Fix $x \in \mathbb{R} \setminus \Omega^*$. Note that $T^+_{[\epsilon_{i+1},\epsilon_i]}\varphi_j(x) = 0$ when $x > c_j + |I_j|$. Moreover, $y - x \geq c_j - x \geq 0$ for any $y \in I_j$. Then,

$$|T^+_{[\epsilon_{i+1},\epsilon_i]}\varphi_j(x)| \leq B_2 \int_{B_i(x)} \frac{|\varphi_j(y)|}{|x-y|}dy \leq B_2|x-c_j|^{-1}\chi_{(-\infty,c_j-2|I_j|]}(x)\int_{B_i(x)}|\varphi_j(y)|dy.$$

Combining this with (20) (with $\delta = \varrho$) implies that

$$\begin{aligned}
& w\Big(\Big\{x \in \mathbb{R} \setminus \Omega^* : \Big(\sum_i \sum_{j\in N_{i,2}} |T^+_{[\epsilon_{i+1},\epsilon_i]}\varphi_j(x)|^\varrho\Big)^{1/\varrho} > \frac{1}{16}\Big\}\Big) \\
& \leq 16^\varrho \int_{\mathbb{R}\setminus\Omega^*} \sum_i \sum_{j\in N_{i,2}} |T^+_{[\epsilon_{i+1},\epsilon_i]}\varphi_j(x)|^\varrho w(x)dx \\
& \leq C(\varrho)\int_{\mathbb{R}\setminus\Omega^*} \sum_j \Big(\sum_i |T^+_{[\epsilon_{i+1},\epsilon_i]}\varphi_j(x)|\Big)^\varrho w(x)dx \\
& \leq C(\varrho)\sum_j \int_{(-\infty,c_j-2|I_j|]} |x-c_j|^{-\varrho}\Big(\sum_i \int_{B_i(x)}|\varphi_j(y)|dy\Big)^\varrho w(x)dx \\
& \leq C(\varrho)\sum_j \int_{(-\infty,c_j-2|I_j|]} \frac{w(x)}{|x-c_j|^\varrho}\Big(\int_{\mathbb{R}}|\varphi_j(y)|dy\Big)^\varrho dx \\
& \leq C(\varrho)\sum_j |I_j|^{\varrho-1}\int_{I_j}\int_{(-\infty,c_j-2|I_j|]} \frac{w(x)}{|x-c_j|^\varrho}dx|f(y)|dy \\
& \leq C(\varrho)\sum_j \int_{I_j}|f(y)|M^-(w)(y)dy \\
& \leq C(\varrho,[w]_{A_1^+})\|f\|_{L^1(\mathbb{R},w(x)dx)},
\end{aligned}$$

which together with (21) and (18) yields (17). Then, (12) follows from (13)–(17). This proves Lemma 3. □

Applying similar arguments used in deriving Lemma 3, we can get the following:

Corollary 1. *Let $K \in \mathrm{OCZK}(B_1, B_2, B_3)$ with support in $(-\infty, 0)$. Let $\varrho > 2$ and $\mathcal{T} = \{T^+_\epsilon\}_{\epsilon>0}$ be given as in Equation (2). Assume that $\|\mathcal{V}_\varrho(\mathcal{T})\|_{L^q(\mathbb{R},dx)\to L^q(\mathbb{R},dx)} < \infty$ for some $q \in (1,\infty)$. Then,*

$$\|\mathcal{V}_\varrho(\mathcal{T})f\|_{L^{1,\infty}(\mathbb{R},dx)} \leq C\|f\|_{L^1(\mathbb{R},dx)}, \quad \forall f \in L^1(\mathbb{R},dx).$$

The following lemma will play a pivotal role in the proof of Theorem 1.

Lemma 4. *Let $m \in \mathbb{N}$, $\varrho > 2$, $b \in \mathrm{BMO}(\mathbb{R})$ and $K \in \mathrm{OCZK}(B_1, B_2, B_3)$ with support in $(-\infty, 0)$. Let $\mathcal{T}^m_b = \{T^{+,b,m}_\epsilon\}_{\epsilon>0}$ and $\mathcal{T} = \{T^+_\epsilon\}_{\epsilon>0}$ be given as in Equations (1) and (2), respectively. Assume that $\|\mathcal{V}_\varrho(\mathcal{T})\|_{L^q(\mathbb{R},dx)\to L^q(\mathbb{R},dx)} < \infty$ for some $q \in (1,\infty)$. Then, for any $r > 1$ and $x \in \mathbb{R}$, it holds that*

$$M^{+,\sharp}(\mathcal{V}_\varrho(\mathcal{T}^m_b)f)(x) \leq C\Big(\sum_{i=0}^{m-1}\|b\|^{m-i}_{\mathrm{BMO}(\mathbb{R})}M^+_r(\mathcal{V}_\varrho(\mathcal{T}^i_b)f)(x) + \|b\|^m_{\mathrm{BMO}(\mathbb{R})}M^+_r f(x)\Big). \tag{22}$$

Proof. We only prove (22) for the case $1 < r < \min\{q, 2\}$, since $M^+_{r_1}f \leq M^+_{r_2}f$ for any $r_2 \geq r_1$. Invoking Corollary 1, we see that $\mathcal{V}_\varrho(\mathcal{T})$ is of weak type (1,1). By the Marcinkiewicz interpolation theorem and our assumption, we have that $\mathcal{V}_\varrho(\mathcal{T})$ is bounded on $L^p(\mathbb{R}, dx)$ for any $1 < p < q$. Fix $x_0 \in \mathbb{R}$ and $h > 0$. We decompose f as $f = f_1 + f_2 + f_3$, where $f_1 = f\chi_{[x_0,x_0+2h]}$ and $f_2 = f\chi_{(x_0+2h,\infty)}$. Let $I = [x_0 - 2h, x_0 + 2h]$. In view of (3), to prove (22), we only prove

$$\begin{aligned}&\frac{1}{h}\int_{x_0}^{x_0+h}|\mathcal{V}_\varrho(\mathcal{T}_b^m)f(y)-\mathcal{V}_\varrho(\mathcal{T}_b^m)((b-b_I)^m f_2)(x_0)|dy\\&\leq C\Big(\sum_{i=0}^{m-1}\|b\|_{\mathrm{BMO}(\mathbb{R})}^{m-i}M_r^+(\mathcal{V}_\varrho(\mathcal{T}_b^i)f)(x)+\|b\|_{\mathrm{BMO}(\mathbb{R})}^m M_r^+f(x)\Big),\end{aligned} \tag{23}$$

where $C>0$ is independent of x_0, h. Using the arguments similar to those used in deriving the inequality (11) in [20], we get

$$T_\epsilon^{+,b,m}f(y)=T_\epsilon^+((b-b_I)^m f)(y)+\sum_{k=0}^{m-1}C_{k,m}(b(y)-b_I)^{m-k}T_\epsilon^{+,b,k}f(y),\ \ \forall y\in\mathbb{R}. \tag{24}$$

Note that $T_\epsilon^{+,b,k}f_3(y)=0$ for any $\epsilon>0$, $0\leq k\leq m-1$ and $y\geq x_0$. (24) leads to

$$\begin{aligned}V(\mathcal{T}_b^m)f(y)&=V(\mathcal{T})((b-b_I)^m f_1)(y)+V(\mathcal{T})((b-b_I)^m f_2)(y)\\&\quad+\sum_{k=0}^{m-1}C_{k,m}(b(y)-b_I)^{m-k}V(\mathcal{T}_b^k)f(y),\ \ \forall y\geq x_0.\end{aligned} \tag{25}$$

We notice from (11) that

$$\begin{aligned}&\frac{1}{h}\int_{x_0}^{x_0+h}|\mathcal{V}_\varrho(\mathcal{T}_b^m)f(y)-\mathcal{V}_\varrho(\mathcal{T}_b^m)((b-b_I)^m f_2)(x_0)|dy\\&=\frac{1}{h}\int_{x_0}^{x_0+h}|\|V(\mathcal{T}_b^m)f(y)\|_{F_\varrho}-\|V(\mathcal{T}_b^m)((b-b_I)^m f_2)(x_0)\|_{F_\varrho}|dy\\&\leq\frac{1}{h}\int_{x_0}^{x_0+h}\|V(\mathcal{T}_b^m)f(y)-V(\mathcal{T}_b^m)((b-b_I)^m f_2)(x_0)\|_{F_\varrho}dy.\end{aligned}$$

This together with (25) and (11) yield that

$$\begin{aligned}&\frac{1}{h}\int_{x_0}^{x_0+h}|\mathcal{V}_\varrho(\mathcal{T}_b^m)f(y)-\mathcal{V}_\varrho(\mathcal{T}_b^m)((b-b_I)^m f_2)(x_0)|dy\\&\leq\frac{1}{h}\int_{x_0}^{x_0+h}\mathcal{V}_\varrho(\mathcal{T})((b-b_I)^m f_1)(y)dy\\&+\sum_{k=0}^{m-1}C_{k,m}\frac{1}{h}\int_{x_0}^{x_0+h}|b(y)-b_I|^{m-k}\mathcal{V}_\varrho(\mathcal{T}_b^k)f(y)dy\\&+\frac{1}{h}\int_{x_0}^{x_0+h}\|V(\mathcal{T})((b-b_I)^m f_2)(y)-V(\mathcal{T})((b-b_I)^m f_2)(x_0)\|_{F_\varrho}dy\\&=:I_1+I_2+I_3.\end{aligned} \tag{26}$$

Observe that, for any $\delta>1$ and $k\in\mathbb{N}$,

$$\begin{aligned}\frac{1}{|2^kI|}\int_{2^kI}|b(z)-b_I|^\delta dz&\leq 2^{\delta-1}\Big(\frac{1}{|2^kI|}\int_{2^kI}|b(z)-b_{2^kI}|^\delta dz+|b_I-b_{2^kI}|^\delta\Big)\\&\leq C(\delta)(k+1)^\delta\|b\|_{\mathrm{BMO}(\mathbb{R})}^\delta.\end{aligned} \tag{27}$$

We set $\rho = \sqrt{r}$. By Hölder's inequality, the L^ρ boundedness for $\mathcal{V}_\varrho(\mathcal{T})$ and (27), we have

$$\begin{aligned} I_1 &\le \Big(\frac{1}{h}\int_{x_0}^{x_0+h} |\mathcal{V}_\varrho(\mathcal{T})((b-b_I)^m f_1)(y)|^\rho dy\Big)^{1/\rho} \\ &\le C(\rho)\Big(\frac{1}{h}\int_{x_0}^{x_0+2h} |(b(y)-b_I)^m f(y)|^\rho dy\Big)^{1/\rho} \\ &\le C(\rho)\Big(\frac{1}{h}\int_{x_0}^{x_0+2h} |f(y)|^r dy\Big)^{1/r}\Big(\frac{1}{|I|}\int_I |b(y)-b_I|^{m\rho\rho'} dy\Big)^{1/\rho\rho'} \\ &\le C(m,r)\|b\|^m_{\mathrm{BMO}(\mathbb{R})} M_r^+ f(x_0) \end{aligned} \tag{28}$$

and

$$\begin{aligned} I_2 &\le \sum_{k=0}^{m-1} C_{k,m}\Big(\frac{1}{h}\int_{x_0}^{x_0+h} |\mathcal{V}_\varrho(\mathcal{T}_b^k)f(y)|^r dy\Big)^{1/r}\Big(\frac{1}{|I|}\int_{|I|} |(b(y)-b_I)^{(m-k)r'} dy\Big)^{1/r'} \\ &\le C(m,r)\sum_{k=0}^{m-1} C_{k,m}\|b\|^{m-k}_{\mathrm{BMO}(\mathbb{R})} M_r^+(\mathcal{V}_\varrho(\mathcal{T}_b^k)f)(x_0). \end{aligned} \tag{29}$$

For I_3, let $y \in [x_0, x_0+h]$ and $\beta = \{\epsilon_i\} \in \Theta$, since

$$\begin{aligned} & T^+_{[\epsilon_{i+1},\epsilon_i]}((b-b_I)^m f_2)(y) - T^+_{[\epsilon_{i+1},\epsilon_i]}((b-b_I)^m f_2)(x_0) \\ =& \int_{\mathbb{R}} [K(y-z)\chi_{(y+\epsilon_{i+1},y+\epsilon_i]}(z) - K(x_0-z)\chi_{(x_0+\epsilon_{i+1},x_0+\epsilon_i]}(z)](b(z)-b_I)^m f_2(z)dz \\ =& \int_{\mathbb{R}} (K(y-z)-K(x_0-z))\chi_{(y+\epsilon_{i+1},y+\epsilon_i]}(z)(b(z)-b_I)^m f_2(z)dz \\ &+ \int_{\mathbb{R}} [K(x_0-z)(\chi_{(y+\epsilon_{i+1},y+\epsilon_i]}(z) - \chi_{(x_0+\epsilon_{i+1},x_0+\epsilon_i]}(z))(b(z)-b_I)^m f_2(z)dz. \end{aligned}$$

It follows that

$$\begin{aligned} & \|V(\mathcal{T})((b-b_I)^m f_2)(y) - V(\mathcal{T})((b-b_I)^m f_2)(x_0)\|_{F_\varrho} \\ \le& \Big\|\Big\{\int_{\mathbb{R}} (K(y-z)-K(x_0-z))\chi_{(y+\epsilon_{i+1},y+\epsilon_i]}(z)(b(z)-b_I)^m f_2(z)dz\Big\}_{i\in\mathbb{N},\beta=\{\epsilon_i\}\in\Theta}\Big\|_{F_\varrho} \\ &+\Big\|\Big\{\int_{\mathbb{R}} K(x_0-z)(\chi_{(y+\epsilon_{i+1},y+\epsilon_i]}(z) - \chi_{(x_0+\epsilon_{i+1},x_0+\epsilon_i]}(z)) \\ &\times(b(z)-b_I)^m f_2(z)dz\Big\}_{i\in\mathbb{N},\beta=\{\epsilon_i\}\in\Theta}\Big\|_{F_\varrho} \\ =:& I_{11} + I_{12}. \end{aligned} \tag{30}$$

Since $|x_0 - z| > 2h \ge 2|x_0 - y|$ for $z > x_0 + 2h$, then $|K(y-z) - K(x_0-z)| \le B_3|x_0-y||x_0-z|^{-2} \le B_3 h|x_0-z|^{-2}$ for any $z > x_0 + 2h$. Note that

$$\|\{\chi_{(y+\epsilon_{i+1},y+\epsilon_i]}(z)\}_{i\in\mathbb{N},\beta=\{\epsilon_i\}\in\Theta}\|_{F_\varrho} \le 1, \quad \forall y \in \mathbb{R}.$$

By Minkowski's inequality, Hölder's inequality and (27) with $\delta = mr'$, we obtain

$$\begin{aligned}
I_{11} &\leq \int_{\mathbb{R}} |K(y-z) - K(x_0 - z)| \| \{\chi_{(y+\epsilon_{i+1}, y+\epsilon_i]}(z)\}_{i\in\mathbb{N}, \beta=\{\epsilon_i\}\in\Theta} \|_{F_\varrho} \\
&\quad \times |(b(z) - b_I)^m f_2(z)| dz \\
&\leq B_3 h \int_{x_0+2h}^{\infty} \frac{|(b(z) - b_I)^m f(z)|}{(z - x_0)^2} dz \\
&\leq B_3 h \sum_{k=1}^{\infty} \int_{x_0+2^k h}^{x_0+2^{k+1}h} \frac{|(b(z) - b_I)^m f(z)|}{(2^k h)^2} dz \\
&\leq 4B_3 \sum_{k=1}^{\infty} 2^{-k} \Big(\frac{1}{2^{k+1}h} \int_{x_0}^{x_0+2^{k+1}h} |f(z)|^r dz \Big)^{1/r} \Big(\frac{1}{|2^k I|} \int_{2^k I} |b(z) - b_I|^{mr'} dz \Big)^{1/r'} \\
&\leq 4B_3 \sum_{k=1}^{\infty} \frac{(k+1)^m}{2^k} \|b\|^m_{\mathrm{BMO}(\mathbb{R})} M_r^+ f(x_0) \leq C(m, r, B_3) \|b\|^m_{\mathrm{BMO}(\mathbb{R})} M_r^+ f(x_0).
\end{aligned} \tag{31}$$

It remains to estimate I_{12}. Fix $\{\epsilon_i\} \in \Theta$. Let $N_1 = \{i \in \mathbb{Z} : \epsilon_i - \epsilon_{i+1} \geq y - x_0\}$ and $N_2 = \{i \in \mathbb{Z} : \epsilon_i - \epsilon_{i+1} < y - x_0\}$. We can write

$$\begin{aligned}
&\sum_{i\in\mathbb{Z}} \Big| \int_{\mathbb{R}} K(x_0 - z)(\chi_{(y+\epsilon_{i+1}, y+\epsilon_i]}(z) - \chi_{(x_0+\epsilon_{i+1}, x_0+\epsilon_i]}(z))(b(z) - b_I)^m f_2(z) dz \Big|^\rho \\
&\leq \sum_{i\in N_1} \Big| \int_{\mathbb{R}} K(x_0 - z)(\chi_{(y+\epsilon_{i+1}, y+\epsilon_i]}(z) - \chi_{(x_0+\epsilon_{i+1}, x_0+\epsilon_i]}(z))(b(z) - b_I)^m f_2(z) dz \Big|^\rho \\
&\quad + \sum_{i\in N_2} \Big| \int_{\mathbb{R}} K(x_0 - z)(\chi_{(y+\epsilon_{i+1}, y+\epsilon_i]}(z) - \chi_{(x_0+\epsilon_{i+1}, x_0+\epsilon_i]}(z))(b(z) - b_I)^m f_2(z) dz \Big|^\rho \\
&=: J_{11} + J_{12}.
\end{aligned} \tag{32}$$

By Hölder's inequality, we obtain

$$\begin{aligned}
J_{11} &\leq B_2^\rho \sum_{i\in N_1} \Big| \int_{\mathbb{R}} \frac{|(b(z) - b_I)^m f_2(z)|}{|x_0 - z|} (\chi_{(x_0+\epsilon_{i+1}, y+\epsilon_{i+1}]}(z) + \chi_{(x_0+\epsilon_i, y+\epsilon_i]}(z)) dz \Big|^\rho \\
&\leq (4B_2)^\rho \sum_{i\in N_1} \Big| \int_{\mathbb{R}} \frac{|(b(z) - b_I)^m f_2(z)|}{|x_0 - z|} \chi_{(x_0+\epsilon_i, y+\epsilon_i]}(z) dz \Big|^\rho \\
&\leq (4B_2)^\rho h^{\rho-1} \sum_{i\in N_1} \int_{\mathbb{R}} \frac{|(b(z) - b_I)^m f_2(z)|^\rho}{|x_0 - z|^\rho} \chi_{(y+\epsilon_{i+1}, y+\epsilon_i]}(z) dz \\
&\leq (4B_2)^\rho h^{\rho-1} \int_{\mathbb{R}} \frac{|(b(z) - b_I)^m f_2(z)|^\rho}{|x_0 - z|^\rho} dz.
\end{aligned} \tag{33}$$

$$\begin{aligned}
J_{12} &\leq B_2^\rho \sum_{i\in N_2} \Big| \int_{\mathbb{R}} \frac{|(b(z) - b_I)^m f_2(z)|}{|x_0 - z|} (\chi_{(y+\epsilon_{i+1}, y+\epsilon_i]}(z) + \chi_{(x_0+\epsilon_{i+1}, x_0+\epsilon_i]}(z)) dz \Big|^\rho \\
&\leq (2B_2)^\rho \sum_{i\in N_2} \Big| \int_{\mathbb{R}} \frac{|(b(z) - b_I)^m f_2(z)|}{|x_0 - z|} (\chi_{(y+\epsilon_{i+1}, y+\epsilon_i]}(z) dz \Big|^\rho \\
&\quad + (2B_2)^\rho \sum_{i\in N_2} \Big| \int_{\mathbb{R}} \frac{|(b(z) - b_I)^m f_2(z)|}{|x_0 - z|} \chi_{(x_0+\epsilon_{i+1}, x_0+\epsilon_i]}(z) dz \Big|^\rho \\
&\leq h^{\rho-1} (2B_2)^\rho \sum_{i\in N_2} \int_{\mathbb{R}} \frac{|(b(z) - b_I)^m f_2(z)|^\rho}{|x_0 - z|^\rho} \chi_{(y+\epsilon_{i+1}, y+\epsilon_i]}(z) dz \\
&\quad + h^{\rho-1} (2B_2)^\rho \sum_{i\in N_2} \int_{\mathbb{R}} \frac{|(b(z) - b_I)^m f_2(z)|^\rho}{|x_0 - z|^\rho} \chi_{(x_0+\epsilon_{i+1}, x_0+\epsilon_i]}(z) dz \\
&\leq 2(2B_2)^\rho h^{\rho-1} \int_{\mathbb{R}} \frac{|(b(z) - b_I)^m f_2(z)|^\rho}{|x_0 - z|^\rho} dz.
\end{aligned} \tag{34}$$

It follows from (32)–(34) that

$$I_{12} \leq C(B_2, r) h^{1-1/\rho} \Big(\int_{\mathbb{R}} \frac{|(b(z) - b_I)^m f_2(z)|^\rho}{|x_0 - z|^\rho} dz \Big)^{1/\rho}. \tag{35}$$

By Hölder's inequality and (27) (with $\delta = m\rho\rho'$), we have

$$\begin{aligned}
&\int_{\mathbb{R}} \frac{|(b(z)-b_I)^m f_2(z)|^\rho}{|x_0-z|^\rho} dz \\
&= \sum_{k=1}^{\infty} \int_{x_0+2^k h}^{x_0+2^{k+1}h} \frac{|(b(z)-b_I)^m f(z)|^\rho}{|x_0-z|^\rho} dz \\
&\leq \sum_{k=1}^{\infty} (2^k h)^{-\rho} \int_{x_0+2^k h}^{x_0+2^{k+1}h} |(b(z)-b_I)^m f(z)|^\rho dz \\
&\leq 4h^{1-\rho} \sum_{k=1}^{\infty} 2^{-k(\rho-1)} \Big(\frac{1}{2^{k+1}h} \int_{x_0}^{x_0+2^{k+1}h} |f(z)|^r dz \Big)^{1/\rho} \\
&\quad \times \Big(\frac{1}{|2^k I|} \int_{2^k I} |b(z)-b_I|^{m\rho\rho'} dz \Big)^{1/\rho'} \\
&\leq 4h^{1-\rho} \|b\|^{m\rho}_{\mathrm{BMO}(\mathbb{R})} \sum_{k=1}^{\infty} \frac{(k+1)^{m\rho}}{2^{k(\rho-1)}} (M_r^+ f(x_0))^\rho.
\end{aligned}$$

This yields directly

$$\int_{\mathbb{R}} \frac{|(b(z)-b_I)^m f_2(z)|^\rho}{|x_0-z|^\rho} dz \leq C(m,r) h^{1-\rho} \|b\|^{m\rho}_{\mathrm{BMO}(\mathbb{R})} (M_r^+ f(x_0))^\rho. \tag{36}$$

Combining (36) with (35) yields (37) together with (30) and (31) implies

$$I_{12} \leq C(m,r,B_2) \|b\|^m_{\mathrm{BMO}(\mathbb{R})} M_r^+ f(x_0), \tag{37}$$

$$I_3 \leq C(m,r,B_2,B_3) \|b\|^m_{\mathrm{BMO}(\mathbb{R})} M_r^+ f(x_0). \tag{38}$$

Combining (38) with (26), (28) and (29) yields (23). This completes the proof. □

We now turn to prove our main results.

Proof of Theorem 1. We first prove (i). For any $w \in A_p^+$ with $1 < p < \infty$, there exists $r \in (1, p)$ such that $w \in A_{p/r}^+$. Then, we have

$$\|M_r^+ f\|_{L^p(\mathbb{R},w(x)dx)} \leq \|M^+|f|^r\|^{1/r}_{L^{p/r}(\mathbb{R},w(x)dx)} \leq C_{p,r} \|f\|_{L^p(\mathbb{R},w(x)dx)}. \tag{39}$$

On the other hand, it was proved in [23] that

$$\|M^+ f\|_{L^p(\mathbb{R},w(x)dx)} \leq C \|M^{+,\sharp} f\|_{L^p(\mathbb{R},w(x)dx)} \tag{40}$$

for $1 < p < \infty$ and $w \in A_\infty^+$. We get from (22), (39) and (40) and that

$$\begin{aligned}
\|\mathcal{V}_\varrho(\mathcal{T})f\|_{L^p(\mathbb{R},w(x)dx)} &\leq \|M^+(\mathcal{V}_\varrho(\mathcal{T})f)\|_{L^p(\mathbb{R},w(x)dx)} \\
&\leq C \|M^{+,\sharp}(\mathcal{V}_\varrho(\mathcal{T})f))\|_{L^p(\mathbb{R},w(x)dx)} \\
&\leq C \|M_r^+ f\|_{L^p(\mathbb{R},w(x)dx)} \leq C \|f\|_{L^p(\mathbb{R},w(x)dx)}.
\end{aligned}$$

This together with Lemma 3 yields Theorem 1 (i).

Applying Lemma 4 and the arguments similar to those used in deriving Theorem 1.3 in [19], we can get Theorem 1 (ii). The details are omitted.

We now prove (iii). For $w^{-1} \in A_1^-$, there exists $r > 1$ such that $w^{-r} \in A_1^-$. Thus, for any $x \in \mathbb{R}$,

$$\begin{aligned}
M_r^+ f(x) w(x) &= w(x) \Big(\sup_{h>0} \frac{1}{h} \int_x^{x+h} (|f(y)|w(y))^r w^{-r}(y) dy \Big)^{1/r} \\
&\leq \|f\|_{L^\infty(\mathbb{R},w(x)dx)} w(x) (M^+(w^{-r})(x))^{1/r} \leq \|w^{-r}\|_{A_1^-} \|f\|_{L^\infty(\mathbb{R},w(x)dx)},
\end{aligned}$$

which together with (23) yield that

$$\begin{aligned}\|\mathcal{V}_\varrho(\mathcal{T})f\|_{\mathrm{BMO}^+(\mathbb{R},w(x)dx)} &= \|M^{+,\sharp}(\mathcal{V}_\varrho(\mathcal{T})f))\|_{L^\infty(\mathbb{R},w(x)dx)}\\ &\le C\|M_r^+ f\|_{L^\infty(\mathbb{R},w(x)dx)} \le C\|f\|_{L^\infty(\mathbb{R},w(x)dx)}\end{aligned}$$

for any $1<r<\infty$. This proves Theorem 1. □

Proof of Theorem 2. We first prove (i). Fix $x_0\in\mathbb{R}$ and $h>0$. It suffices to show that

$$\Big(\frac{1}{w(x_0-h,x_0)}\int_{x_0}^{x_0+h}|\mathcal{V}_\varrho(\mathcal{T}_b^m)f(x)|^p dx\Big)^{1/p}\le C\|b\|^m_{\mathrm{BMO}(\mathbb{R})}h^\beta\|f\|_{L^{p,\beta,+}(w)}, \tag{41}$$

where $C>0$ is independent of x_0, h. Let $f_1=f\chi_{[x_0,x_0+2h)}$, $f_2=f\chi_{[x_0+2h,\infty)}$ and $f_3=f-f_1-f_2$. Let $I=[x_0-2h,x_0+2h]$. Note that $T_\epsilon^{+,b,m}f_3(x)=0$ for any $\epsilon>0$ and $x\ge x_0$. It follows that $\mathcal{V}_\varrho(\mathcal{T}_b^m)f_3(x)=0$ for all $x\ge x_0$. Thus, we can write

$$\begin{aligned}&\Big(\frac{1}{w(x_0-h,x_0)}\int_{x_0}^{x_0+h}|\mathcal{V}_\varrho(\mathcal{T}_b^m)f(x)|^p dx\Big)^{1/p}\\ \le&\Big(\frac{1}{w(x_0-h,x_0)}\int_{x_0}^{x_0+h}|\mathcal{V}_\varrho(\mathcal{T}_b^m)f_1(x)|^p dx\Big)^{1/p}\\ &+\Big(\frac{1}{w(x_0-h,x_0)}\int_{x_0}^{x_0+h}|\mathcal{V}_\varrho(\mathcal{T}_b^m)f_2(x)|^p dx\Big)^{1/p}=:S_1+S_2.\end{aligned}\tag{42}$$

Invoking Lemma 1 (i) and Theorem 1 (ii), there exists $C>0$ independent of x_0, h, such that

$$\begin{aligned}S_1&\le C\|b\|^m_{\mathrm{BMO}(\mathbb{R})}\Big(\frac{1}{w(x_0-h,x_0)}\int_{x_0}^{x_0+2h}|f(x)|^p dx\Big)^{1/p}\\ &\le C\|b\|^m_{\mathrm{BMO}(\mathbb{R})}\Big(\frac{w(x_0-2h,x_0)}{w(x_0-h,x_0)}\frac{1}{w(x_0-2h,x_0)}\int_{x_0}^{x_0+2h}|f(x)|^p dx\Big)^{1/p}\\ &\le C\|b\|^m_{\mathrm{BMO}(\mathbb{R})}h^\beta\|f\|_{L^{p,\beta,+}(w)}.\end{aligned}\tag{43}$$

Applying Lemma 1 (ii), there exists $C>0$ independent of x_0, h such that

$$\begin{aligned}&\Big(\frac{1}{w(x_0-h,x_0)}\int_{x_0+2^kh}^{x_0+2^{k+1}h}|f(z)|^p dz\Big)^{1/p}\\ \le&\Big(\frac{w(x_0-h-2^{k+2}h,x_0-h)}{w(x_0-h,x_0)}\frac{1}{w(x_0-h-2^{k+2}h,x_0-h)}\int_{x_0-h}^{x_0-h+2^{k+2}h}|f(z)|^p dz\Big)^{1/p}\\ \le& C2^{(k+2)(1+\beta)}h^\beta\|f\|_{L^{p,\beta,+}(w)}.\end{aligned}\tag{44}$$

One can easily check that $|x-z|>|z-x_0|/2$ for $x\in[x_0,x_0+h]$ and $z\in[x_0+2h,\infty)$. Fix $x\in[x_0,x_0+h]$. Then, by (11) and Minkowski's inequality, we have

$$\begin{aligned}\mathcal{V}_\varrho(\mathcal{T}_b^m)f_2(x)&=\|V(\mathcal{T}_b^m)f_2(x)\|_{F_\varrho}\\ &\le\Big\|\Big\{\int_{\epsilon_{i+1}<z-x\le\epsilon_i}K(x-z)(b(x)-b(z))^m f_2(z)dz\Big\}_{i\in\mathbb{N},\beta=\{\epsilon_i\}\in\Theta}\Big\|_{F_\varrho}\\ &\le\int_{\mathbb{R}}|K(x-z)(b(x)-b(z))^m f_2(z)|\Big\|\Big\{\chi_{\epsilon_{i+1}<z-x\le\epsilon_i}\Big\}_{i\in\mathbb{N},\beta=\{\epsilon_i\}\in\Theta}\Big\|_{F_\varrho}dz\\ &\le C\int_{\mathbb{R}}\frac{|f_2(z)(b(x)-b(z))^m|}{|z-x_0|}dz,\end{aligned}\tag{45}$$

where $C > 0$ is independent of x_0, h. It is clear that

$$\int_{\mathbb{R}} \frac{|f_2(z)(b(x)-b(z))^m|}{|z-x_0|}dz = \sum_{k=1}^{\infty}\int_{x_0+2^kh}^{x_0+2^{k+1}h} \frac{|f(z)(b(x)-b(z))^m|}{|z-x_0|}dz.$$

Fix $k \geq 1$. By Hölder's inequality, we obtain

$$\begin{aligned}
&\int_{x_0+2^kh}^{x_0+2^{k+1}h} \frac{|f(z)(b(x)-b(z))^m|}{|z-x_0|}dz \\
&\leq 2^m(2^kh)^{-1}\Big(\int_{x_0+2^kh}^{x_0+2^{k+1}h}|f(z)||b(x)-b_{2^kI}|^m dz + \int_{x_0+2^kh}^{x_0+2^{k+1}h}|f(z)||b(z)-b_{2^kI}|^m dz\Big) \\
&\leq 2^m(2^kh)^{-1/p}|b(x)-b_{2^kI}|^m\Big(\int_{x_0+2^kh}^{x_0+2^{k+1}h}|f(z)|^p dz\Big)^{1/p} \\
&\quad +2^m(2^kh)^{-1}\Big(\int_{x_0+2^kh}^{x_0+2^{k+1}h}|f(z)|^p dz\Big)^{1/p}\Big(\int_{x_0+2^kh}^{x_0+2^{k+1}h}|b(z)-b_{2^kI}|^{mp'}dz\Big)^{1/p'}.
\end{aligned}$$

This together with (27) and (44) yields that

$$\begin{aligned}
&\int_{x_0+2^kh}^{x_0+2^{k+1}h} \frac{|f(z)(b(x)-b(z))^m|}{|z-x_0|}dz \\
&\leq C2^{k(1+\beta)}h^\beta\|f\|_{L^{p,\beta,+}(w)}(2^kh)^{-1/p}w((x_0-h,x_0))^{1/p}(|b(x)-b_{2^kI}|^m+\|b\|^m_{\mathrm{BMO}(\mathbb{R})}).
\end{aligned} \tag{46}$$

Here, $C > 0$ is independent of x_0, h. By (45) and (46) and Hölder's inequality, we have

$$\begin{aligned}
S_2 &\leq Ch^\beta\|f\|_{L^{p,\beta,+}(w)}\sum_{k=1}^{\infty}2^{k(1+\beta)}(2^kh)^{-1/p} \\
&\quad\times\Big(\int_{x_0}^{x_0+h}|(|b(x)-b_{2^kI}|^m+\|b\|^m_{\mathrm{BMO}(\mathbb{R})})|^p dx\Big)^{1/p} \\
&\leq Ch^\beta\|f\|_{L^{p,\beta,+}(w)}\sum_{k=1}^{\infty}2^{k(1+\beta)}(2^kh)^{-1/p} \\
&\quad\times\Big(\int_{x_0}^{x_0+h}(2^m|b(x)-b_I|^m+2^m|b_I-b_{2^kI}|^m+\|b\|^m_{\mathrm{BMO}(\mathbb{R})})^p dx\Big)^{1/p} \\
&\leq C\|b\|^m_{\mathrm{BMO}(\mathbb{R})}h^\beta\|f\|_{L^{p,\beta,+}(w)}\sum_{k=1}^{\infty}\frac{(k+1)^m}{2^{(1/p-1-\beta)k}} \\
&\leq C\|b\|^m_{\mathrm{BMO}(\mathbb{R})}h^\beta\|f\|_{L^{p,\beta,+}(w)}.
\end{aligned} \tag{47}$$

Here, $C > 0$ is independent of x_0, h. In the last inequality of (47), we have used the condition $1/p > 1+\beta$. (47) together with (42) and (43) yield (41).

Next, we prove (ii). Let $f_1 = f\chi_{[x_0,x_0+2h)}$, $f_2 = f\chi_{[x_0+2h,\infty)}$ and $f_3 = f - f_1 - f_2$. Let $I = [x_0-2h, x_0+2h]$. By (4), we want to show that

$$\Big(\frac{1}{w(x_0-h,x_0)}\int_{x_0}^{x_0+h}|\mathcal{V}_\varrho(\mathcal{T})f(x)-\mathcal{V}_\varrho(\mathcal{T})f_2(x_0)|^p dx\Big)^{1/p} \leq Ch^\beta\|f\|_{L^{p,\beta,+}(w)}, \tag{48}$$

where $C > 0$ independent of x_0, h. Using (11) and Minkowski's inequality, one has

$$\begin{aligned}
&|\mathcal{V}_\varrho(\mathcal{T})f(x)-\mathcal{V}_\varrho(\mathcal{T})f_2(x_0)| \\
&= |\|V(\mathcal{T})f(x)\|_{F_\varrho}-\|V(\mathcal{T})f_2(x_0)\|_{F_\varrho}| \\
&\leq \|V(\mathcal{T})f(x)-V(\mathcal{T})f_2(x_0)\|_{F_\varrho} \leq |\mathcal{V}_\varrho(\mathcal{T})f_1(x)| + \|V(\mathcal{T})f_2(x)-V(\mathcal{T})f_2(x_0)\|_{F_\varrho}.
\end{aligned}$$

This together with Minkowski's inequality again yield that

$$\begin{aligned}
&\Big(\frac{1}{w(x_0-h,x_0)}\int_{x_0}^{x_0+h}|\mathcal{V}_\varrho(\mathcal{T})f(x)-\mathcal{V}_\varrho(\mathcal{T})f_2(x_0)|^p dx\Big)^{1/p}\\
&\leq \Big(\frac{1}{w(x_0-h,x_0)}\int_{x_0}^{x_0+h}|\mathcal{V}_\varrho(\mathcal{T})f_1(x)|^p dx\Big)^{1/p}\\
&+\Big(\frac{1}{w(x_0-h,x_0)}\int_{x_0}^{x_0+h}\|V(\mathcal{T})f_2(x)-V(\mathcal{T})f_2(x_0)\|_{F_\varrho}^p dx\Big)^{1/p}.
\end{aligned}\tag{49}$$

We get from (43) (with $m=0$) that

$$\Big(\frac{1}{w(x_0-h,x_0)}\int_{x_0}^{x_0+h}|\mathcal{V}_\varrho(\mathcal{T})f_1(x)|^p dx\Big)^{1/p}\leq Ch^\beta\|f\|_{L^{p,\beta,+}(w)},\tag{50}$$

where $C>0$ is independent of x_0, h. Fix $x\in[x_0,x_0+h]$. (30), (31) and (35) (with $m=0$) imply that

$$\begin{aligned}
&\|V(\mathcal{T})f_2(x)-V(\mathcal{T})f_2(x_0)\|_{F_\varrho}\\
&\leq B_3h\int_{\mathbb{R}}\frac{|f_2(z)|}{|z-x_0|^2}dz+C(B_2,p)h^{1-1/p}\Big(\int_{\mathbb{R}}\frac{|f_2(z)|^p}{|x_0-z|^p}dz\Big)^{1/p}.
\end{aligned}$$

It follows that

$$\begin{aligned}
&\Big(\frac{1}{w(x_0-h,x_0)}\int_{x_0}^{x_0+h}\|V(\mathcal{T})f_2(x)-V(\mathcal{T})f_2(x_0)\|_{F_\varrho}^p dx\Big)^{1/p}\\
&\leq \frac{h^{1+1/p}}{w(x_0-h,x_0)^{1/p}}\int_{x_0+2h}^{\infty}\frac{|f(z)|}{(z-x_0)^2}dz+h\Big(\frac{1}{w(x_0-h,x_0)}\int_{\mathbb{R}}\frac{|f_2(z)|^p}{|x_0-z|^p}dz\Big)^{1/p}\\
&=: V_1+V_2.
\end{aligned}\tag{51}$$

By (44) and Hölder's inequality, there exists $C>0$ independent of x_0, h, such that

$$\begin{aligned}
V_1 &\leq \frac{h^{1+1/p}}{w(x_0-h,x_0)^{1/p}}\sum_{k=1}^{\infty}(2^kh)^{-2}\int_{x_0+2^kh}^{x_0+2^{k+1}h}|f(z)|dz\\
&\leq \sum_{k=1}^{\infty}2^{k(-2+1/p')}\Big(\frac{1}{w(x_0-h,x_0)}\int_{x_0+2^kh}^{x_0+2^{k+1}h}|f(z)|^pdz\Big)^{1/p}\\
&\leq C\sum_{k=1}^{\infty}2^{k(-2+1/p')}2^{k(1+\beta)}h^\beta\|f\|_{L^{p,\beta,+}(w)}\\
&\leq C\sum_{k=1}^{\infty}2^{k(\beta-1/p)}h^\beta\|f\|_{L^{p,\beta,+}(w)}\leq Ch^\beta\|f\|_{L^{p,\beta,+}(w)}.
\end{aligned}\tag{52}$$

$$\begin{aligned}
V_2 &\leq \Big(\sum_{k=1}^{\infty}2^{-kp}\frac{1}{w(x_0-h,x_0)}\int_{x_0+2^kh}^{x_0+2^{k+1}h}\frac{|f(z)|^p}{(z-x_0)^p}dz\Big)^{1/p}\\
&\leq C\Big(\sum_{k=1}^{\infty}2^{-kp}2^{k(1+\beta)p}h^{\beta p}\|f\|_{L^{p,\beta,+}(w)}^p\Big)^{1/p}\\
&\leq C\Big(\sum_{k=1}^{\infty}2^{k\beta p}\Big)^{1/p}h^\beta\|f\|_{L^{p,\beta,+}(w)}\leq Ch^\beta\|f\|_{L^{p,\beta,+}(w)}.
\end{aligned}\tag{53}$$

(53) together with (49)–(52) yields (48). This finishes the proof of Theorem 2. □

4. λ-Jump Operators and the Number of Up-Crossing

This section is devoted to study the λ-jump operators and the number of up-crossing associated with the operators sequence $\{T_\epsilon^{+,b,m}\}_{\epsilon>0}$, which give certain quantitative information on the convergence of the above families of operators.

Definition 4. *Given a family of bounded operators $\mathcal{T} = \{T_\epsilon\}_{\epsilon>0}$ acting between spaces of functions, the λ-jump operator associated with $\mathcal{T}$ applied to a function f at a point x is defined by*

$$\begin{aligned}\Lambda_\lambda(\mathcal{T})f(x) \quad &:= \sup\{n: \text{ there exist } s_1 < t_1 \le s_2 < t_2 < \cdots \le s_n < t_n \\ &\quad \text{such that } |T_{s_i}f(x) - T_{t_i}f(x)| > \lambda\}.\end{aligned}$$

For $0 < \alpha < \gamma$, the number of up-crossing associated with $\mathcal{T}$ applied to a function f at a point x is defined by

$$\begin{aligned}N_{\alpha,\gamma}(\mathcal{T})f(x) \quad &:= \sup\{n: \text{ there exist } s_1 < t_1 < s_2 < t_2 < \dots < s_n < t_n \\ &\quad \text{such that } T_{s_i}f(x) < \alpha,\ T_{t_i}f(x) > \gamma\}.\end{aligned}$$

It was shown in [11] that, if the λ-jump operators is finite a.e. for each choice of $\lambda > 0$, then we must have a.e. convergence of our family of operators. Moreover,

$$\lambda(\Lambda_\lambda(\mathcal{T})f(x))^{1/\varrho} \le \mathcal{V}_\varrho(\mathcal{T})f(x) \text{ and } N_{\alpha,\lambda}(\mathcal{T})f(x) \le \Lambda_{\lambda-\alpha}(\mathcal{T})f(x),\ \forall \lambda > \alpha > 0. \tag{54}$$

By Theorem 1 (ii) and Theorem 2 and (54), we can get the following result.

Theorem 3. *Let $m \in \mathbb{N}$, $\varrho > 2$, $b \in \mathrm{BMO}(\mathbb{R})$ and $K \in \mathrm{OCZK}(B_1, B_2, B_3)$ with support in $(-\infty, 0)$. Let $\mathcal{T}_b^m = \{T_\epsilon^{+,b,m}\}_{\epsilon>0}$ and $\mathcal{T} = \{T_\epsilon^+\}_{\epsilon>0}$ be given as in (1.1) and (1.2), respectively. Let $\lambda > \alpha > 0$. Assume that $\|\mathcal{V}_\varrho(\mathcal{T})\|_{L^q(\mathbb{R},dx)\to L^q(\mathbb{R},dx)} < \infty$ for some $q \in (1,\infty)$. Then,*

(i) for any $1 < p < \infty$, $w \in A_p^+$ and $f \in L^p(\mathbb{R}, w(x)dx)$,

$$\|(\Lambda_\lambda(\mathcal{T}_b^m)f)^{1/\varrho}\|_{L^p(\mathbb{R},w(x)dx)} \le \frac{C(p,\varrho)}{\lambda}\|b\|^m_{\mathrm{BMO}(\mathbb{R})}\|f\|_{L^p(\mathbb{R},w(x)dx)};$$

$$\|(N_{\alpha,\lambda}(\mathcal{T}_b^m)f)^{1/\varrho}\|_{L^p(\mathbb{R},w(x)dx)} \le \frac{C(p,\varrho)}{\lambda-\alpha}\|b\|^m_{\mathrm{BMO}(\mathbb{R})}\|f\|_{L^p(\mathbb{R},w(x)dx)};$$

(ii) for any $1 < p < 1/(\beta+1)$, $-1/p \le \beta < 0$, $w \in A_p^+$ and $f \in L^{p,\beta,+}(w)$,

$$\|(\Lambda_\lambda(\mathcal{T}_b^m)f)^{1/\varrho}\|_{L^{p,\beta,+}(w)} \le \frac{C(p,\varrho)}{\lambda}\|b\|^m_{\mathrm{BMO}(\mathbb{R})}\|f\|_{L^{p,\beta,+}(w)};$$

$$\|(N_{\alpha,\lambda}(\mathcal{T}_b^m)f)^{1/\varrho}\|_{L^{p,\beta,+}(w)} \le \frac{C(p,\varrho)}{\lambda-\alpha}\|b\|^m_{\mathrm{BMO}(\mathbb{R})}\|f\|_{L^{p,\beta,+}(w)}.$$

5. Conclusions and Further Comments

It should be pointed out that our main results represent one-sided extensions of the main results in [19,28]. Combining with the two-sided case, the one-sided case is often more complex. Our main results not only enrich the variation inequalities for singular integrals and related commutators, but also explore some one-sided techniques to serve our aim (for example, see Lemma 1). In fact, it is unknown whether the variation operators for one-sided singular integrals are bounded on $L^p(\mathbb{R})$, which will be our forthcoming objective of research. On the other hand, some new one-sided methods and techniques can be explored to apply other one-sided operators.

Author Contributions: Formal analysis, Z.F.; writing–original draft preparation, F.L.; writing–review and editing, S.J. and S.-K.O.

Funding: This work was funded by the NNSF of China (grant Nos.11701333,11671185) and the SP-OYSTTT-CMSS (grant No. Sxy2016K01).

Conflicts of Interest: All of authors in this article declare no conflict of interest. All of funders in this article support the article's publication.

References

1. Aimar, H.; Forzani, L.; Martín-Reyes, F.J. On weighted inequalities for singular integrals. *Proc. Am. Math. Soc.* **1997**, *125*, 2057–2064. [CrossRef]
2. Bourgain, J. Pointwise ergodic theorems for arithmetic sets. *Inst. Hautes Études Sci. Publ. Math.* **1989**, *69*, 5–45. [CrossRef]
3. Campbell, J.T.; Jones, R.L.; Reinhdd, K.; Wierdl, M. Oscillation and variation for the Hilbert transform. *Duke Math. J.* **2000**, *105*, 59–83.
4. Campbell, J.T.; Jones, R.L.; Reinhdd, K.; Wierdl, M. Oscillation and variation for singular integrals in higher dimensions. *Trans. Am. Math. Soc.* **2003**, *355*, 2115–2137. [CrossRef]
5. Crescimbeni, R.; Martín-Reyes, F.J.; Torre, A.L.; Torrea, J.L. The ρ-variation of the Hermitian Riesz transform. *Acta Math. Sin. Engl. Ser.* **2010**, *26*, 1827–1838. [CrossRef]
6. Ding, Y.; Hong, G.; Liu, H. Jump and variational inequalities for rough operators. *J. Fourier Anal. Appl.* **2017**, *23*, 679–711. [CrossRef]
7. Fu, Z.; Lu, S.; Sato, S.; Shi, S. On weighted weak type norm inequalities for one-sided oscillatory singular integrals. *Stud. Math.* **2011**, *207*, 137–150. [CrossRef]
8. Fu, Z.; Lu, S.; Shi, S.; Pan, Y. Some one-sided estimates for oscillatory singular integrals. *Nonlinear Anal.* **2014**, *108*, 144–160. [CrossRef]
9. Gillespie, T.A.; Torrea, J.L. Dimension free estimates for the oscillation of Riesz transforms. *Israel J. Math.* **2004**, *141*, 125–144. [CrossRef]
10. Harboure, E.; Macías, R.A.; Menárguez, M.T.; Torrea, J.L. Oscillation and variation for the Gaussian Riesz transforms and Poisson integral. *Proc. R. Soc. Edinb.* **2005**, *135A*, 85–104. [CrossRef]
11. Jones, R.L. Variation inequalities for singular integrals and related operators. *Contemp. Math.* **2006**, *411*, 89–121.
12. Jones, R.L.; Kaufman, R.; Rosenblatt, J.; Wierdl, M. Oscillation in ergodic theory. *Ergod. Theory Dyn. Syst.* **1998**, *18*, 889–935. [CrossRef]
13. Jones, R.L.; Rosenblatt, J.; Wierdl, M. Oscillation in ergodic theory: Higher dimensional results. *Israel J. Math.* **2003**, *135*, 1–27. [CrossRef]
14. Jones, R.L.; Seeger, A.; Wright, J. Strong variational and jump inequalities in harmonic analysis. *Trans. Am. Math. Soc.* **2008**, *360*, 6711–6742. [CrossRef]
15. Lépingle, D. La variation d'ordre p des semi-martingales. *Zeitschrift für Wahrscheinlichkeitstheori Verwandte Gebiete* **1976**, *36*, 295–316. [CrossRef]
16. Liu, F. On the regularity of one-sided fractional maximal functions. *Math. Slovaca* **2018**, *68*, 1097–1112. [CrossRef]
17. Liu, F.; Mao, S. On the regularity of the one-sided Hardy–Littlewood functions. *Czechoslov. Math. J.* **2017**, *67*, 219–234. [CrossRef]
18. Liu, F.; Xu, L. Regularity of one-sided multilinear fractional maximal functions. *Open Math.* **2018**, *16*, 1556–1572. [CrossRef]
19. Liu, F.; Wu, H. A criterion on oscillation and variation for the commutators of singular integral operators. *Forum Math.* **2015**, *27*, 77–97. [CrossRef]
20. Lorente, M.; Riveros, M.S. Weighted inequalities for commutators of one-sided singular integrals. *Comment. Math. Univ. Carolinae* **2002**, *43*, 83–101.
21. Ma, T.; Torrea, J.L.; Xu, Q. Weighted variation inequalities for differential operators and singular integrals. *J. Funct. Anal.* **2015**, *268*, 376–416. [CrossRef]
22. Ma, T.; Torrea, J.L.; Xu, Q. Weighted variation inequalities for differential operators and singular integrals in higher dimensions. *Sci. China Math.* **2017**, *60*, 1419–1442. [CrossRef]
23. Martín-Reyes, F.J.; de la Torre, A. One-sided BMO spaces. *J. Lond. Math. Soc.* **1994**, *49*, 529–542. [CrossRef]
24. Le Merdy, C.; Xu, Q. Strong q-variation inequalities for analytic semigroups. *Ann. Inst. Fourier* **2012**, *62*, 2069–2097. [CrossRef]
25. Pisier, G.; Xu, Q. The strong p-variation of martingales and orthogonal series. *Probab. Theory Relat. Fields* **1988**, *77*, 497–514. [CrossRef]
26. Sawyer, E. Weighted inequalities for the one-sided Hardy–Littlewood maximal functions. *Trans. Am. Math. Soc.* **1986**, *291*, 53–61. [CrossRef]

27. Shi, S.G.; Fu, Z.W. Estimates of some operators on one-sided weighted Morrey spaces. *Abstr. Appl. Anal.* **2013**, *2013*, 829218. [CrossRef]
28. Zhang, J.; Wu, H. Oscillation and variation inequalities for singular integrals and commutators on weighted Morrey spaces. *Front. Math. China* **2016**, *11*, 423–447. [CrossRef]

mathematics

MDPI

Article

On the Generalization for Some Power-Exponential-Trigonometric Inequalities

Aníbal Coronel [1,*], Peter Kórus [2], Esperanza Lozada [1] and Elias Irazoqui [1]

[1] Departamento de Ciencias Básicas, Facultad de Ciencias, Universidad del Bío-Bío, Campus Fernando May, 3780000 Chillán, Chile; elozada@udec.cl (E.L.); eliasirazoqui@gmail.com (E.I.)

[2] Department of Mathematics, Juhász Gyula Faculty of Education, University of Szeged, Hattyas utca 10, H-6725 Szeged, Hungary; korpet@jgypk.szte.hu

* Correspondence: acoronel@ubiobio.cl; Tel.: +56-42-2463259

Received: 31 August 2019; Accepted: 26 September 2019; Published: 17 October 2019

Abstract: In this paper, we introduce and prove several generalized algebraic-trigonometric inequalities by considering negative exponents in the inequalities.

Keywords: power inequalities; exponential inequalities; trigonometric inequalities

1. Introduction

In recent years, an increasing amount of attention has been paid to the study of power-exponential inequalities [1–10]. A review of some problems and historical landmarks are given in [2,11]. In particular, in order to contextualize, we recall that the basic problem of comparing a^b and b^a for all positive real numbers a and b was presented in [12–14]. Increasing in algebraic difficulty, the comparison of $a^a + b^b$ and $a^b + b^a$ was studied independently by Laub–Ilani and Zeikii–Cirtoaje–Berndt, see [15–18], respectively. The result is the fact that the inequality

$$a^a + b^b \geq a^b + b^a, \quad a, b \in [0, \infty[\tag{1}$$

holds. An extension of (1) was proposed, analyzed and proved by Matejíčka, Cîrtoaje and Coronel-Huancas in [2,17,19] obtaining the inequality

$$a^{ra} + b^{rb} \geq a^{rb} + b^{ra}, \quad a, b \in [0, \infty[, \quad r \in [0, e[. \tag{2}$$

More recently, other extensions and generalizations of (1) were introduced, proved and conjectured by Özban in [11], where, in particular, the author proved the following inequalities:

$$\begin{aligned}
&(\sin x)^{\sin x} + (\sin y)^{\sin y} > (\sin x)^{\sin y} + (\sin y)^{\sin x}, \quad 0 < x < y < \pi/2, \\
&(\cos x)^{\cos x} + (\cos y)^{\cos y} > (\cos x)^{\cos y} + (\cos y)^{\cos x}, \quad 0 < x < y < \pi/2, \\
&(\cos x)^{\sin x} + (\cos y)^{\sin y} < (\cos x)^{\sin y} + (\cos y)^{\sin x}, \quad 0 < x < y \leq 1, \\
&(\cos x)^{x} + (\cos y)^{y} < (\cos x)^{y} + (\cos y)^{x}, \quad 0 < x < y \leq \pi/2, \\
&(\sin x)^{x} + (\sin y)^{y} > (\sin x)^{y} + (\sin y)^{x}, \quad 0 < x < y \leq \pi/2, \\
&x^{\cos x} + y^{\cos y} < x^{\cos y} + y^{\cos x}, \quad 0 < x < y, \ 1 \leq y \leq \pi/2, \\
&x^{\sin x} + y^{\sin y} > x^{\sin y} + y^{\sin x}, \quad 0 < x < y \leq \pi/2.
\end{aligned} \tag{3}$$

In order to extend or generalize (2) and (3), it seems natural to ask some questions: What happens with the inequality (2) when $r \in \mathbb{R} - [0, e[$? and what happens with the inequalities in (3) if we include a negative power r? We note that the powers in question exist, since the basis of powers in (2) and (3) are positive. Indeed, in this article, we study (2) for $r \in]-\infty, 0[$ and establish reverse inequalities for some cases. Moreover, we study the generalization of the inequalities in (3) with negative power r.

The main results of the paper are the following theorems:

Theorem 1. *Let the function $\varphi_\alpha : \mathbb{R} \to \mathbb{R}$ be defined by $\varphi_\alpha(m) = m\alpha^m$ for each $\alpha > 1$ and consider the following sets:*

$$\begin{aligned} A_{old} &= \Big\{(a,b,r) \in \mathbb{R}^3 \quad : \quad a \geq 0, \quad b \geq 0, \quad r \in [0, e[\Big\}, \\ A^d_{new} &= \Big\{(a,b,r) \in \mathbb{R}^3 \quad : \quad a > 1, \quad b > 1, \quad r < 0, \quad \varphi_b(rb) > \varphi_b(ra)\Big\} \\ &\quad \bigcup \Big\{(a,b,r) \in \mathbb{R}^3 \quad : \quad a > 1, \quad b > 1, \quad r < 0, \quad \varphi_b(rb) < \varphi_b(ra), \quad a^{rb} < \overline{\gamma}\Big\}, \\ A^r_{new} &= \Big\{(a,b,r) \in \mathbb{R}^3 \quad : \quad 0 \leq a \leq 1, \quad 0 \leq b \leq 1, \quad r < 0\Big\} \\ &\quad \bigcup \Big\{(a,b,r) \in \mathbb{R}^3 \quad : \quad a > 1, \quad b > 1, \quad r < 0, \quad \varphi_b(rb) < \varphi_b(ra), \quad a^{rb} > \overline{\gamma}\Big\}, \end{aligned} \tag{4}$$

where $\overline{\gamma} \in]0,1[$ is such that $\overline{\gamma} \neq b^{rb}$ and $(\overline{\gamma})^{a/b} - \overline{\gamma} - b^{ra} + b^{rb} = 0$. Then, the following inequalities

$$a^{ra} + b^{rb} \geq a^{rb} + b^{ra}, \quad (a,b,r) \in A_{old} \cup A^d_{new}, \tag{5}$$

$$a^{ra} + b^{rb} \leq a^{rb} + b^{ra}, \quad (a,b,r) \in A^r_{new} \tag{6}$$

are satisfied.

Remark 1. *The inclusion of the notation $\overline{\gamma}$ is related with the fact that the argumentation of the proof is based on the properties of function $f(t) = (t)^s - t - \gamma^s + \gamma$ with $t = a^{rb}$ $s = a/b$ and $\gamma = b^{rb}$. In particular, we observe that, if $0 < t < \gamma < 1$, there are two solutions of $f(t) = 0$ on the interval $]0,1[$; one solution is clearly γ and the other solution is difficult to get explicitly and is denoted by $\overline{\gamma}$.*

Theorem 2. *If $x, y \in (0, \pi/2)$ and $r < 0$, then*

$$(\sin x)^{r \sin x} + (\sin y)^{r \sin y} \leq (\sin x)^{r \sin y} + (\sin y)^{r \sin x}, \tag{7}$$

$$(\cos x)^{r \cos x} + (\cos y)^{r \cos y} \leq (\cos x)^{r \cos y} + (\cos y)^{r \cos x}, \tag{8}$$

$$(\cos x)^{r \sin x} + (\cos y)^{r \sin y} \geq (\cos x)^{r \sin y} + (\cos y)^{r \sin x}. \tag{9}$$

Theorem 3. *If $x, y \in (0, \pi/2)$ and $r < 0$, then*

$$(\cos x)^{rx} + (\cos y)^{ry} \geq (\cos x)^{ry} + (\cos y)^{rx}, \tag{10}$$

$$(\sin x)^{rx} + (\sin y)^{ry} \leq (\sin x)^{ry} + (\sin y)^{rx}. \tag{11}$$

Theorem 4. *If $x, y \in (0, \pi/2)$, $\min\{x, y\} \in (0, 1]$ and $r < 0$, then*

$$x^{r \cos x} + y^{r \cos y} \geq x^{r \cos y} + y^{r \cos x}, \tag{12}$$

$$x^{r \sin x} + y^{r \sin y} \leq x^{r \sin y} + y^{r \sin x}. \tag{13}$$

The rest of the paper is dedicated to the proof of Theorems 1–4.

2. Proofs of Main Results

2.1. *Proof of Theorem 1*

For completeness and self-contained structure of the proof, we recall the notation and a result given in [1]. Indeed, let us consider $s \in \mathbb{R}^+$ and we define the functions f and g from $\mathbb{R}^+$ to $\mathbb{R}$ by the relations

$$f(t) = t^s - t - \gamma^s + \gamma,$$
$$g(t) = \begin{cases} e^{-\ln(t)/(t-1)}, & \text{for } t \notin \{0,1\}, \\ e^{-1}, & \text{for } t = 1, \\ 0, & \text{for } t = 0. \end{cases}$$

Then, the following properties are satisfied: $f(\gamma) = 0$ and $f(0) = f(1) = -\gamma^s + \gamma$; if $s > 1$ (resp. $s < 1$), f is strictly increasing (resp. decreasing) on $]g(s), \infty[$ and strictly decreasing (resp. increasing) on $]0, g(s)[$; and g is continuous on $\mathbb{R}^+ \cup \{0\}$, strictly increasing on $\mathbb{R}^+$, $y = 1$ is a horizontal asymptote of $y = g(t)$, and the range of g is $[0,1]$. Moreover, if we consider the function $\xi : \mathbb{R}^+ \to \mathbb{R}\, \xi(m) = -m^s + m$ and φ_α defined in the enunciate of the theorem, we observe that the following following assertions are satisfied: $\xi(0) = \xi(1) = 0$; if $s > 1$ (resp. $s < 1$) w has a maximum at $g(s)$ (resp. minimum at $g(s)$); $\varphi_\alpha(0) = 0$; φ_α has a minimum at $m^* = -1/\ln(\alpha)$; φ_α has a inflection point at $m^{**} = -2/\ln(\alpha)$; $y = 0$ is a left horizontal asymptote of φ_α and the range of g is $[\varphi_\alpha(m^*), \infty[$ with $\varphi_\alpha(m^*) < 0$.

Let us consider $t = a^{rb}$, $\gamma = b^{rb}$, and $s = a/b$ and we observe that

$$f(t) = (a^{rb})^{a/b} - a^{rb} - (b^{rb})^{a/b} + b^{rb} = a^{ra} - a^{rb} - b^{ra} + b^{rb}. \tag{14}$$

Then, the proofs of (5) and (6) are reduced to analyze the sign of $f(t)$ for $t \in [0, \gamma]$. Indeed, without loss of generality and by the symmetric form of the inequalities in (5) and (6), we assume that $0 \le b < a$ (i.e., $s = a/b > 1$) and consider three cases:

(i) Let a, b such that $1 > a > b \ge 0$. Then, for $r < 0$, we note that $1 < a^r < b^r$ or equivalently we have that $1 < t < \gamma$. Moreover, observing that $s > 1$ and $g(s) < 1$, by the strictly increasing behavior of f on $[g(s), \infty)$, we deduce that $f(g(s)) < f(1) < f(t) < f(\gamma) = 0$. Thus, from (14) and $f(t) < 0$, we follow that the inequality $a^{ra} + b^{rb} < a^{rb} + b^{ra}$ is satisfied.

(ii) Let a, b such that $a > 1 > b \ge 0$. In this case, we have that $a^r < 1 < b^r$ or equivalently $t < 1 < \gamma$. We note that $s > 1$ implies the strictly decreasing behavior of f on $[0, g(s)]$ and the strictly increasing behavior of f on $[g(s), \infty[$. Moreover, observing that $g(s) \in [0,1]$, we deduce that $f(t) < f(1) = -\gamma^s + \gamma := \xi(\gamma)$ for any $t < 1 < \gamma$. Now, by the fact that ξ is decreasing on $[g(s), \infty[$, we have that $\xi(\gamma) < \xi(1) = 0$ for any $\gamma > 1$. Thus, $f(t) < \xi(\gamma) < 0$ for $t < 1 < \gamma$ and, from (14), the inequality $a^{ra} + b^{rb} < a^{rb} + b^{ra}$ is satisfied.

(iii) Let a, b such that $a > b > 1$. Similarly to cases (i) and (ii), we have that $s > 1$ and $0 < a^r < 1 < b^r < 1$ or equivalently $0 < t < \gamma < 1$. Here, we distinguish two subcases: $\gamma \le g(s)$ and $g(s) < \gamma < 1$. First, if $\gamma \le g(s)$, we have that f is strictly decreasing on $[0, \gamma]$ and consequently $f(t) \ge f(\gamma) = 0$ for $t \in [0, \gamma]$. Second, if $g(s) < \gamma < 1$, by the fact that $f(0) = \xi(\gamma) > 0 = f(\gamma) > f(g(s))$, we have that there exists $\overline{\gamma} \in [0, g(s)[$ such that $f(\overline{\gamma}) = 0$. Then, $f(t) \ge f(\overline{\gamma}) = 0$ for $t \in [0, \overline{\gamma}]$ and $f(t) \le f(\gamma) = f(\overline{\gamma}) = 0$ for $t \in [\overline{\gamma}, \gamma]$. Thus, from both subcases, we conclude that the inequality $a^{ra} + b^{rb} < a^{rb} + b^{ra}$ is satisfied for $t \in [\overline{\gamma}, \gamma]$ with $\gamma \in]g(s), 1[$ and the inequality $a^{ra} + b^{rb} > a^{rb} + b^{ra}$ is satisfied for $t \in [0, \overline{\gamma}]$ with $\gamma \in]g(s), 1[$ or for $t \in [0, \gamma]$ with $\gamma \in]0, g(s)]$.

On the other hand, by the definition of γ, s, g and φ_b, we observe that $\gamma < g(s)$ (resp. $\gamma > g(s)$) is equivalent to $\varphi_b(rb) > \varphi_b(ra)$ (resp. $\varphi_b(rb) < \varphi_b(ra)$). Moreover, the relation $t > \overline{\gamma}$ (resp. $t < \overline{\gamma}$) is equivalent to $a^{rb} > \overline{\gamma}$ (resp. $a^{rb} < \overline{\gamma}$). Thus, the subcases can be characterized in terms of the function φ_b and $a^{rb} > \overline{\gamma}$ or $a^{rb} < \overline{\gamma}$.

Hence, translating (i), (ii) and (iii) to the corresponding notation in (4) and observing that the set A_{old} is the set for the inequality in (2), we conclude the proof the theorem.

2.2. Proof of Theorem 2

Since $\sin t, \cos t > 0$ for $t \in (0, \pi/2)$, Theorem 1 immediately implies inequalities (7) and (8). To prove (9), we define

$$f(t) = (\cos t)^{r \sin t} + (\cos y)^{r \sin y} - (\cos t)^{r \sin y} - (\cos y)^{r \sin t}$$

for y is fixed and arbitrarily selected such that $y \in (0, \pi/2)$ and $0 < t \leq y$. We note that $f(y) = 0$, then the result follows if f is decreasing. Indeed, to see this, we write

$$f'(t) = r\left[g(t)\cos t + \frac{\sin t}{\cos t} h(t)\right],$$

where

$$\begin{aligned} g(t) &= (\cos t)^{r \sin t} \ln(\cos t) - (\cos y)^{r \sin t} \ln(\cos y), \\ h(t) &= (\cos t)^{r \sin y} \sin y - (\cos t)^{r \sin t} \sin t. \end{aligned}$$

Now, since $r < 0$, it is enough to show that $g(t), h(t) > 0$. For g, we have that

$$\begin{aligned} g(t) &= -\int_t^y \frac{d}{ds}(\cos s)^{r \sin t} \ln(\cos s) \\ &= \int_t^y ((\cos s)^{r \sin t - 1} \sin s)(1 + r \sin t \ln(\cos s))\, ds > 0 \end{aligned}$$

and, similarly for h, we deduce that

$$\begin{aligned} h(t) &= \int_t^y \frac{d}{ds}(\cos t)^{r \sin s} \sin s \\ &= \int_t^y ((\cos t)^{r \sin s} \cos s)(1 + r \sin s \ln(\cos t))\, ds > 0. \end{aligned}$$

2.3. Proof of Theorem 3

Set $0 < t \leq y < \pi/2$ and $r < 0$ arbitrarily. Along the proofs, we will use that $\sin s, \cos s > 0$ for $s \in (0, \pi/2)$.

In order to prove (10), let us consider $f_1(t) = (\cos t)^{rt} + (\cos y)^{ry} - (\cos t)^{ry} - (\cos y)^{rt}$. Observing that $f_1(y) = 0$, it is enough to show that f_1 is decreasing. Indeed, the decreasing behavior of f_1 follows immediately since

$$f_1'(t) = r\left[g_1(t) + \frac{\sin t}{\cos t} h_1(t)\right],$$

where

$$\begin{aligned} g_1(t) &= (\cos t)^{rt} \ln(\cos t) - (\cos y)^{rt} \ln(\cos y) = -\int_t^y \frac{d}{ds}(\cos s)^{rt} \ln(\cos s) \\ &= \int_t^y ((\cos s)^{rt-1} \sin s)(1 + rt \ln(\cos s))\, ds > 0 \end{aligned}$$

and

$$h_1(t) = y(\cos t)^{ry} - t(\cos t)^{rt} = \int_t^y \frac{d}{ds} s(\cos t)^{rs}$$
$$= \int_t^y (\cos t)^{rs}(1 + rs \ln(\cos t))\, ds > 0.$$

We prove (11) by analogous arguments to the proof of (10). Indeed, let us introduce the notation $f_2(t) = (\sin t)^{ry} + (\sin y)^{rt} - (\sin t)^{rt} - (\sin y)^{ry}$. We observe that

$$f_2'(t) = r\left[g_2(t) + \frac{\cos t}{\sin t} h_2(t)\right] < 0,$$

since

$$g_2(t) = (\sin y)^{rt} \ln(\sin y) - (\sin t)^{rt} \ln(\sin t) = \int_t^y \frac{d}{ds} (\sin s)^{rt} \ln(\sin s)$$
$$= \int_t^y ((\sin s)^{rt-1} \cos s)(1 + rt \ln(\sin s))\, ds > 0$$

and

$$h_2(t) = y(\sin t)^{ry} - t(\sin t)^{rt} = \int_t^y \frac{d}{ds} s(\sin t)^{rs}$$
$$= \int_t^y (\sin t)^{rs}(1 + rs \ln(\sin t))\, ds > 0.$$

Thus, (11) is a consequence of the decreasing behavior of f_2 and the fact that $f_2(y) = 0$.

2.4. *Proof of Theorem 4*

We set $0 < x \leq y < \pi/2$ with $x \leq 1$ and $r < 0$ arbitrarily selected. Then, by the fact that $\cos x \geq \cos y > 0$, we deduce the following estimate:

$$x^{r\cos x} - x^{r\cos y} = x^{r\cos y}(x^{r(\cos x - \cos y)} - 1)$$
$$\geq y^{r\cos y}(y^{r(\cos x - \cos y)} - 1) = y^{r\cos x} - y^{r\cos y},$$

which implies (12). Similarly, using the fact that $\sin y \geq \sin x > 0$ implies that

$$x^{r\sin y} - x^{r\sin x} = x^{r\sin x}(x^{r(\sin y - \sin x)} - 1)$$
$$\geq y^{r\sin x}(y^{r(\sin y - \sin x)} - 1) = y^{r\sin y} - y^{r\sin x},$$

and we get the proof of (13).

Author Contributions: A.C. and P.K. worked together in the initial formulation of the mathematical results. E.L. and E.I. helped with the analysis and original draft presntation. All the authors provided critical feedback and helped shape the research, analysis, and manuscript.

Funding: A.C. is thankful for the support from the research projects DIUBB GI 172409/C and DIUBB 183309 4/R.

Acknowledgments: A.C., E.L. and E.I. are grateful for the suggestions from the colleges at Ciencias Básicas of Universidad del Bío-Bío.

Conflicts of Interest: The authors declare that they have no competing interests.

References

1. Coronel, A.; Huancas, F. On the inequality $a^{2a} + b^{2b} + c^{2c} \geq a^{2b} + b^{2c} + c^{2a}$. *Aust. J. Math. Anal. Appl.* **2012**, *9*, 3.
2. Coronel, A.; Huancas, F. Proof of three power-exponential inequalitie. *J. Inequal. Appl.* **2014**, *2014*, 509. [CrossRef]
3. Cîrtoaje, V. Proofs of three open inequalities with power-exponential functions. *J. Nonlinear Sci. Appl.* **2011**, *4*, 130–137. [CrossRef]
4. Matejíčka, L. Some remarks on Cîrtoaje's conjecture. *J. Inequal. Appl.* **2016**, *159*, 269. [CrossRef]
5. Matejíčka, L. On the Cîrtoaje's conjecture. *J. Inequal. Appl.* **2016**, *159*, 152. [CrossRef]
6. Matejíčka, L. Next, generalization of Cîrtoaje's inequality. *J. Inequal. Appl.* **2017**, *159*, 159. [CrossRef] [PubMed]
7. Mitrinović, D.S. *Analytic Inequalities*; In Cooperation with P. M. Vasić. Die Grundlehren der Mathematischen Wissenschaften, Band 165; Springer: New York, NY, USA; Berlin, Germany, 1970.
8. Miyagi, M.; Nishizawa, Y. A short proof of an open inequality with power-exponential functions. *Aust. J. Math. Anal. Appl.* **2014**, *11*, 6.
9. Miyagi, M.; Nishizawa, Y. Extension of an inequality with power exponential functions. *Tamkang J. Math.* **2015**, *46*, 427–433.
10. Miyagi, M.; Nishizawa, Y. A stronger inequality of Cîrtoaje's one with power exponential functions. *J. Nonlinear Sci. Appl.* **2015**, *8*, 224–230. [CrossRef]
11. Özban, A.Y. New algebraic-trigonometric inequalities of Laub-Ilani type. *Bull. Aust. Math. Soc.* **2017**, *96*, 87–97. [CrossRef]
12. Bullen, P.S. *A Dictionary of Inequalities*; Volume 97 of Pitman Monographs and Surveys in Pure and Applied Mathematics; Longman: Harlow, UK, 1998.
13. Luo, J.; Wen, J.J. A power-mean discriminance of comparing a^b and b^a. In *Research Inequalities*; Yand, X.-Z., Ed.; People's Press of Tibet: Lhasa, China, 2000; pp. 83–88.
14. Qi, F.; Debnath, L. Inequalities for power-exponential functions. *J. Inequal. Pure Appl. Math.* **2000**, *1*, 15.
15. Laub, M. Problem E3116. *Am. Math. Mon.* **1985**, *92*, 666. [CrossRef]
16. Laub, M.; Ilani, I. A subtle inequality. *Am. Math. Mon.* **1990**, *97*, 65–67. [CrossRef]
17. Cîrtoaje, V. On some inequalities with power-exponential functions. *J. Inequal. Pure Appl. Math.* **2009**, *10*, 21.
18. Zeikii, A.; Cirtoaje, V.; Berndt, W. Mathlinks. Forum. Available online: http://www.mathlinks.ro/Forum/viewtopic.php?t=118722 (accessed on 11 November 2006).
19. Matejicka, L. Solution of one conjecture on inequalities with power-exponential functions. *J. Inequal. Pure Appl. Math.* **2009**, *10*, 72.

www.ingramcontent.com/pod-product-compliance
Lightning Source LLC
Chambersburg PA
CBHW051850210326
41597CB00033B/5838
* 9 7 8 3 0 3 9 2 8 0 6 2 9 *

MDPI
St. Alban-Anlage 66
4052 Basel
Switzerland
Tel. +41 61 683 77 34
Fax +41 61 302 89 18
www.mdpi.com

Mathematics Editorial Office
E-mail: mathematics@mdpi.com
www.mdpi.com/journal/mathematics